Gibbs Measures in Machine Learning

Gibbs Measures in Machine Learning

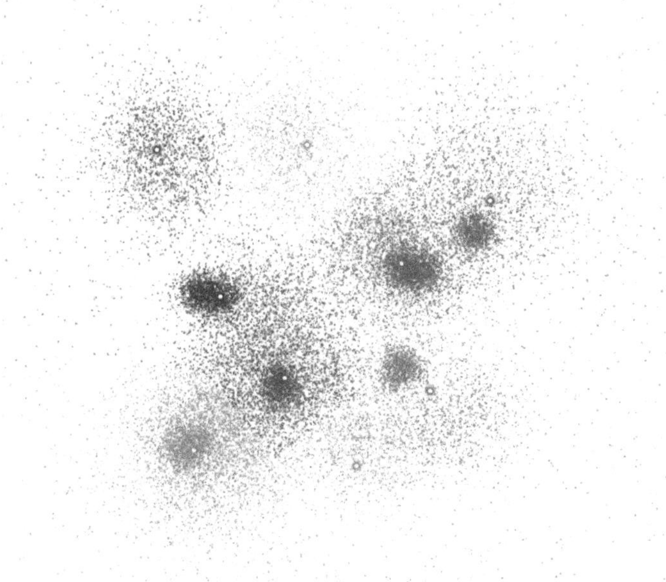

Laziz U Abdullaev
National University of Singapore, Singapore

Utkir A Rozikov
V I Romanovskiy Institute of Mathematics, Uzbekistan

World Scientific

NEW JERSEY • LONDON • SINGAPORE • BEIJING • SHANGHAI • HONG KONG • TAIPEI • CHENNAI • TOKYO

Published by

World Scientific Publishing Co. Pte. Ltd.
5 Toh Tuck Link, Singapore 596224
USA office: 27 Warren Street, Suite 401-402, Hackensack, NJ 07601
UK office: 57 Shelton Street, Covent Garden, London WC2H 9HE

Library of Congress Control Number: 2025038141

British Library Cataloguing-in-Publication Data
A catalogue record for this book is available from the British Library.

GIBBS MEASURES IN MACHINE LEARNING

ISBN 978-981-98-1456-5 (hardcover)
ISBN 978-981-98-1457-2 (ebook for institutions)
ISBN 978-981-98-1458-9 (ebook for individuals)

For any available supplementary material, please visit
https://www.worldscientific.com/worldscibooks/10.1142/14350#t=suppl

Desk Editors: Nambirajan Karuppiah/Lai Fun Kwong

Typeset by Stallion Press
Email: enquiries@stallionpress.com

To our family

Preface

In the rapidly evolving landscape of machine learning and artificial intelligence, the mathematical foundations that underpin our most sophisticated models often trace back to fundamental concepts from statistical physics. Among these, Gibbs measures stand out as a particularly powerful and versatile framework that has found applications across virtually every domain of modern machine learning, from classical statistical methods to cutting-edge deep learning architectures.

This book presents a comprehensive exploration of how Gibbs measures have become integral to machine learning methodology. We begin with foundational concepts, establishing the mathematical framework of Gibbs measures and demonstrating their natural appearance in machine learning contexts from the ubiquitous softmax function to the attention mechanisms and others that power modern transformer architectures. The connection between energy functions and probability distributions provides an elegant way to encode complex dependencies between variables, making Gibbs measures particularly suited for modeling structured data and sequential processes.

Our journey progresses through the computational aspects of working with Gibbs measures, including Markov chain Monte Carlo methods and interacting particle systems that enable practical implementation of these theoretical constructs. We explore how Gibbs sampling has become a cornerstone of Bayesian computation, while interacting particle systems offer new perspectives on dynamic learning processes.

The book dedicates substantial attention to parameter estimation within the Gibbs framework, contrasting parametric and non-parametric approaches. We examine how these measures naturally arise in Bayesian inference and hierarchical modeling, providing principled ways to handle uncertainty and incorporate prior knowledge. The treatment of Ising models with hidden Markov structure illustrates how classical statistical physics models continue to inform contemporary machine learning practice.

Unsupervised learning presents another rich domain where Gibbs measures excel, particularly in clustering applications and dimensionality reduction techniques. We explore how these measures can address fairness considerations in clustering and their role in modern generative models, including normalizing flows that have become essential tools for density estimation and sample generation.

The applications extend to structured prediction and graphical models, where the ability of Gibbs measures to capture complex dependencies makes them indispensable for tasks involving sequential data, spatial relationships, and multi-label classification. Through practical examples including lattice-based models and the Potts-Ising framework, we demonstrate how theoretical insights translate into effective algorithms.

Perhaps most remarkably, Gibbs measures have found new relevance in the era of deep learning. We explore their role in understanding neural network training dynamics through the lens of statistical mechanics, their appearance in attention mechanisms that enable transformer architectures to process sequential data effectively, and their connection to dense associative memories. The statistical mechanical perspective on deep learning provides profound insights into generalization, optimization landscapes, and the remarkable effectiveness of these models.

The final chapters examine emerging frontiers where Gibbs measures continue to push the boundaries of what is possible in machine learning. From their role in Large Language Models where temperature parameters control generation diversity to their application in federated learning and quantum-enhanced algorithms, these measures remain at the forefront of methodological innovation. We explore their integration with probabilistic programming frameworks and their extensions to infinite-dimensional settings through Bayesian nonparametrics.

This book systematically explores the foundational role of Gibbs measures in machine learning, highlighting how tools from statistical mechanics inform modern algorithms. It also serves as a guide for pure mathematicians from relevant fields to trace the relevance of their work in contemporary machine learning, and as a resource for practitioners seeking to understand the theoretical principles underlying the models they use.

Laziz U. Abdullaev and Utkir A. Rozikov

About the Authors

Laziz U. Abdullaev is a PhD candidate in mathematics and machine learning at the National University of Singapore (NUS), where he focuses on developing theoretical foundations for attention mechanisms, mixture-of-experts architectures, and recurrent models to name a few. His research draws from classical domains such as statistics, image processing, and graphical models to reinterpret the workings of modern neural architectures, particularly Large Language Models (LLMs).

He is the co-author of several peer-reviewed papers at top machine learning conferences, including *ICLR*, *NeurIPS*, with recent work exploring mixture-of-experts models, attention mechanisms, and safety alignment in LLMs.

Before joining NUS, Abdullaev received his undergraduate degree in mathematics from the National University of Uzbekistan and was a recipient of multiple international scholarships and awards.

Utkir A. Rozikov (Academician) earned his PhD in 1995 and his Doctor of Sciences degree in 2001 from the V.I. Romanovskiy Institute of Mathematics in Tashkent. Since 2019, Rozikov has served as the Deputy Director of the Institute of Mathematics.

Rozikov is a globally recognized expert in probability, mathematical physics, algebra and analysis, with a specialization in evolution algebras, dynamical systems and statistical mechanics on graphs. His significant contributions include in-depth analyses of Gibbs measures in key models (Ising, Potts, SOS, HC) of statistical mechanics on trees. He has introduced innovative tools for studying Gibbs measures on graphs, including group representation theory, information flows, node-weighted random walks, contour methods on trees, and non-linear analysis.

Rozikov is an exceptionally prolific author, having published over 180 papers since 1995 in prestigious journals. His work includes 12 papers in the renowned *Journal of Statistical Physics*, 27 papers in *Theoretical and Mathematical Physics*, *Lett. Math. Phys.*, *Comm. Math. Phys.*, *J. Stat. Mech. Theory Exp.*, *J. Math. Anal. Appl.*, among others. Notably, he has published extensive review papers in *Rev. Math. Phys.* in 2013 and 2021. Additionally, Rozikov is the author of five research monographs, two of which are related to Gibbs measures.

Contents

Chapter 1

Introduction to Gibbs Measures and Their Relevance in Machine Learning

This chapter provides a comprehensive introduction to Gibbs measures and their emerging significance in the field of machine learning. We begin by reviewing foundational concepts in machine learning, including both deterministic and probabilistic views of predictors. The connection between statistical physics and machine learning is then explored, laying the groundwork for the formal introduction of Gibbs measures. Essential mathematical tools such as measurable spaces, graphs, configuration spaces, and Markov chains are presented to support a rigorous understanding of the Gibbs measure framework. The chapter then demonstrates how Gibbs measures naturally arise in machine learning applications, with detailed discussions on their relationship to the Softmax function, Bayesian inference, and models on infinite graphs. We also examine the role of Gibbs measures in modern architectures like attention mechanisms and transformers.

For those new to the field of machine learning, we recommend the book [94], which covers the mathematical foundations underlying many machine learning methods. In addition, the book demonstrates how these mathematical concepts can be applied to design algorithms that effectively address machine learning tasks.

1.1 Machine learning

During the past decade, machine learning (ML) has become a central focus in both the public domain and scientific research [139]. Rapid advances in artificial neural networks training have contributed significantly to this surge in attention [234]. ML is a subfield of artificial intelligence (AI) designed to automatically extract meaningful information from data.

It encompasses three main branches:
- supervised learning,
- unsupervised learning, and
- reinforcement learning,

each of which addresses different challenges in the development of algorithms that can autonomously learn complex functions, distributions, and policies from data.

Supervised learning is the most widely studied and employed of these branches, focusing on problems like classification and regression. In supervised learning, given an input x, such as an image, the goal is to predict some label or quantitative value $y = f(x)$, with the function f being learned from a training dataset composed of input and ground-truth output pairs (x, y).

Unsupervised learning, on the other hand, is concerned with learning the underlying distribution $p(x)$ of the training data itself. This could involve extracting meaningful features from the data or generating new, realistic instances that resemble the training set.

Reinforcement learning aims to learn a conditional probability $p(a|s)$, where a represents an action (such as a move in a maze) taken in a given state s (the position within the maze), all in pursuit of achieving a specific goal, such as escaping the maze.

Following [94], let us briefly give the problem that a machine learning algorithm is devoted to solve. There are two different senses in "machine learning algorithm": training and prediction. Data, model, and learning are the three major components of a machine learning system.

The core **question in machine learning** is, "What defines a good model?" Note that the term "model" has many nuances which will be clear below. Defining "good" objectively is not straightforward, but a key principle is that good models perform well on unseen data. Below we shall provide an overview of the concept of machine learning, with a particular focus on supervised learning, which will be explored in greater depth later in the book.

In a supervised learning problem, we have a label y_n (e.g., the salary) associated with each example x_n (e.g., the age). The label y_n has various other names, including target, response variable, and annotation. A dataset is written as a set of example-label pairs $\{(x_1, y_1), \ldots, (x_n, y_n), \ldots, (x_N, y_N)\}$.

If we have data in an appropriate vector representation, then we can construct a predictive function (also known as a predictor). A predictor may be a deterministic function or a probabilistic model.

1.1.1 *Predictor as a function*

A function predictor produces an output (prediction) given a particular input example. Assuming that the output is a single number, i.e., a real-valued scalar, this can be written as

$$f : \mathbf{x} \in \mathbb{R}^D \to f(\mathbf{x}) \in \mathbb{R},$$

where the input vector $\mathbf{x}$ is D-dimensional and the function f returns a real number. A simple example of a predictor function is a linear function:

$$f(\mathbf{x}) = \boldsymbol{\theta}^\top \mathbf{x} + \theta_0,$$

for unknown (learnable) parameters $\boldsymbol{\theta}$ and θ_0.

The goal of learning in this case is to find (learn) a model and its corresponding parameters based on some training dataset such that the resulting predictor will perform well on unseen data. Namely, assume we are given N examples $\mathbf{x}_n \in \mathbb{R}^D$ and corresponding scalar labels $y_n \in \mathbb{R}$. In supervised learning, for instance, with pairs $(\mathbf{x}_1, y_1), \ldots, (\mathbf{x}_N, y_N)$, one aims to estimate a predictor $f(\cdot, \boldsymbol{\theta}) : \mathbb{R}^D \to \mathbb{R}$, parametrized by $\boldsymbol{\theta}$, that can capture the underlying input-output relation. To this end, the goal is to find a good parameter $\boldsymbol{\theta}^*$ such that the predictor fits the data well, meaning

$$f(\mathbf{x}_n, \boldsymbol{\theta}^*) \approx y_n \quad \text{for all } n = 1, \ldots, N.$$

To define what it means to fit the data well, introduce a loss function $\ell(y_n, \hat{y}_n)$ that takes the ground truth label and the corresponding prediction as input and produces a non-negative number (the loss) representing the error in this prediction. The goal is to find a good parameter vector $\boldsymbol{\theta}^*$ that *minimizes the average loss* across N training examples.

A common assumption in machine learning is that the set of examples $\{(\mathbf{x}_1, y_1), \ldots, (\mathbf{x}_N, y_N)\}$ is independent and identically distributed (i.i.d.). Independence implies that two data points $(\mathbf{x}_i, y_i)$ and $(\mathbf{x}_j, y_j)$ do not statistically depend on each other, making the empirical mean a good estimate of the population mean. Thus, we can use the empirical mean of the loss on the training data.

For a given training set $\{(\mathbf{x}_1, y_1), \ldots, (\mathbf{x}_N, y_N)\}$, we introduce the sample matrix $\mathbf{X} := [\mathbf{x}_1, \ldots, \mathbf{x}_N]^\top \in \mathbb{R}^{N \times D}$ and the label vector $\mathbf{y} := [y_1, \ldots, y_N]^\top \in \mathbb{R}^N$. Using this matrix notation, the average loss is given by

$$R_{\text{emp}}(f, \mathbf{X}, \mathbf{y}) = \frac{1}{N} \sum_{n=1}^{N} \ell(y_n, \hat{y}_n),$$

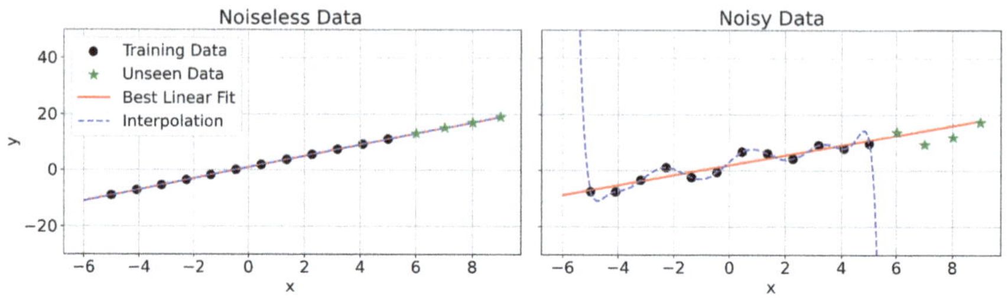

Fig. 1.1: Interpolation versus Parametric Fit on a Linear Data Relation. Interpolation forces the function to pass through all sample values no matter the date quality, while properly chosen parametric function (linear in this case) captures the essential data pattern well under clean (left figure) **and** noisy (right figure) data environments. This is important to be able to predict on the unseen input data.

where $\hat{y}_n = f(\mathbf{x}_n, \boldsymbol{\theta})$.

This equation is referred to as the empirical risk, which depends on three arguments: the predictor f, the input matrix $\mathbf{X}$, and labels $\mathbf{y}$. This general strategy for learning is called *empirical risk minimization* (ERM) .

Another interesting problem is to seek a predictor that performs well (i.e., has low risk) on unseen test data. More formally, one aims to find a predictor f (with fixed parameters) that minimizes the expected risk:

$$R_{\text{true}}(f) = \mathbb{E}_{\mathbf{x},y}[\ell(y, f(\mathbf{x}))],$$

where y is the label, and $f(\mathbf{x})$ is the prediction based on the example $\mathbf{x}$. The notation $R_{\text{true}}(f)$ represents the true risk, which assumes access to an infinite amount of data. The expectation is taken over the (infinite) set of all possible data and labels.

Two practical questions arise from minimizing expected risk:

(1) How should we modify our training procedure to generalize well?
(2) How can we estimate the expected risk from finite data?

Does exact interpolation work? In statistical learning, parametric predictors $f(\mathbf{x}; \boldsymbol{\theta})$ are generally preferred over strict interpolation because they are more suitable to modeling the underlying patterns in the data rather than just fitting the observations perfectly. Interpolation methods, by design, are required to pass through all training samples, ensuring zero training error. However, this often results in overly complex, uninformative functions that are highly sensitive to small perturbations in the data—an

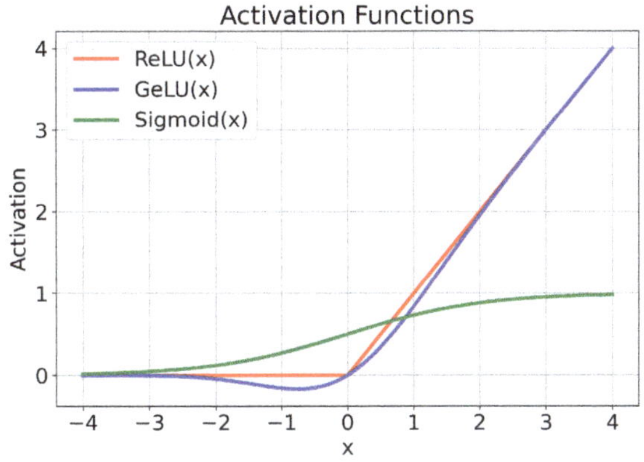

Fig. 1.2: Common nonlinear activation functions: $\mathrm{ReLU}(x) = \max(0, x)$; $\mathrm{GeLU}(x) = \frac{x}{2}\left(1 + \mathrm{erf}(\frac{x}{\sqrt{2}})\right)$; and $\mathrm{Sigmoid}(x) = \frac{1}{1+\exp(-x)}$.

issue that becomes particularly problematic in the presence of noise or outliers. As shown in Figure 1.1, exact interpolation can cause a dramatic blow-up in predictions outside the region covered by the training samples. In contrast, parametric models aim to approximate the true functional relationship between inputs $\mathbf{x}$ and outputs y by selecting f from a parameterized function class and optimizing the parameters $\boldsymbol{\theta}$ to minimize a loss function. This emphasis on generalization, rather than strict adherence to the training data, allows for a more accurate recovery of the underlying data-generating process (notice that the linear fit performs relatively well on unseen data in Figure 1.1). Consequently, in machine learning tasks involving real-world (and potentially noisy) data, exact interpolation methods fail to provide one of the most important qualities of a prediction function: generalization.

How to choose the parametric model? Although we have seen how a simple linear predictor can easily outperform direct interpolation, such predictor is limited to linear data relations. Another important example of widely used class of parametric predictors that balances expressiveness and generalization is *neural networks*, especially those using ReLU (Rectified Linear Unit) activation functions or its smooth variants (see Figure 1.2). The simplest example of this is a 1-layer ReLU network, defined as:

$$f(\mathbf{x}) = \boldsymbol{w}^\top \phi(\mathbf{A}\mathbf{x} + \mathbf{b}) \in \mathbb{R},$$

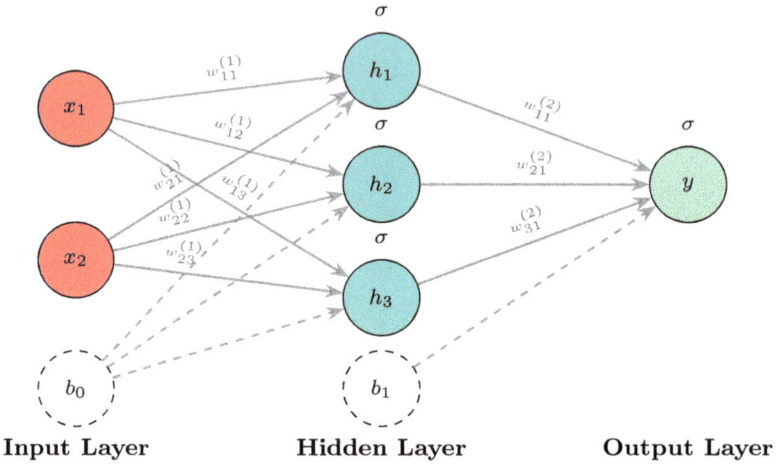

Fig. 1.3: Neural network showing weights, biases, and activation functions

where $\mathbf{A} \in \mathbb{R}^{D' \times D}$ and $\boldsymbol{w} \in \mathbb{R}^{D'}$ are learnable parameters, $\mathbf{b} \in \mathbb{R}^{D'}$ is a bias vector, and $\phi(\cdot)$ denotes elementwise application of the ReLU function, $\phi(z) = \max(0, z)$.

This architecture introduces nonlinearity through elementwise nonlinear functions (e.g. ReLU) while remaining relatively simple to analyze and train. The function f is still a composition of linear operations and pointwise nonlinearities, yet it can approximate complex functions given enough hidden units D'.

Remark. (Deep Neural Networks) By adding another layer–a second affine transform followed by another ReLU–you get a 2-layer ReLU network:

$$f(\mathbf{x}) = \boldsymbol{v}^\top \phi(\mathbf{A}_2 \phi(\mathbf{A}_1 \mathbf{x} + \mathbf{b}_1) + \mathbf{b}_2) \in \mathbb{R},$$

with new set of parameters $\mathbf{A}_1, \mathbf{A}_2, \mathbf{b}_1, \mathbf{b}_2$, and output weights $\boldsymbol{v}$. With two or more layers, the network gains the capacity to model highly nonlinear relationships that cannot be captured with simple linear or even 1-layer models. Extending this idea leads to the notion of *deep neural networks*: compositions of many layers of linear transformations followed by nonlinear activations. These models are capable of learning rich hierarchical representations of data, especially when trained with large datasets. **Deep learning**, a subfield of machine learning, refers to training such deep models, and has proven especially powerful in tasks like image recognition, natural language processing, and speech understanding. From a functional perspective, deep learning can be viewed as learning highly expressive function approximators through gradient-based optimization over large parameter spaces.

1.1.2 *Predictor as a probabilistic model*

While it might seem tempting to think of ERM as "probability-free", this is not entirely accurate. ERM operates under the assumption of an unknown probability distribution $p(\mathbf{x}, y)$, which governs data generation. However, unlike traditional statistical methods, ERM does not rely on a specific choice of this distribution. We make the following points to make this distinction clear:

- *Standard statistical methods*: Require explicit knowledge of $p(x, y)$.
- *ERM*: Assumes the existence of a data-generating distribution but does not require specifying it; it allows labels y to be non-deterministic without assuming a particular noise model.

In contrast to ERM, parameter estimation explicitly models uncertainty using probability distributions. This allows us to address uncertainties both in the observation process and in the predictor parameters. In this setting, the likelihood function plays a role analogous to the loss function in ERM.

Maximum likelihood estimation focuses on selecting parameters that best fit the data by maximizing the likelihood function. Specifically: for a random variable x and a family of probability densities $p(x|\theta)$, parameterized by θ, the negative log-likelihood is the key optimization target given by

$$L_x(\theta) = -\log p(x \mid \theta).$$

Here, the notation $L_x(\theta)$ highlights that θ is a variable parameter, while the data x is fixed. Often, for simplicity, we omit the explicit reference to x and write the function as $L(\theta)$, assuming that the context makes the dependence on x clear.

The probability density $p(x \mid \theta)$ represents a model for the uncertainty in the data x given a fixed parameter setting θ. For a specific dataset, the likelihood function provides a way to evaluate different parameter settings θ, allowing us to identify the parameters that are most likely to have generated the data.

Alternatively, if we consider x as fixed (since it has been observed), and vary θ, the negative log-likelihood $L(\theta)$ quantifies how plausible a particular parameter setting θ is, given the observations x. Based on this perspective, the *maximum likelihood estimator* (MLE) identifies the parameter θ

that maximizes this plausibility. MLE's framework connects probabilistic foundations with practical model fitting, offering a structured approach to parameter inference.

As mentioned above, in the supervised learning setting, we are given pairs of data points $(x_1, y_1), \ldots, (x_N, y_N)$, where $x_n \in \mathbb{R}^D$ represents the input features, and $y_n \in \mathbb{R}$ represents the corresponding labels. The goal is to construct a predictor that maps x_n to y_n (or a value close to it).

This can be formulated probabilistically by specifying the conditional probability distribution of the labels y_n given the examples x_n, for a particular parameter setting θ:

$$p(y_n \mid x_n, \theta).$$

This formulation provides a probabilistic framework for supervised learning, where the goal is to optimize θ based on the observed data.

Assume that the set of examples $(x_1, y_1), \ldots, (x_N, y_N)$ are independent and identically distributed (i.i.d.). The term "independent" implies that the likelihood of the entire dataset, $Y = \{y_1, \ldots, y_N\}$ and $X = \{x_1, \ldots, x_N\}$, can be expressed as the product of the likelihoods of individual examples:

$$p(Y \mid X, \theta) = \prod_{n=1}^{N} p(y_n \mid x_n, \theta),$$

where $p(y_n \mid x_n, \theta)$ represents a specific distribution (e.g., Gaussian). The phrase "identically distributed" signifies that all terms in the product share the same form of distribution and parameters.

In machine learning, one frequently considers the negative log-likelihood, given by:

$$L(\theta) = -\log p(Y \mid X, \theta) = -\sum_{n=1}^{N} \log p(y_n \mid x_n, \theta). \tag{1.1}$$

To determine the parameter vector θ that best explains the data $(x_1, y_1), \ldots, (x_N, y_N)$, **the goal is** to minimize the negative logarithmic probability $L(\theta)$ with respect to θ.

1.1.3 *Statistical physics and machine learning*

In recent years, the development of deep neural networks (DNNs), which take advantage of high-performance graphical processing units (GPUs), has propelled machine learning tasks, particularly image classification, to outperform human-level performance in some areas. Moreover, unsupervised learning has also seen impressive progress with the advent of generative models such as generative adversarial networks (GANs) [139], [161]. GANs have demonstrated the ability to generate highly realistic images that are often indistinguishable from the real ones [208].

Despite these remarkable successes, theoretical understanding in deep learning remains a challenging and evolving field. Although substantial progress has been made in understanding certain aspects of machine learning [263], [32], [197], [145], the connection between the choice of neural network architecture and the complexity of the data it processes is not yet fully understood. One of the core questions in statistical learning theory [372], [373] is the consistency of machine learning models. In this framework, a model is considered consistent if, as the size of the training data increases, the risk (or error) associated with the model's predictions approaches the minimal possible value. This can be expressed mathematically as:

$$R(\theta) = \int Q(z, \theta) dF(z)$$

where $Q(z, \theta)$ represents the cost function for a given model, and $R(\theta)$ is the risk associated with the model's parameters θ. The true risk is approximated by the empirical risk, computed from the training data:

$$R_{\text{emp}}(\theta) = \frac{1}{n} \sum_{i=1}^{n} Q(z_i, \theta)$$

The principle of empirical risk minimization (ERM) is based on minimizing the empirical risk to approximate the true risk. Consistency of a model means that as the training dataset grows, the empirical risk converges to the true risk. However, this ideal is often disrupted by overfitting, where the model becomes excessively tailored to the training data, failing to generalize to new data. Intuitively, consistency is achieved if the model's complexity (its resolution) is appropriately matched to the size of the training dataset.

The theoretical concept of the Vapnik-Chervonenkis (VC) entropy offers a measure of the complexity of a set of functions $\mathcal{H}$ and their capacity to fit a dataset. It is defined as:

$$H(\mathcal{H}) = \mathbb{E}_z \left[\log \left(\sum_{i=1}^{n} h(z_i) \right) \right]$$

This quantity characterizes the number of distinct realizations that the functional space $\mathcal{H}$ can achieve on a set of points $(z_1, \dots, z_n)$. Consistency is guaranteed asymptotically if the VC entropy vanishes as the number of data points n increases.

If the model's resolution is too high, the VC entropy grows with the number of data points, potentially leading to overfitting. In practice, a balance between the resolution and the data size must be maintained to ensure consistency. The growth function, $G(\mathcal{H})$, provides a practical way to analyze this balance. The VC dimension h represents the maximal number of data points that can be shattered (separated) by a set of functions, providing insight into the model's capacity. Typically, a model with a VC dimension around 20 is considered to avoid overfitting.

From a statistical physics perspective [250], [269], machine learning can be reformulated to better understand its underlying mechanisms. One of the most common approaches is to view the problem in a Bayesian framework, considering the empirical risk as an energy function within a Gibbs distribution. The posterior distribution $P_{\text{posterior}}(\theta)$ of the model parameters can be written as:

$$P_{\text{posterior}}(\theta) = P_{\text{prior}}(\theta) \exp\left(-\frac{E(\theta)}{\beta}\right) \exp\left(-\frac{R_{\text{emp}}(\theta)}{\beta}\right)$$

where β is an inverse temperature parameter related to model noise. This framework allows for a stochastic interpretation of the model's training process as governed by a Langevin equation [336]:

$$\frac{d\theta}{dt} = -\nabla E(\theta) + \sqrt{2\beta}\,\eta(t)$$

Here, $\eta(t)$ represents a white noise term. The training data can be seen as a "quenched disorder," akin to a physical system in equilibrium, which can be averaged over to obtain useful insights into the model's behavior.

Historically, statistical physics has played a critical role in the development of neural networks. In the 1980s, the Hopfield model of associative memory [193] spurred significant theoretical work, drawing from spin-glass theory [272]. The steady-state behavior of the Hopfield network is described by a Gibbs measure, with an associated Hamiltonian on spin configurations of the system.

The overlaps between spin configurations provide order parameters that describe the retrieval phase, helping to understand the model's memory capabilities. The mean-field theory of the Hopfield model, solved using replica techniques [11], provided foundational insights into the capacity of neural networks to store patterns, an issue closely related to the computation of the VC entropy.

In recent developments, mean-field methods originally applied in complex systems—such as the TAP equations [366] and cavity method [272] —have found applications in machine learning. In these contexts, a Bayesian framework with a teacher-student setup [143] is often employed. A deeper understanding of Bayes-optimality, especially in the context of error-correcting codes, has provided connections to statistical physics, with the Nishimori line [291] serving as a useful concept for analyzing the behavior of machine learning models. This interdisciplinary approach continues to unlock new insights into the fundamental workings of machine learning systems. Below we will mainly consider applications of the theory of Gibbs measures to machine learning.

1.2 Introduction to Gibbs measures

1.2.1 *Measurable space and measure*

When a mathematician is asked by a customer to study a system or object, such as a dataset in machine learning, they typically follow these steps:

- *Represent the system*: Represent the system as a collection of elements within a set X.
- *Determine relevant properties*: Identify the properties of X that are of interest to the customer.
- *Define Interactions*: Define the interactions between elements of X that are crucial for studying the required properties.
- *Measure the system*[1]: Assign numbers to the properties of the system that are of interest to the customer.
- *Interpret the measurements*: Provide interpretations of the measurements in a way that is understandable to the customer.

Recall that a *system* is a set of interacting or interdependent components, members, or points.

Remark 1. *We note that a measure is a cornerstone of most natural sciences, technology, economics, and quantitative research in other social sciences. In this book we mainly present application of the (Gibbs) measure theory in machine learning.*

[1] In general, a mathematician may apply various mathematical techniques, such as statistical analysis, optimization to study the system or object.

Assume we have determined which properties of the set X are interested to the customer. Then we consider such subsets of X which are relevant to these properties. Let

$$\tau_0 = \{A \subset X : A - \text{relevant to an interested property}\}$$

be the set of subsets of X. We want to measure each elements of τ_0. But to make our analysis simpler, we need to assume that $\mu(A) \geq 0$ and if $\mu(A)$, $\mu(B)$ is known then we can define $\mu(A \cup B)$ too. This aim is mathematically formulated as follows.

Let X be an arbitrary set.

Definition 1. *A non-empty set τ of subsets of X is called a ring if*

$$\forall A, B \in \tau \Longrightarrow A \cup B, A \cap B, A \setminus B \in \tau.$$

Recall $A \triangle B = (A \setminus B) \cup (B \setminus A)$. Therefore, $A \triangle B \in \tau$.

For $A \subset X$ denote $\overline{A} = X \setminus A$. Since $A \cup B = \overline{\overline{A} \cap \overline{B}}$, $A \cap B = \overline{\overline{A} \cup \overline{B}}$, in Definition 1 one can omit condition $A \cup B \in \tau$ or $A \cap B \in \tau$.

Definition 2. *If τ is a ring with $X \in \tau$ then τ is called an algebra.*

From this definition it follows that if τ is an algebra then $X \setminus X = \emptyset \in \tau$.

We observe that a ring or an algebra can be defined on any set regardless of the nature of its elements. For instance, it could be a set containing various types of data (as in Machine learning).

Example 1. *Each of the following sets is ring and algebra:*

$$(1)\, \tau = \{\emptyset,\ X\}, \quad (2)\, \tau = \{A,\ \overline{A} = X \setminus A,\ \emptyset,\ X\}, \quad (3)\, \tau = \{A : A \in X\}.$$

(4) *The set τ_{fin} of all finite subsets of the set $\mathbb{R}$ of real numbers is a ring. But τ_{fin} is not an algebra, because $\mathbb{R} \notin \tau_{\text{fin}}$*

(5) *Consider the set*

$$X = \{apple, book, car, diamond\}.$$

On this set we define the following algebra

$$\tau_1 = \{\emptyset, \{book, diamond\}, \{apple, car\}, X\}.$$

But the following is not a ring

$$\tau_2 = \{\emptyset, \{apple, book\}, \{diamond\}, X\}.$$

The following is straightforward

Theorem 1. *Let I be an arbitrary non-empty set and τ_α, $\alpha \in I$ are rings (algebras), then their intersection $\tau = \bigcap_{\alpha \in I} \tau_\alpha$ is a ring (algebra).*

In this theorem, the intersection of rings (or algebras) is considered as the intersection of sets.

Theorem 2. *Let X be a set and τ_0 a system composed of some subsets of X. There exists a system τ such that:*

(1) τ is a ring;
(2) $\tau_0 \in \tau$;
(3) Among the rings satisfying (2), τ is the smallest (minimal).

Proof. Using τ_0, we construct the following set: $Y = \bigcup_{A \in \tau_0} A$. The system $\mathcal{P}(Y)$ composed of all subsets of this set Y will be a ring. Hence, $\tau = \mathcal{P}(Y)$ satisfies the first two conditions of the theorem. Suppose there are several rings that satisfy the first and second conditions of the theorem. We take their intersection as τ. □

Definition 3. *The τ in the above theorem is called the minimal ring generated by the system τ_0, and it is denoted as $\tau = \tau(\tau_0)$.*

Definition 4. *Let τ be an algebra formed from the subsets of the set X. A function $\mu : \tau \to [0, +\infty)$ that satisfies*

$$\mu(A \cup B) = \mu(A) + \mu(B), \quad \forall A, B \in \tau, \; A \cap B = \emptyset$$

is called a finitely-additive measure. If $\mu(X) < \infty$, the finitely-additive measure is called finite. If $\mu(X) = 1$, μ is called a probability measure.

Definition 5. *For a system τ formed from subsets of the set X, if*

1. $X \in \tau$;
2. $A \in \tau \Rightarrow \overline{A} = X \setminus A \in \tau$;
3. $A_1, A_2, \ldots \in \tau \Rightarrow \bigcup_{n=1}^{\infty} A_n \in \tau$

then the system τ is called a σ-algebra.

Definition 6. *If τ is a σ-algebra, the pair (X, τ) is called a measurable space. Every set in the system τ is called a measurable set.*

Definition 7. *If μ given on a measurable space (X, τ), satisfies*

$$\mu\left(\bigcup_{n=1}^{\infty} A_n\right) = \sum_{n=1}^{\infty} \mu(A_n), \quad \forall A_1, A_2, \ldots, A_n, \ldots \in \tau, \; A_i \cap A_j = \emptyset, \; i \neq j,$$

then it is called a σ-additive measure.

Definition 8. *A finitely-additive measure μ is called σ-finite if $X = \bigcup_{n=1}^{\infty} X_n$ and for all n, $\mu(X_n) < \infty$. If $\mu(X) = 1$, a σ-additive measure is called a probability measure.*

Denote by $\mathcal{M}_1(X)$ the set of all probability measures defined on the set X.

Definition 9. *The triplet (X, τ, μ) is called a measure space, where*
(a) X is an arbitrary set,
(b) τ is a σ-algebra (formed from subsets of X),
(c) μ is a σ-additive measure.

Measure space $(\mathbb{R}, \mathcal{B}(\mathbb{R}))$: Consider the following system of subsets of $\mathbb{R} = (-\infty, \infty)$:

$$\tau_0 = \{A \subset \mathbb{R} : A = \bigcup_{i=1}^{n} (a_i, b_i],\ n < \infty\}.$$

Clearly, this system (where $(a, \infty]$ is understood as $(a, +\infty)$ and $\emptyset$ is also considered in this system) is an algebra. However, τ_0 is not a σ-algebra, because if we take $A_n = (0, 1 - \frac{1}{n}] \in \tau_0$, then $\bigcup_n A_n = (0, 1) \notin \tau_0$.

The minimal σ-algebra containing τ_0 is called the Borel σ-algebra and is denoted by $\mathcal{B} = \mathcal{B}(\mathbb{R})$. Each element of $\mathcal{B}$ is called a Borel set.

This σ-algebra plays an important role in analysis and probability theory.

If we take the system $J = \{(a, b] : \forall a, b \in \mathbb{R}\}$ and consider the minimal σ-algebra containing J as $\sigma(J)$, then $\sigma(J) = \mathcal{B}$. That is, the Borel σ-algebra can also be generated from J.

Clearly,

$$(a, b) = \bigcup_{n=1}^{\infty} (a, b - \frac{1}{n}],\ [a, b] = \bigcap_{n=1}^{\infty} (a - \frac{1}{n}, b],\ a < b,\ \{a\} = \bigcap_{n=1}^{\infty} (a - \frac{1}{n}, a].$$

Thus, the Borel σ-algebra also includes the sets (a, b), $[a, b]$, $[a, b)$, $(-\infty, b)$, $(-\infty, b]$, (a, ∞). Therefore, other sets besides the sets of the form $(a, b]$ can be used to generate the Borel σ-algebra.

The $\mathcal{B}$ σ-algebra is very rich. It is difficult to find sets that do not belong to it. In the following example, we will construct a set that does not belong to $\mathcal{B}$:

Example of an unmeasurable set with respect to the Borel σ-algebra.

One of the classic examples (see e.g. [220], [363]) of a set that is not measurable with respect to the Borel σ-algebra is the Vitali set. Here is a brief explanation of how the Vitali set is constructed and why it is not Borel measurable.

Construction of the Vitali set:

(1) Equivalence relation: Consider the unit interval $[0, 1)$ on the real line. Define an equivalence relation $\sim$ on $[0, 1)$ such that two numbers x and y are equivalent if and only if their difference $x - y$ is a rational number, i.e., $x \sim y \iff x - y \in \mathbb{Q}$.
(2) Equivalence classes: This equivalence relation partitions $[0, 1)$ into disjoint equivalence classes, where each class consists of numbers that differ by a rational number.
(3) Vitali set: From each equivalence class, select exactly one representative. The collection of these representatives forms the Vitali set V.

Why the Vitali set is not Borel measurable:

(1) Non-measurability proof sketch: Assume for contradiction that the Vitali set V is Borel measurable. Then the Lebesgue measure of V, denoted by $\mu(V)$, must be well-defined.
(2) Translation invariance: For each rational number $q \in [0, 1) \cap \mathbb{Q}$, consider the translation $V + q = \{v + q \mid v \in V\}$. Each translated set $V + q$ is also a subset of $[0, 1)$.
(3) Disjoint translates: The sets $V + q$ for different $q \in [0, 1) \cap \mathbb{Q}$ are disjoint. There are countably many such rational translations because the rational numbers are countable.
(4) Covering the interval: The union of all these disjoint translations $\bigcup_{q \in [0,1) \cap \mathbb{Q}} (V + q)$ covers the interval $[0, 1)$.
(5) Measure contradiction: Since μ is countably additive, we have:

$$\mu\left(\bigcup_{q \in [0,1) \cap \mathbb{Q}} (V + q) \right) = \sum_{q \in [0,1) \cap \mathbb{Q}} \mu(V + q).$$

If $\mu(V)$ were positive, this sum would be infinite, which contradicts the fact that $\mu([0, 1)) = 1$. If $\mu(V)$ were zero, the sum would be zero, which again contradicts $\mu([0, 1)) = 1$.

Thus, no consistent measure can be assigned to the Vitali set, implying that it is not measurable with respect to the Borel σ-algebra.

Weak convergence of measures describes a way in which a sequence of probability measures converges to another probability measure:

Definition 10. *A sequence of probability measures (μ_n) on a measurable space $(X, \mathcal{B})$ is said to converge weakly to a probability measure μ if for* *every bounded continuous function* $f : X \to \mathbb{R}$*, we have:*

$$\int_X f\, d\mu_n \to \int_X f\, d\mu$$

as $n \to \infty$*.*

Weak convergence of (μ_n) to μ is often denoted as:

$$\mu_n \xrightarrow{w} \mu, \quad \text{or} \quad \mu_n \Rightarrow \mu.$$

The book [40] provides a comprehensive treatment of weak convergence of measures. In particular, the following results are well known:

1. *Portmanteau theorem*: There are several equivalent definitions of weak convergence, such as:

- For every bounded continuous function f, $\int_X f\, d\mu_n \to \int_X f\, d\mu$.
- For every open set $G \subset X$, $\liminf_{n\to\infty} \mu_n(G) \geq \mu(G)$.
- For every closed set $F \subset X$, $\limsup_{n\to\infty} \mu_n(F) \leq \mu(F)$.
- For every Borel set $A \subset X$ with $\mu(\partial A) = 0$, $\mu_n(A) \to \mu(A)$.

2. *Prokhorov's theorem*: This theorem provides a criterion for relative compactness in the space of probability measures. It states that a family of probability measures on a complete separable metric space is relatively compact (i.e., every sequence has a weakly convergent subsequence) if and only if it is tight.

3. *Tightness*: A sequence of measures (μ_n) is tight if for every $\epsilon > 0$, there exists a compact set $K \subset X$ such that $\mu_n(K) > 1 - \epsilon$ for all n.

In the probability theory a weak convergence is used in the study of convergence of random variables. If (X_n) is a sequence of random variables with corresponding distributions (μ_n), then X_n converges in distribution to a random variable X if $\mu_n \Rightarrow \mu$, where μ is the distribution of X.

In statistics, weak convergence is used to describe the convergence of empirical distributions to the true distribution.

Example 2. *Consider the sequence of measures* μ_n *on* $\mathbb{R}$ *defined by* $\mu_n = \delta_{1/n}$*, where* $\delta_{1/n}$ *is the Dirac measure centered at* $1/n$*. For any continuous bounded function* f*,*

$$\int_{\mathbb{R}} f\, d\mu_n = f(1/n).$$

As $n \to \infty$*,* $f(1/n) \to f(0)$*. Hence,* $\mu_n \xrightarrow{w} \delta_0$*, where* δ_0 *is the Dirac measure centered at 0.*

1.2.2 *Graphs*

A *graph* is a pair $G = (V, L)$, where V is a set whose elements are called vertices, and $L \subset V \times V$ is a set of two-sets of vertices, whose elements are called edges. The edge connecting $x, y \in V$ is denoted by $l = \langle x, y \rangle$.

A *path* in a graph is a subgraph in which the vertices can be listed in an order $v_1, v_2, \ldots, v_n$ such that the edges are the $\langle v_i, v_{i+1} \rangle$ where $i = 1, 2, \ldots, n-1$.

The distance $d(x, y), x, y \in V$ on the graph is the number of edges of the shortest path from x to y:

$$d(x, y) = \min\{d \mid \exists x = x_0, x_1, \ldots, x_{d-1}, x_d = y \in V$$

$$\text{such that } \langle x_0, x_1 \rangle, \ldots, \langle x_{d-1}, x_d \rangle\}.$$

A (non-empty) set $\Lambda \subset V$ is called *connected* if for any $x, y \in \Lambda$ the path connecting x and y lies in Λ. We denote $\Lambda^c = V \setminus \Lambda$ and the *boundary* of Λ by

$$\partial\Lambda := \{x \in \Lambda^c \colon \exists y \in \Lambda,\ d(x, y) = 1\},$$

and we write $\bar{\Lambda} = \Lambda \cup \partial\Lambda$. The subset of edges in Λ is denoted

$$L_\Lambda := \{\langle x, y \rangle \in L \colon x, y \in \Lambda\}.$$

An unordered pair of vertices $\{x, y\}$ is called connected if a path leads from x to y.

A *connected graph* is an undirected graph in which every unordered pair of vertices in the graph is connected. Otherwise, it is called a disconnected graph.

A *lattice graph* is a graph whose drawing, embedded in some Euclidean space $\mathbb{R}^n$, forms a regular tiling. This implies that the group of bijective transformations that send the graph to itself is a lattice in the group-theoretical sense.

The lattice $\mathbb{Z}^n$: the n-dimensional integer lattice (or cubic lattice), which is the lattice in the Euclidean space $\mathbb{R}^n$ whose lattice points are n-tuples of integers. The set $\mathbb{Z}^n$ is an additive group with respect to adding the corresponding components of the vectors.

The Cayley tree Γ^k: The Cayley tree Γ^k of order $k \geq 1$ is an infinite tree, i.e., a graph without cycles, such that exactly $k + 1$ edges originate from each vertex. Let $\Gamma^k = (V, L, i)$, where V is the set of vertices Γ^k, L the set of edges and i is the incidence function setting each edge $l \in L$ into correspondence with its endpoints $x, y \in V$. If $i(l) = \{x, y\}$, then the vertices x and y are called the *nearest neighbors*, denoted by $l = \langle x, y \rangle$.

Fix a vertex $x^0 \in V$, interpreted as the *root* of the tree. We say that $y \in V$ is a *direct successor* of $x \in V$ if x is the penultimate vertex on the unique path leading from the root x^0 to the vertex y; that is, $d(x^0, y) = d(x^0, x) + 1$ and $d(x, y) = 1$. The set of all direct successors of $x \in V$ is denoted by $S(x)$.

For a fixed $x^0 \in V$ we set $W_n = \{x \in V \mid d(x, x^0) = n\}$,

$$V_n = \{x \in V \mid d(x, x^0) \le n\}, \quad L_n = \{l = \langle x, y\rangle \in L \mid x, y \in V_n\}. \tag{1.2}$$

For $x \in W_n$ the set $S(x)$ then has the form

$$S(x) = \{y \in W_{n+1} : \langle x, y\rangle\}. \tag{1.3}$$

For any $x \in V$ denote

$$W_m(x) = \{y \in V : d(x, y) = m\}, \quad m \ge 1.$$

Note that the sequence of balls (V_n) $(n \ge 0)$ is *cofinal* (see [149, Section 1.2, page 17]), that is, any finite subset $\Lambda \subset V$ is contained in some V_n.

1.2.3 *Configuration space*

Take a finite set $\Phi = \{1, 2, ..., q\}$, $q \ge 2$. This is the set of spin values. For $A \subseteq V$ a spin *configuration* σ_A on A is defined as a function

$$x \in A \to \sigma_A(x) \in \Phi.$$

The set of all configurations coincides with $\Omega_A = \Phi^A$. The cardinality of Ω_A is $|\Omega_A| = |\Phi|^{|A|} = q^{|A|}$.

We denote $\Omega = \Omega_V$ and $\sigma = \sigma_V$. Since V is a countable set, Ω is an uncountable set.

In the set $\Omega = \Phi^V$ a notion of convergence for sequences of configurations (i.e. topology) is defined as follows [136, page 262]:

Definition 11. *A sequence $\sigma^{(n)} \in \Omega$ converges to $\sigma \in \Omega$ (written as $\sigma^{(n)} \to \sigma$) if*

$$\lim_{n\to\infty} \sigma^{(n)}(x) = \sigma(x), \quad x \in V.$$

Recall sequentially compactness: for every sequence $\{\sigma^{(n)}\}_{n\ge1} \subset \Omega$, there exists $\sigma^* \in \Omega$ and a subsequence $\{n_m\}_{m\ge1}$ such that $\sigma^{(n_m)} \to \sigma^*$ when $m \to \infty$.

Since Φ is finite the following fact is known

Proposition 1. *[136, page 262] With the above notion of convergence, Ω is compact.*

Definition 12. *A function $f : \Omega \to \mathbb{R}$ is called continuous if $\sigma^{(n)} \to \sigma$ implies $f(\sigma^{(n)}) \to f(\sigma)$. The set of all continuous functions on Ω is denoted by $C(\Omega)$.*

1.2.4 *Markov chains*

Following [58], [339], [344] we briefly give basic notations and necessary properties of Markov chains.

A stochastic process is a collection of random variables $X_t(\omega)$ on a probability space Ω, indexed by a time parameter $t \in I$ from some index set I, and taking values in a common state space $\mathcal{S}$.

If the time index set $I = \mathbb{N}_0 = \{0, 1, 2, \dots\}$, then such a process $(X_n)_{n\geq 0}$ is called a discrete time stochastic process. The random variable X_n gives the state (position) at time n. The state space $\mathcal{S}$ will always be a discrete set, either finite or countably infinite.

The fundamental assumption on the stochastic processes $(X_n)_{n\geq 0}$ is the so-called Markov property, which defined in the following way: At each time n, the future positions X_t (for $t > n$) of the process only depend on the position at time n and not on the positions of X_s for $s < n$.

A stochastic process $(X_n)_{n\geq 0}$ with discrete state space $\mathcal{S}$ that has the Markov property is called a Markov chain.

Consider a discrete (at most countable) state space $\mathcal{S}$. Fix a natural number n and consider the direct product space

$$\Omega = \mathcal{S} \times \mathcal{S} \times \cdots \times \mathcal{S} \quad \text{with } (n+1) \text{ factors } \mathcal{S}$$

which is denoted by $\Omega = \mathcal{S}^{n+1}$, i.e.,

$$\Omega = \{(x_0, x_1, \dots, x_n) \mid x_i \in \mathcal{S} \text{ for } 0 \leq i \leq n\}.$$

Let $I = \{0, 1, \dots, n\}$ be the time set and let $p : \mathcal{S} \times \mathcal{S} \to [0, 1]$ be a function with the property

$$\sum_{b \in \mathcal{S}} p(a, b) = 1 \quad \text{for all } a \in \mathcal{S}. \tag{1.4}$$

Let $\mathcal{F} = \mathcal{P}(\Omega)$ be the set of all subsets of Ω. The set $\mathcal{F}$ consists of all sets of the form

$$F = F_0 \times F_1 \times \cdots \times F_n \quad \text{with } F_0, F_1, \dots, F_n \subseteq \mathcal{S}.$$

$(\Omega, \mathcal{F})$ is a σ-algebra.

Let π_0 be a probability distribution on $\mathcal{S}$, which we call the initial distribution.

We construct a probability $\mathbb{P}$ on $(\Omega, \mathcal{F})$ with the use of p in the following way: For all $\omega = (x_0, x_1, \dots, x_n) \in \Omega$, define

$$\mathbb{P}(\omega) = \pi_0(x_0)\, p(x_0, x_1)\, p(x_1, x_2) \cdots p(x_{n-1}, x_n), \tag{1.5}$$

and from this for all $F \in \mathcal{F}$, since Ω is discrete,

$$\mathbb{P}(F) = \sum_{\omega \in F} \mathbb{P}(\omega).$$

By (1.4) and (1.5) it follows that

$$\sum_{\omega \in \Omega} \mathbb{P}(\omega) = 1.$$

Thus $(\Omega, \mathcal{F}, \mathbb{P})$ defines a probability space.

Now, consider the coordinate random variables $X_0, X_1, \ldots, X_n$ on $(\Omega, \mathcal{F}, \mathbb{P})$ defined by

$$X_i(\omega) = x_i \qquad \text{for } \omega = (x_0, x_1, \ldots, x_n) \in \Omega, 0 \leq i \leq n.$$

As a consequence of this construction, conditional probabilities involving the coordinate random variables X_i have the following important property, called the Markov property:

Proposition 2. *(Markov property). Let $(\Omega, \mathcal{F}, \mathbb{P})$ be the probability space as defined above and $(X_k)_{0 \leq k \leq n}$ the random vector whose components are the coordinate random variables as defined above. Then for all $0 \leq i \leq n-1$ and for all $x_0, x_1, \ldots, x_{i+1} \in \mathcal{S}$*

$$\begin{aligned} \mathbb{P}\left(X_{i+1} = x_{i+1} \mid X_0 = x_0, \ldots, X_i = x_i\right) \\ = \mathbb{P}\left(X_{i+1} = x_{i+1} \mid X_i = x_i\right) = p\left(x_i, x_{i+1}\right) \end{aligned} \tag{1.6}$$

Proof. Using (1.5) we get

$$\begin{aligned} &\mathbb{P}\left(X_{i+1} = x_{i+1} \mid X_0 = x_0, \ldots, X_i = x_i\right) \\ &= \frac{\mathbb{P}\left(X_0 = x_0, X_1 = x_1, \ldots, X_{i+1} = x_{i+1}\right)}{\mathbb{P}\left(X_0 = x_0, X_1 = x_1, \ldots, X_i = x_i\right)} \\ &= \frac{\pi_0\left(x_0\right) p\left(x_0, x_1\right) p\left(x_1, x_2\right) \cdots p\left(x_i, x_{i+1}\right)}{\pi_0\left(x_0\right) p\left(x_0, x_1\right) p\left(x_1, x_2\right) \cdots p\left(x_{i-1}, x_i\right)} \\ &= p\left(x_i, x_{i+1}\right). \end{aligned}$$

Moreover,

$$\begin{aligned} &\mathbb{P}\left(X_{i+1} = x_{i+1} \mid X_i = x_i\right) \\ &= \frac{\mathbb{P}\left(X_i = x_i, X_{i+1} = x_{i+1}\right)}{\mathbb{P}\left(X_i = x_i\right)} \\ &= \frac{\sum_{x_0, \ldots, x_{i-1} \in \mathcal{S}} \pi_0\left(x_0\right) p\left(x_0, x_1\right) \cdots p\left(x_{i-1}, x_i\right) p\left(x_i, x_{i+1}\right)}{\sum_{x_0, \ldots, x_{i-1} \in \mathcal{S}} \pi_0\left(x_0\right) p\left(x_0, x_1\right) \cdots p\left(x_{i-1}, x_i\right)} \\ &= \frac{\mathbb{P}\left(X_i = x_i\right) p\left(x_i, x_{i+1}\right)}{\mathbb{P}\left(X_i = x_i\right)} = p\left(x_i, x_{i+1}\right). \end{aligned}$$

This completes the proof of (1.6). □

Conversely, let us now assume that we are given a probability space $(\Omega, \mathcal{F}, \mathbb{P})$, a discrete set $\mathcal{S}$, and a sequence of random variables $X_0, X_1, \ldots, X_n$ with $X_i : \Omega \to \mathcal{S}$ for $0 \leq i \leq n$. Let X_0 is chosen by π_0 and assume that there exists a function $p : \mathcal{S} \times \mathcal{S} \to [0, 1]$ such that for all $0 \leq i \leq n-1$ the Markov property (1.6) holds. Then by sequential conditioning one can show that for all $\omega = (x_0, x_1, \ldots, x_n) \in \mathcal{S}^{n+1}$ we have

$$\mathbb{P}(X_0 = x_0, \ldots, X_n = x_n) = \pi_0(x_0)\, p(x_0, x_1) \cdots p(x_{n-1}, x_n) \tag{1.7}$$

which describes the joint distribution of the random vector $(X_k)_{0 \leq k \leq n}$.

Recall the conditional probabilities

$$p(x, y) = \mathbb{P}(X_1 = y \mid X_0 = x) \quad \text{for} \quad x, y \in \mathcal{S}. \tag{1.8}$$

The collection $(X_k)_{0 \leq k \leq n}$ constructed above is called a homogeneous Markov chain (HMC) of length n with state space $\mathcal{S}$, one-step transition probabilities $p(x, y)$, $x, y \in \mathcal{S}$, and initial distribution π_0.

For $k \geq 1$ and $i + k \leq m \leq n$ the Markov property (1.6) implies that

$$\begin{aligned}\mathbb{P}(X_{i+k} &= x_{i+k}, \ldots, X_m = x_m \mid X_0 = x_0, X_1 = x_1, \ldots, X_i = x_i) \\ &= \mathbb{P}(X_{i+k} = x_{i+k}, \ldots, X_m = x_m \mid X_i = x_i).\end{aligned}$$

Usually, one says that a Markov chain is given, if the following are fixed:
- a state space $\mathcal{S}$,
- an initial distribution π_0 on $\mathcal{S}$,
-a set of one-step transition probabilities $p(x, y)$, $x, y \in \mathcal{S}$ (see (1.8)), that model the probabilistic properties of the process in consideration.

Then this information allows us to compute all finite-dimensional joint distributions of $(X_k)_{0 \leq k \leq n}$, for the process (see (1.7). However, a HMC $(X_n)_{n \geq 0}$ for the process (1.5) is defined for all $n \in \mathbb{N}$.

The matrix $\mathbf{P} = (p(x, y))_{x, y \in \mathcal{S}}$, given by (1.8), is the *transition matrix* of the HMC. Since its entries are probabilities, and since a transition from any state x must be to *some* state, it follows that

$$p(x, y) \geq 0, \quad \sum_{y \in \mathcal{S}} p(x, y) = 1$$

for all states x, y. A matrix $\mathbf{P}$ indexed by $\mathcal{S}$ and satisfying the above properties is called a *stochastic matrix.*

Many HMCs receive a natural description in terms of a recurrence equation driven by white noise [58], which comes from signal theory and refers to the driving sequence $\{Z_n\}_{n \geq 1}$ given in the following theorem.

Theorem 3. HMCs Driven by White Noise. *Let $\{Z_n\}_{n\geq 1}$ be an i.i.d sequence of random variables with values in an arbitrary space F. Let E be a countable space, and $f : E \times F \to E$ be some function. Let X_0 be a random variable with values in E, independent of $\{Z_n\}_{n\geq 1}$. The recurrence equation*

$$X_{n+1} = f(X_n, Z_{n+1}) \tag{1.9}$$

then defines an HMC.

Proof. By iteration of recurrence (1.9) one can show that for all $n \geq 1$, there is a function g_n such that

$$X_n = g_n(X_0, Z_1, \ldots, Z_n),$$

consequently,

$$\mathbb{P}(X_{n+1} = j \mid X_n = i, X_{n-1} = i_{n-1}, \ldots, X_0 = i_0) =$$

$$\mathbb{P}(f(i, Z_{n+1}) = j \mid X_n = i, X_{n-1} = i_{n-1}, \ldots, X_0 = i_0) = \mathbb{P}(f(i, Z_{n+1}) = j),$$

since the event $\{X_0 = i_0, \ldots, X_{n-1} = i_{n-1}, X_n = i\}$ is expressible in terms of $X_0, Z_1, \ldots, Z_n$ and is therefore independent of Z_{n+1}. Similarly,

$$\mathbb{P}(X_{n+1} = j \mid X_n = i) = \mathbb{P}(f(i, Z_{n+1}) = j).$$

Thus we get HMC, since the right-hand side of the last equality does not depend on n. Namely

$$p(i, j) = \mathbb{P}(f(i, Z_1) = j).$$

□

Definition 13. *A probability distribution π satisfying*

$$\pi^T = \pi^T \mathbf{P} \tag{1.10}$$

is called a stationary distribution of the transition matrix $\mathbf{P}$, or of the corresponding HMC.

The balance (or fixed point) equation (1.10) means that for all states i, the following equality holds

$$\pi(i) = \sum_{j \in \mathcal{S}} \pi(j) p(j, i).$$

From (1.10) by iterating we obtain $\pi^T = \pi^T \mathbf{P}^n$ for all $n \geq 0$. If the initial distribution $\nu = \pi$, then $\nu_n = \nu^T \mathbf{P}^n = \pi$ for all $n \geq 0$. Thus, if a chain is started with a stationary distribution, it keeps the same distribution forever. Moreover,

$$\mathbb{P}\left(X_n = i_0, X_{n+1} = i_1, \ldots, X_{n+k} = i_k\right) = \pi\left(i_0\right) p_{i_0 i_1} \cdots p_{i_k, i_k}$$

does not depend on n. In this sense the chain is stationary. In this case the chain is in a stationary regime or in equilibrium state.

The balance equation $\pi^T \mathbf{P} = \pi^T$, together with the requirement that π be a probability vector, i.e., $\pi^T \mathbf{1} = 1$ (where $\mathbf{1}$ is a column vector with all its entries equal to 1), constitute when $\mathcal{S}$ is finite, $|\mathcal{S}| + 1$ equations for $|\mathcal{S}|$ unknown variables. One of the equations is superfluous given the constraint $\pi^T \mathbf{1} = 1$.

Reversed HMC.

The time-reversal and time-reversibility are very useful in the theory of Markov chains, and especially in Monte Carlo simulation.

Let $\{X_n\}_{n \geq 0}$ be an HMC with transition matrix $\mathbf{P}$ and admitting a stationary distribution π such that $\pi(i) > 0$ for all states i. Define the matrix $\mathbf{Q}$, indexed by $\mathcal{S}$, as

$$\pi(i) q(i, j) = \pi(j) p(j, i) \tag{1.11}$$

The matrix $\mathbf{Q}$ is stochastic. Indeed,

$$\sum_{j \in \mathcal{S}} q(i, j) = \sum_{j \in \mathcal{S}} \frac{\pi(j)}{\pi(i)} p(j, i) = \frac{1}{\pi(i)} \sum_{j \in \mathcal{S}} \pi(j) p(j, i) = \frac{\pi(i)}{\pi(i)} = 1,$$

where the third equality uses the balance equations.

Interpretation: Let the initial distribution of $\{X_n\}$ be π, in which case for each $n \geq 0$ and all $i \in \mathcal{S}$, we have

$$\mathbb{P}\left(X_n = i\right) = \pi(i). \tag{1.12}$$

Then, from Bayes's formula we get

$$\mathbb{P}\left(X_n = j \mid X_{n+1} = i\right) = \frac{\mathbb{P}\left(X_{n+1} = i \mid X_n = j\right) \mathbb{P}\left(X_n = j\right)}{\mathbb{P}\left(X_{n+1} = i\right)}$$

that by (1.11) and (1.12) gives

$$\mathbb{P}\left(X_n = j \mid X_{n+1} = i\right) = q(j, i).$$

We see that $\mathbf{Q}$ is the transition matrix of the initial chain when time is reversed.

Theorem 4. (Reversal test.) *Let* $\mathbf{P}$ *be a stochastic matrix indexed by a countable set* $\mathcal{S}$*, and let* π *be a probability distribution on* $\mathcal{S}$*. Let* $\mathbf{Q}$ *be a stochastic matrix such that for all* $i, j \in \mathcal{S}$ *the equality (1.11) is satisfied.*

Then π *is a stationary distribution of* $\mathbf{P}$.

Proof. For fixed $i \in \mathcal{S}$, sum equalities (1.11) with respect to $j \in \mathcal{S}$ to obtain

$$\sum_{j\in\mathcal{S}} \pi(i)q(i,j) = \sum_{j\in\mathcal{S}} \pi(j)p(j,i).$$

The left-hand side is equal to $\pi(i)\sum_{j\in\mathcal{S}} q(i,j) = \pi(i)$, and therefore, for all $i \in \mathcal{S}$,

$$\pi(i) = \sum_{j\in\mathcal{S}} \pi(j)p(j,i).$$

□

Definition 14. (Reversible Chain.) *One calls reversible a stationary Markov chain with initial distribution π (a stationary distribution) assumed positive if for all $i, j \in \mathcal{S}$,*

$$\pi(i)p(i,j) = \pi(j)p(j,i). \tag{1.13}$$

In this scenario, $q(i,j) = p(i,j)$, and thus the chain and its time-reversed counterpart are statistically identical. This is due to the fact that the distribution of a homogeneous Markov chain is entirely determined by its initial distribution and transition matrix. Equations (1.13) are referred to as the detailed balance equations.

The following is an immediate corollary of Theorem 4.

Corollary 1. *Let P be a transition matrix on the countable state space $\mathcal{S}$, and let π be some probability distribution on $\mathcal{S}$. If for all $i, j \in \mathcal{S}$, the detailed balance equations (1.13) are satisfied, then π is a stationary distribution of P.*

The nth power of the transition matrix $\mathbf{P}$ given by (1.8), denoted as $\mathbf{P}^n = (p_n(x,y))_{x,y\in\mathcal{S}}$.

In this book we also need notations irreducible, aperiodic, and ergodic Markov chains, which ensure long-term stability and predictability of chains behavior:

Definition 15. *A Markov chain is irreducible if it is possible to get from any state to any other state (not necessarily in one step). Formally, for any states i and j, there exists a positive integer n such that $p_n(i,j) > 0$.*

A Markov chain is aperiodic if there is no fixed cycle that the chain follows. In other words, the chain does not get "stuck" in a cyclic pattern.

Formally, the period of a state i is the greatest common divisor of the set of all possible lengths of paths that return to i. A Markov chain is aperiodic if the period of every state is 1.

Definition 16. *A Markov chain is ergodic if it is both irreducible and aperiodic.*

Definition 17. *Let μ and ν be two probability distributions on a common finite or countable state space $\mathcal{S}$. The total variation distance between μ and ν, denoted by $\|\mu - \nu\|_{TV}$ (or $\|\mu - \nu\|$), is defined as:*

$$\|\mu - \nu\| = \frac{1}{2}\sum_{x\in\mathcal{S}} |\mu(x) - \nu(x)|.$$

The total variation distance can be interpreted as the maximum difference in probabilities that the two distributions assign to the same event. Alternatively, it represents the largest possible difference in probabilities assigned to any single event by the two distributions.

The following theorem guarantees the convergence of an ergodic Markov chain to its stationary distribution:

Theorem 5. *(*Ergodic Theorem*) Let $\{X_n\}_{n\geq 0}$ be an irreducible, aperiodic, and positive recurrent Markov chain with state space $\mathcal{S}$ and transition matrix $\mathbf{P}$. Let π be the unique stationary distribution of the Markov chain, satisfying $\pi\mathbf{P} = \pi$. Then, for any initial distribution μ on $\mathcal{S}$,*

$$\lim_{n\to\infty} \|\mu\mathbf{P}^n - \pi\| = 0,$$

where $\|\cdot\|$ denotes the total variation distance.

Proof. See e.g. [339, Chapter 8]. □

Thus the properties of irreducibility, aperiodicity, and ergodicity are crucial for ensuring that the Markov chain has a well-defined long-term behavior and can be used to make reliable predictions about the system it models.

1.2.5 *Gibbs measure*

Gibbs measures are probability measures that arise in the study of systems with many interacting components. These measures are defined on a space of configurations, where each configuration represents a possible state of the system. The probability of a particular configuration is determined by a local energy function, which assigns an energy value to each component of the system.

1.2.5.1 *Gibbs measures on finite graphs*

In the context of finite graphs, a state can represent a configuration of the system, such as spins in a spin system on the graph.

A *finite graph* $G = (V, E)$ consists of a set of vertices V and edges E. Each vertex $v \in V$ can be assigned a state (or spin) from a finite set. The energy of the system is typically defined in terms of interactions between neighboring vertices (connected by edges).

The *configuration space* Ω consists of all possible assignments of states to the vertices of the graph. If each vertex can take on values from a set S, then $\Omega = S^V$. It is clear that if number of elements $|S|$ of the set S is finite and $|V|$ is finite then Ω also is fine set, i.e.,

$$|\Omega| = |S|^{|V|}.$$

The *Hamiltonian* H is a function that assigns an energy to each configuration.

For example, in the Ising model, the Hamiltonian is given by:

$$H(\sigma) = -\sum_{\{i,j\}\in E} J_{ij}\sigma_i\sigma_j - \sum_{i\in V} h_i\sigma_i,$$

where $\sigma \in \Omega = \{-1, 1\}^V$ is a configuration, $J_{ij} \in \mathbb{R}$ are interaction strengths, and $h_i \in \mathbb{R}$ are external magnetic fields.

For $q \in \mathbb{N}$, the q-state Potts model is given by Hamiltonian

$$H(\sigma) = -\sum_{\{i,j\}\in E} J_{ij}\delta_{\sigma_i,\sigma_j} - \sum_{i\in V} h_i\delta_{1,\sigma_i},$$

where $\sigma \in \Omega = \{1, 2, \ldots, q\}^V$ is a configuration, J_{ij} are interaction strengths, and h_i are external magnetic fields, δ is the Kronecker delta:

$$\delta_{a,b} = \begin{cases} 0, & \text{if } a \neq b \\ 1, & \text{if } a = b \end{cases}$$

The Gibbs measure μ assigns probabilities to configurations according to their energy. It is defined as:

$$\mu(\sigma) = \frac{1}{Z}e^{-\beta H(\sigma)}, \tag{1.14}$$

where

- β is the inverse temperature, i.e., $\beta = 1/T$, $T > 0$ is temperature.
- $H(\sigma)$ is the Hamiltonian,
- Z is the partition function, which normalizes the measure:

$$Z = \sum_{\sigma\in\Omega} e^{-\beta H(\sigma)}.$$

For a finite graph G and a finite set S, the following properties of Gibbs measure are obvious:

1. *Normalization*: The Gibbs measure is a probability measure, so $\mu(\sigma) > 0$ and $\sum_{\sigma\in\Omega}\mu(\sigma) = 1$.
2. *The formula (1.5) for a Markov chain is a specific form of a Gibbs measure.* Consider a Markov chain with $p(x_i, x_j) > 0$ for all $x_i, x_j \in \mathcal{S}$. For $\Lambda = \{k, k+1, \ldots, k+m\}$ the probability of any configuration $\varphi(\Lambda) = (\varphi(k), \varphi(k+1), \ldots, \varphi(k+m)) \in \mathcal{S}^{m+1}$ is

$$\pi_{\varphi(k)}p(\varphi(k), \varphi(k+1))p(\varphi(k+1), \varphi(k+2))\cdots p(\varphi(k+m-1), \varphi(k+m))$$
$$= \exp\left\{\ln \pi_{\varphi(k)} + \sum_{i=k}^{k+m-1} \ln p(\varphi(i), \varphi(i+1))\right\}.$$

Introduce Hamiltonian

$$H(\varphi) = -\sum_{i=-\infty}^{\infty} \ln \pi_{\varphi(i)\varphi(i+1)}.$$

This Hamiltonian is called the Hamiltonian of the Markov chain [343].
3. *Maximum entropy principle* (MEP): In terms of randomness, the outcomes of a random experiment whose probability distribution is uniform are the least predictable.
Let Ω be an arbitrary finite set of microstates. One convenient way of quantifying the unpredictability of the outcomes of a probability distribution μ on Ω is to use the notion of entropy introduced by Shannon:

Definition 18. *The (Shannon) entropy of $\mu \in \mathcal{M}_1(\Omega)$ is defined by*

$$S(\mu) = -\sum_{\omega\in\Omega} \mu(\omega)\log\mu(\omega).$$

Recall that the uniform distribution ν_0 on Ω is defined by:

$$\nu_0(\omega) = \frac{1}{|\Omega|}, \quad \forall\omega\in\Omega.$$

Lemma 1. *The uniform distribution ν_0 on Ω, is the unique probability distribution at which S attains its maximum:*

$$S(\nu_0) = \sup_{\mu\in\mathcal{M}_1(\Omega)} S(\mu).$$

Proof. Let's show why the uniform distribution maximizes the Shannon entropy. Denote $n = |\Omega|$. Let's calculate the entropy $S(\nu_0)$:

$$S(\nu_0) = -\sum_{\omega\in\Omega} \nu_0(\omega)\log\nu_0(\omega).$$

Since $\nu_0(\omega) = \frac{1}{n}$ for all $\omega \in \Omega$, this becomes:

$$S(\nu_0) = -\sum_{\omega \in \Omega} \frac{1}{n} \log \frac{1}{n}.$$

Since there are n terms in the sum, this simplifies to:

$$S(\nu_0) = -n \left(\frac{1}{n} \log \frac{1}{n} \right) = -\log \frac{1}{n} = \log n.$$

To see that the uniform distribution maximizes entropy, we need to show that $S(\mu) \leq \log n$ for any probability distribution μ on Ω.
Consider the Kullback-Leibler divergence (also called relative entropy) $D(\mu \parallel \nu_0)$ between μ and ν_0:

$$D(\mu \parallel \nu_0) = \sum_{\omega \in \Omega} \mu(\omega) \log \frac{\mu(\omega)}{\nu_0(\omega)}.$$

Since $\nu_0(\omega) = \frac{1}{n}$, we have:

$$D(\mu \parallel \nu_0) = \sum_{\omega \in \Omega} \mu(\omega) \log (\mu(\omega) n) = \sum_{\omega \in \Omega} \mu(\omega) \log \mu(\omega) + \sum_{\omega \in \Omega} \mu(\omega) \log n.$$

Since $\sum_{\omega \in \Omega} \mu(\omega) = 1$, this simplifies to:

$$D(\mu \parallel \nu_0) = \sum_{\omega \in \Omega} \mu(\omega) \log \mu(\omega) + \log n.$$

Thus, we have:

$$D(\mu \parallel \nu_0) = -S(\mu) + \log n.$$

Note that the Kullback-Leibler divergence $D(\mu \parallel \nu_0)$ is always non-negative[2]:

$$D(\mu \parallel \nu_0) \geq 0.$$

Therefore,

$$-S(\mu) + \log n \geq 0,$$

which implies:

$$S(\mu) \leq \log n.$$

Attaining maximum entropy: The equality $S(\mu) = \log n$ is achieved if and only if $D(\mu \parallel \nu_0) = 0$, which occurs if and only if $\mu = \nu_0$.
Therefore, the uniform distribution ν_0 is the unique distribution that maximizes the entropy, and its entropy is $\log n$:

$$S(\nu_0) = \log n = \sup_{\mu \in \mathcal{M}_1(\Omega)} S(\mu).$$

Thus, the lemma is proved. □

[2] https://en.wikipedia.org/wiki/Kullback-Leibler_divergence.

Since it is concave, with a unique maximum at ν_0, Shannon's entropy provides a method to quantify the deviation of a distribution from uniform one. Furthermore, Shannon entropy is the sole function that accomplishes this task under appropriate conditions. Therefore, we can employ it to identify, from a given set of probability distributions, the one that exhibits the highest degree of uniformity.

Namely, assume that we have a set of probability distributions $\mathcal{M}_1'(\Omega) \subset \mathcal{M}_1(\Omega)$ representing the set of all distributions compatible with the information at our disposal. Then, the one that best describes our state of knowledge is the one maximizing the Shannon entropy; this way of selecting a distribution is called the MEP.

This principle states that the probability distribution which maximizes the entropy subject to given constraints is the most unbiased estimate we can make given the information we have. In this context, we are interested in finding the probability distribution $\mu \in \mathcal{M}_1'(\Omega)$ that maximizes the entropy under the constraint that the expected value of the Hamiltonian $\mathcal{H}$ equals a fixed value U.

We need to maximize the entropy S of the probability distribution μ:

$$S(\mu) = -\sum_{\omega\in\Omega} \mu(\omega) \log \mu(\omega),$$

subject to the constraint:

$$\langle \mathcal{H} \rangle_\mu = \sum_{\omega\in\Omega} \mu(\omega)\mathcal{H}(\omega) = U,$$

and the normalization condition:

$$\sum_{\omega\in\Omega} \mu(\omega) = 1.$$

To solve this constrained optimization problem, we use the method of Lagrange multipliers. Introduce Lagrange multipliers α and β for the normalization condition and the energy constraint, respectively. Define the Lagrangian:

$$\mathcal{L}(\mu, \alpha, \beta) = -\sum_{\omega\in\Omega} \mu(\omega) \log \mu(\omega)$$

$$+\alpha \left(\sum_{\omega\in\Omega} \mu(\omega) - 1 \right) + \beta \left(\sum_{\omega\in\Omega} \mu(\omega)\mathcal{H}(\omega) - U \right).$$

Taking the derivative of $\mathcal{L}$ with respect to $\mu(\omega)$ and setting it to zero gives:

$$\frac{\partial \mathcal{L}}{\partial \mu(\omega)} = -\log \mu(\omega) - 1 + \alpha + \beta\mathcal{H}(\omega) = 0.$$

Solving for $\mu(\omega)$:

$$\log \mu(\omega) = \alpha - 1 + \beta\mathcal{H}(\omega),$$

$$\mu(\omega) = \exp(\alpha - 1)\exp(\beta\mathcal{H}(\omega)).$$

Let $\exp(\alpha - 1) = Z_\beta^{-1}$, where Z_β is a normalizing factor (the partition function):

$$\mu(\omega) = Z_\beta^{-1}\exp(-\beta\mathcal{H}(\omega)).$$

The partition function Z_β ensures that $\mu(\omega)$ is a valid probability distribution:

$$Z_\beta = \sum_{\omega\in\Omega}\exp(-\beta\mathcal{H}(\omega)).$$

Thus, the unique solution to the problem, satisfying both the entropy maximization and the constraint on the expected energy, is the Gibbs measure:

$$\mu_\beta(\omega) = Z_\beta^{-1}\exp(-\beta\mathcal{H}(\omega)).$$

To verify the uniqueness, note that the function $\exp(-\beta\mathcal{H}(\omega))$ is strictly positive and strictly convex in terms of β. The partition function Z_β acts as a normalizing constant ensuring the distribution sums to 1. Given the constraints and the form of the entropy function, the optimization problem is strictly convex, implying a unique maximum.
Thus, the probability distribution $\mu_\beta(\omega)$ is uniquely determined by the constraints, making the Gibbs measure the unique solution to the problem.

4. *Markov type property.* The Gibbs measure satisfies a form of the Markov property, where the conditional probability of a state at a vertex, given the states of the neighboring vertices, depends only on the states of those neighbors. In the other words, for systems with finite-range interactions (for example, nearest-neighbor interactions), the Gibbs measure satisfies

$$\mu(\sigma_k = s \mid \sigma_j,\, j \neq k) = \mu(\sigma_k = s \mid \sigma_j,\, j \in N_k),$$

where N_k is a neighborhood of the site k. That is, the probability at site k depends only on the spins in a finite neighborhood. This last equation is in the form of a local Markov property.
The converse is also true: any positive probability distribution (nonzero density everywhere) having the Markov property can be represented as a Gibbs measure for an appropriate energy function (see [304] for detailed presentation of this statement and its proof).

5. *Applications*:

- Statistical physics: Gibbs measures model physical systems, such as magnetic materials, where the states represent spins of particles.
- Graphical models: In machine learning and statistics, Gibbs measures are used in graphical models to represent dependencies between random variables.
- Combinatorics and optimization: They are used to study combinatorial structures and to design algorithms for optimization problems, such as the max-cut problem.

Thus Gibbs measures on finite graphs provide a framework for understanding the distribution of states in complex systems with interactions defined by the graph's structure. They are essential in various fields, from physics to machine learning, due to their ability to model dependencies and interactions in a structured way.

Remark 2. *In some sources, the Gibbs measure is referred to as the exponential family (see, for example, [42, Section 3.4]). As noted in [42], many standard probability distributions discussed in [42, Chapter 3] are specific examples of Gibbs measures. Therefore, Gibbs measures share numerous important properties, and discussing these properties in a general framework can be highly illuminating.*

1.2.5.2 *Gibbs measures on countable graphs*

Consider a graph $G = (V, E)$, where V is countable. The Gibbs measure μ on G can be described in terms of local interactions between vertices.

Given a configuration space $\Omega = S^V$, where S is a finite set of possible states for each vertex, as the finite graph case the Hamiltonian H encodes the energy associated with each configuration $\sigma \in \Omega$.

A Gibbs measure on a countable graph $G = (V, E)$ (where V is the set of vertices and E is the set of edges) is a probability measure on the set of configurations $\Omega = S^V$ (for some finite set S) that satisfies the Dobrushin-Lanford-Ruelle (DLR) conditions. These conditions ensure that the probability of a configuration in a given region, conditional on the configuration outside that region, is determined by the local Hamiltonian.

Formal definition is given by the following steps:

1. *Configuration Space*: Let $\Omega = S^V$ be the set of all possible configurations of spins on the graph, where S is a finite set. Let $A \subset V$ be a subset. On this set fix a configuration φ_A then, the following set is called

a cylinder set with base A:

$$\mathcal{S}(\varphi_A) = \{\sigma \in \Omega : \sigma\big|_A = \varphi_A\},$$

that is the set of all configurations from Ω which on the set A coincide with φ_A.

The cylinder σ-algebra is the one that is generated[3] by cylinder sets.

The σ-algebra generated by cylinders with base contained in A is denoted by $\mathcal{F}_A$, this consists of all the events that depend only on the spins inside A.

Denote by $\mathcal{M}_1(\Omega, \mathcal{F})$ the set of all probability measures on the measurable space $(\Omega, \mathcal{F})$. It is known (see [136, Lemma 6.7]) that probability measures on a measurable space $(\Omega, \mathcal{F})$ are uniquely determined by their values on cylinders. Moreover, if $\mu, \nu \in \mathcal{M}_1(\Omega, \mathcal{F})$, then $\mu = \nu$ iff $\mu(\mathcal{S}) = \nu(\mathcal{S})$ for all cylinders $\mathcal{S}$.

2. *Hamiltonian*: A Hamiltonian H is a function that assigns an energy to each configuration. For a configuration $\sigma \in \Omega$, the Hamiltonian $H(\sigma)$ typically takes the form:

$$H(\sigma) = \sum_{A \subset V} \Phi_A(\sigma_A),$$

where Φ_A are interaction potentials depending on the spins in subset A.

3. *Gibbs Measure*: A probability measure μ on Ω is a Gibbs measure with respect to the Hamiltonian H if, for every finite subset $\Lambda \subset V$ and every configuration σ outside Λ (denoted by σ_{Λ^c}), the conditional probability μ satisfies:

$$\mu(\sigma_\Lambda | \sigma_{\Lambda^c}) = \frac{1}{Z_\Lambda(\sigma_{\Lambda^c})} \exp\left(-\beta H_\Lambda(\sigma_\Lambda | \sigma_{\Lambda^c})\right),$$

where

- β is the inverse temperature (a positive constant).

- $H_\Lambda(\sigma_\Lambda | \sigma_{\Lambda^c})$ is the conditional Hamiltonian, which depends on the spins in Λ given the configuration outside Λ.

- $Z_\Lambda(\sigma_{\Lambda^c})$ is the normalizing constant (partition function) ensuring the probabilities sum to 1:

$$Z_\Lambda(\sigma_{\Lambda^c}) = \sum_{\sigma_\Lambda} \exp\left(-\beta H_\Lambda(\sigma_\Lambda | \sigma_{\Lambda^c})\right).$$

---

[3]Let F be an arbitrary family of subsets of Ω. Then there exists a unique smallest σ-algebra which contains every set in F (even though F may or may not itself be a σ-algebra). It is, in fact, the intersection of all σ-algebras containing F. This σ-algebra is called the σ-algebra generated by F.

The Gibbs measure describes a probability distribution over configurations such that the conditional probability of the configuration in any finite region, given the configuration outside that region, follows a Boltzmann distribution determined by the Hamiltonian. This means the measure respects the local interactions and dependencies defined by the Hamiltonian, ensuring that local changes in configuration are influenced by the surrounding environment in a manner consistent with the physical model.

In essence, Gibbs measures capture the equilibrium states of a system where the probability of each state is proportional to the exponential of the negative Hamiltonian (energy) of that state, reflecting the system's tendency to favor lower energy configurations, particularly at lower temperatures.

1.2.5.3 *Existence and uniqueness of Gibbs measures*

Dobrushin's conditions.

Dobrushin's conditions provide a criterion for the uniqueness of Gibbs measures on countable graphs. They are based on the concept of strong mixing conditions and can be stated as follows:

- There exists a constant $\alpha < 1$ such that for any pair of vertices i and j, the total influence of the configuration at j on the conditional probability at i is bounded by α.

Mathematically, let μ_Λ be the Gibbs measure restricted to a finite subset $\Lambda \subset V$ with boundary condition τ. Dobrushin's condition is:

$$\sum_{j \neq i} \sup_{\tau, \tau'} |\mu_\Lambda(\sigma_i = \sigma | \sigma_{\Lambda^c} = \tau) - \mu_\Lambda(\sigma_j = \sigma | \sigma_{\Lambda^c} = \tau')| \leq \alpha.$$

The DLR (Dobrushin-Lanford-Ruelle) equations.

The DLR equations characterize Gibbs measures through consistency conditions of local specifications:

$$\mu(A|\mathcal{F}_{\Lambda^c}) = \mu_\Lambda(A|\mathcal{F}_{\Lambda^c})$$

for all finite $\Lambda \subset V$ and measurable sets $A \subset \Omega$. These equations ensure that the global measure is consistent with local conditional probabilities.

Phase transitions and the uniqueness problem.

Phase transitions occur when there are multiple Gibbs measures for the same Hamiltonian at a given temperature, corresponding to different macroscopic states. The uniqueness problem seeks to determine under what conditions the Gibbs measure is unique.

Key results include:

- High-temperature uniqueness: For sufficiently high temperatures (small β), the Gibbs measure is unique.

- Low-temperature non-uniqueness: For low temperatures (large β), multiple Gibbs measures may exist, indicating a phase transition.

Convergence of Gibbs measures.

Convergence properties of Gibbs measures are crucial for understanding their stability and phase transitions. A sequence of Gibbs measures $\{\mu_n\}$ on increasing finite subsets $\Lambda_n \subset V$ is said to converge to a Gibbs measure μ if:

$$\mu_n \xrightarrow{w} \mu,$$

where $\xrightarrow{w}$ denotes weak convergence of measures.

Stability under perturbations.

Stability of Gibbs measures under perturbations is a significant aspect, especially when considering small changes in the interaction potentials or external fields. A Gibbs measure μ is stable under a perturbation if the perturbed measure μ' remains close to μ in a suitable sense.

Role of boundary conditions.

Boundary conditions play a vital role in determining the properties of Gibbs measures, especially in countable graphs where boundary effects can propagate throughout the system. Different boundary conditions can lead to different Gibbs measures, particularly in the presence of phase transitions.

Relationship between Gibbs Measures and Markov Random Fields.

Gibbs measures and Markov random fields (MRFs) are closely related. A Gibbs measure can be viewed as a probability distribution that satisfies the Markov property with respect to the graph structure. The Hammersley-Clifford theorem formalizes this relationship, stating that a positive probability distribution is a Gibbs measure if and only if it is a Markov random field.

Hammersley-Clifford Theorem.

The Hammersley-Clifford theorem, also known as the Markov random field theorem, is a fundamental result[4] in the theory of probabilistic graphical models. It provides a crucial link between the global properties of a probability distribution and its local properties in terms of conditional independence. Here's a detailed explanation:

[4] http://www.stat.yale.edu/~pollard/Courses/251.spring04/Handouts/Hammersley-Clifford.pdf.

The Hammersley-Clifford theorem states that a positive probability distribution P on a finite set of random variables $\mathbf{X} = \{X_1, X_2, \ldots, X_n\}$ can be factorized according to the cliques of an undirected graph[5] G if and only if P satisfies the Markov properties with respect to G.

Before giving the theorem let us give key concepts on an undirected graph:

- Markov random field (MRF): An MRF is an undirected graph $G = (V, E)$ where each node represents a random variable, and edges represent direct probabilistic interactions between the variables.

- Pairwise Markov property: Two non-adjacent nodes are conditionally independent given all other nodes.

- Local Markov property: A node is conditionally independent of all other nodes given its neighbors.

- Global Markov property: Two sets of nodes are conditionally independent given a separating set in the graph.

-Cliques: A clique (or complete subgraph) is a subset of nodes such that every two distinct nodes are adjacent.

Theorem 6. *(Hammersley-Clifford) For a positive probability distribution P on $\mathbf{X}$ associated with an undirected graph G:*

- *P is a Gibbs random field, i.e., it can be written as a product of potential functions over the cliques of G:*

$$P(\mathbf{X}) = \frac{1}{Z} \prod_{C \in \mathcal{C}} \psi_C(X_C)$$

  where $\mathcal{C}$ is the set of cliques in G, ψ_C are non-negative potential functions, and Z is a normalization constant.
- *P satisfies the Markov properties with respect to G if and only if it can be factorized as described.*

Implications:

1. Factorization: The theorem provides a way to decompose a joint probability distribution into a product of simpler functions (potential functions) associated with the cliques of the graph.

2. Conditional Independence: The theorem establishes that the structure of the graph G encodes conditional independencies in the distribution P.

[5] An undirected graph is a set of objects called vertices (or nodes) connected by edges, where the edges have no direction. This means the connection between two vertices is bidirectional, indicating a mutual relationship.

3. Computational Efficiency: This factorization can simplify the computation of the probability distribution, especially for large systems, as operations can be localized to smaller subsets (cliques). Applications:

- Statistical Physics: The theorem is used in the study of Gibbs distributions in statistical mechanics.

- Computer Vision: MRFs are used for image segmentation and denoising.

- Machine Learning: The theorem underpins the use of graphical models for probabilistic reasoning and inference in various machine learning tasks.

The Hammersley-Clifford theorem thus provides a foundational understanding of how probabilistic dependencies can be represented and manipulated using graphical models.

Examples of MRFs application:

- Image processing: MRFs are used in image segmentation and denoising.

- Spatial statistics: MRFs model spatial dependencies in geographical data.

- Machine learning: MRFs are used in graphical models for representing dependencies among variables.

Criteria for phase transitions on countable graphs.

Criteria for phase transitions involve conditions on the interaction potentials and external fields. For example, in the Ising model, phase transitions are characterized by the existence of multiple Gibbs measures at low temperatures.

Rigorous results include theorems on the existence of phase transitions in specific models. For example, it is known that the Ising and Potts models on a Bethe lattice (Cayley trees) exhibit a phase transition (see [149], [304], [323], [324]).

1.3 How Gibbs measure appears in machine learning?

1.3.1 *As applications of statistical physics methods in machine learning*

Following [139] (see also the references therein), we give theoretical foundations of machine learning (ML) where the Gibbs measure is an important measure.

Let us recall first the main problems of ML and then give relevance of them to Gibbs measures. ML, a subfield of artificial intelligence (AI), focuses on the automated extraction of information from data.

The discipline is broadly divided into three main areas: supervised learning, unsupervised learning, and reinforcement learning. Each of these areas addresses distinct challenges in learning from data:

1. Supervised Learning: This involves learning a mapping from input data x (e.g., an image) to output labels (e.g. classifying the input image as an image of a cat) or quantitative values $y = f(x)$, based on a training dataset of labeled pairs (x, y). The goal is typically to solve classification or regression problems.

2. Unsupervised Learning: Here, the objective is to model the underlying distribution $p(x)$ of the training data. This allows extracting meaningful features or generating new, realistic data samples.

3. Reinforcement Learning: This area focuses on learning a policy $a|s$, which represents the probability of taking an action a (e.g., a move in a maze) given a state s (the position in the maze). The ultimate aim is to optimize decisions to achieve a specific goal, such as escaping the maze.

The statistical foundations of learning address the consistency of machine learning models in general and elegant terms. Given an underlying distribution $F(z)$ from which a set $z = \{z_1, \ldots, z_n\}$ of training data is sampled, and a set of models with an associated cost function $Q(z)$ (e.g., a classification score or the negative log-likelihood $\log(p(z))$), one seeks to minimize the following risk functional:

$$R(\theta) = \int Q(z; \theta)\, dF(z), \tag{1.15}$$

where θ represents the parameters of the model. In practice, this true risk is approximated by the empirical risk estimated from the training data:

$$R_{\text{emp}}(\theta) = \frac{1}{n} \sum_{i=1}^{n} Q(z_i; \theta). \tag{1.16}$$

Machine learning, particularly in the context of supervised learning, relies on the *empirical risk minimization* (ERM) principle. Consistency, in this context, means that the true risk $R(\theta)$ of the model minimizing $R_{\text{emp}}(\theta)$ is close to the optimal risk R^*. Failure to achieve this often leads to overfitting. Intuitively, the model will be consistent if its complexity is appropriately adapted to the amount of data, allowing some form of the law of large numbers to apply over the functional space Θ.

This idea is formalized by introducing the Vapnik–Chervonenkis (VC) entropy:

$$H_n(\Theta) = \mathbb{E}_{z_1, \ldots, z_n \sim F(z)} \left[\log G_n(\Theta, \{z_1, \ldots, z_n\})\right], \tag{1.17}$$

where $G_n(\Theta, \{z_1, \dots, z_n\})$ measures the number of distinct dichotomies (labelings) that the functional space Θ can realize over the set of data points $\{z_1, \dots, z_n\}$. Consistency is ensured asymptotically when:

$$\lim_{n\to\infty} \frac{H_n(\Theta)}{n} = 0. \tag{1.18}$$

In contrast, a model with excessively high complexity corresponds to $H_n(\Theta) = O(n)$ (for a binary classification problem).

A more practical tool is the growth function $G_n(\Theta)$, defined as:

$$G_n(\Theta) = \max_{z_1,\dots,z_n} \log G_n(\Theta, \{z_1, \dots, z_n\}), \tag{1.19}$$

which bounds the VC entropy. The VC dimension, denoted as h, is the largest value for which $G_n(\Theta) = 2^n$ for some set of n points. For Θ with VC dimension h, $G_n(\Theta)$ grows slower than $n \log 2$ for $n > h$, reflecting the functional space's limited capacity. In practice, a ratio of $h/n \ll 1$ typically avoids overfitting.

Machine learning can also be reframed using a language familiar to physicists (see [139], [234]) and references therein), revealing the relevance of statistical physics. A common approach relates machine learning to statistical physics in a Bayesian framework by interpreting the empirical risk as an energy function of a Gibbs measure over candidate models:

$$P_{\text{posterior}}(\theta) \propto P_{\text{prior}}(\theta) \exp\left(-\beta R_{\text{emp}}(\theta)\right), \tag{1.20}$$

where P_{prior} and $P_{\text{posterior}}$ represent the prior and posterior distributions of the models after considering the training data $\{z_i\}$, and β (inverse temperature) relates to the output noise of the model. This distribution can be viewed as the equilibrium Gibbs measure associated with a stochastic learning dynamics, expressed as a Langevin equation [336]:

$$d\theta_t = -\nabla_\theta R_{\text{emp}}(\theta)\, dt + \sqrt{2/\beta}\, dW_t, \tag{1.21}$$

where dW_t represents a white noise term. In this framework, the training data acts as quenched disorder, requiring averaging over the stochasticity.

The role of statistical physics in neural networks: The Hopfield model of associative memory [336] stimulated extensive theoretical studies, coinciding with advances in spin-glass theory [272]. The steady-state properties of the Hopfield model can be described using the Gibbs measure based on the Hamiltonian:

$$H[s] = -\frac{1}{N} \sum_{i,j=1}^{N} \sum_{k=1}^{P} \xi_i^{(k)} \xi_j^{(k)} s_i s_j, \tag{1.22}$$

where $\{\xi_i^{(k)}\}_{i=1,\dots,N}^{k=1,\dots,P}$ represents a set of P patterns to be memorized, encoded into a random spin model of N variables. A natural set of order parameters, called the overlaps, is defined as:

$$m^{(k)} = \frac{1}{N}\sum_{i=1}^{N} \mathbb{E}_s\left[\xi_i^{(k)} s_i\right], \quad k = 1, \dots, P, \tag{1.23}$$

where the expectation $\mathbb{E}_s$ involves both thermal averages over the spin configurations $\{s_i\}$ and averages with respect to the quenched disorder $\{\xi_i^{(k)}\}$.

The mean-field theory of the Hopfield model was developed (see [272] and references therein). The model exhibits three distinct phases, separated by two transition temperatures:

- T_g, which separates the paramagnetic phase from the spin-glass phase, and
- T_c, which separates the spin-glass phase from the ferromagnetic (retrieval) phase.

A key question relates to the storage capacity of neural networks, specifically the number of patterns that can be stored as a function of the number of neurons N. This has been extensively studied for models such as the perceptron under various assumptions (see [139] and references therein), revealing connections to the computation of VC entropy. While statistical learning theory provides bounds on the true risk, resulting in algorithms like Support Vector Machines (SVMs), statistical physics methods often provide asymptotic estimates of the learning curve (i.e., the risk as a function of the number of examples).

Layered networks have also been analyzed in the thermodynamic limit [104,273] using mean-field techniques to investigate mechanisms of information storage and the efficiency of learning algorithms based on associative memory.

1.3.2 *Relation between Softmax and Gibbs measure*

Let us begin with a definition of the so-called softmax function:

Definition 19. *For a vector* $\mathbf{z} = (z_1, z_2, \dots, z_n) \in \mathbb{R}^n$, $n \geq 1$ *the softmax function* $s : \mathbb{R}^n \to (0,1)^n$ *is defined as:*

$$s(\mathbf{z})_i := softmax(z_i) = \frac{e^{z_i}}{\sum_{j=1}^{n} e^{z_j}}.$$

This softmax function (see [162, Chapter 6] and [41, Chapter 4] for in detail discussions on the softmax function) is commonly used in machine learning, particularly in various multiclass classification methods, such as multinomial logistic regression, multiclass linear discriminant analysis, naive Bayes classifiers, and widely adopted artificial neural networks such as Transformers [374].

Remark 3. *The softmax function can be viewed as a special case of the Gibbs measure (see* (1.14)*). Because in statistical mechanics, the softargmax function is known as the Boltzmann distribution (or Gibbs measure): the index set* $\{1, 2, \dots n\}$ *are the microstates of the system; the inputs* z_i *are the energies of the states.*

Here are the key mathematical properties of the softmax function:

1. The softmax function outputs a strongly positive probability distribution:

$$0 < s(\mathbf{z})_i \leq 1, \ \sum_{i=1}^{n} s(\mathbf{z})_i = 1.$$

2. The gradient of the softmax function with respect to the input vector $\mathbf{z}$ is:

$$\frac{\partial s(\mathbf{z})_i}{\partial z_j} = s(\mathbf{z})_i(\delta_{ij} - s(\mathbf{z})_j),$$

where δ_{ij} is the Kronecker delta. This property is crucial in backpropagation during neural network training.

3. The softmax function is invariant to translation by a constant. That is, adding a constant c to every component of $\mathbf{z}$ does not change the output:

$$s(\mathbf{z} + c) = s(\mathbf{z}).$$

This property is useful for numerical stability, as it allows shifting the inputs to prevent overflow or underflow when exponentiating.

4. The derivative of the log-softmax function, which is often used in loss functions like cross-entropy, simplifies to:

$$\frac{\partial \log s(\mathbf{z})_i}{\partial z_j} = \delta_{ij} - s(\mathbf{z})_j.$$

Since the softmax function is bounded between 0 and 1, it suitable for representing probabilities in classification tasks.

Example 3. *Given a vector* $\mathbf{z} = (1, 2, 3)$*, the softmax function computes the probabilities as follows:*

1. *Compute the exponentials:*

$$e^{\mathbf{z}} = (e^1, e^2, e^3) \approx (2.718, 7.389, 20.086).$$

2. *Sum of exponentials:*

$$\sum_{i=1}^{3} e^{z_i} \approx 2.718 + 7.389 + 20.086 = 30.193.$$

3. *Normalize:*

$$s(\mathbf{z})_i \approx \frac{e^{z_i}}{30.193},$$

$$s(\mathbf{z}) \approx \left(\frac{2.718}{30.193}, \frac{7.389}{30.193}, \frac{20.086}{30.193} \right),$$

$$s(\mathbf{z}) \approx (0.090, 0.245, 0.665).$$

These properties and the example illustrate the softmax function's ability to transform a vector of real numbers into a probability distribution.

For example[6], in multinomial logistic regression and linear discriminant analysis, the input to the function is the result of n distinct linear functions, and the predicted probability for the jth class given a sample vector $\mathbf{x}$ and a weighting vector $\mathbf{w}$ is:

$$P(y = j \mid \mathbf{x}) = \frac{e^{\mathbf{x}^\top \mathbf{w}_j}}{\sum_{k=1}^{n} e^{\mathbf{x}^\top \mathbf{w}_k}}.$$

This can be seen as the composition of n linear functions

$$\mathbf{x} \mapsto \mathbf{x}^\top \mathbf{w}_1, \ldots, \mathbf{x} \mapsto \mathbf{x}^\top \mathbf{w}_n$$

and the softmax function (where $\mathbf{x}^\top \mathbf{w}$ denotes the inner product of $\mathbf{x}$ and $\mathbf{w}$). The operation is equivalent to applying a linear operator defined by $\mathbf{w}$ to vectors $\mathbf{x}$, thus transforming the original, probably high-dimensional, input to vectors in a n-dimensional space $\mathbb{R}^n$.

The softmax function in reinforcement learning[7] is used to convert action values into probabilities for selecting actions. Here the softmax function is defined as:

$$P_t(a) = \frac{\exp(q_t(a)/\tau)}{\sum_{i=1}^{n} \exp(q_t(i)/\tau)},$$

[6] https://en.wikipedia.org/wiki/Softmax_function.

[7] https://en.wikipedia.org/wiki/Reinforcement_learning.

where $P_t(a)$ is the probability of selecting action a at time t, $q_t(a)$ is the action value, which represents the expected reward of following action a, and τ is a temperature parameter that controls the randomness of the action selection.

Thermodynamics can be described as follows:

High temperature ($\tau \to \infty$): All actions have nearly the same probability of being selected. The action selection is more random, meaning the policy explores more.

Low temperature ($\tau \to 0^+$): The probability distribution becomes more peaked, and the action with the highest expected reward is more likely to be selected. The policy is more greedy, exploiting the best-known action.

We note that the use of temperature parameter (τ) is analogous to its role in statistical mechanics, particularly in Gibbs measure theory, where it influences the distribution of energy states.

1.3.3 *Word2Vec*

In this subsection we give another application of softmax function in a natural language processing (NLP) technique, Word2Vec, that is used to obtain vector representations of words [275], effectively enabling models to learn patterns in natural language.

These vectors capture the meaning of a word based on its surrounding words. The Word2Vec algorithm estimates these representations by modeling text in a large corpus. Once trained, the model can detect synonymous words or suggest additional words for a partial sentence. Following [159], [320] we give mathematical explanations of Word2Vec.

A corpus C is a sequence of words. Denote by V the set of all words in the corpus C. Each word $w \in V$ is represented by a vector $\mathbf{v}_w \in \mathbb{R}^n$.

In a skip-gram model the vector of a word should be close to the vectors of its neighboring words.

In Continuous Bag of Words (CBOW) model the vector-sum of a word's neighbors should be close to the vector of the word itself.

Closeness is originally measured using the softmax function.

Suppose we want each word in the corpus to be predicted by every other word in a span of r words. The neighbor set of 0 is

$$N = \{-r, -r+1, \dots, -1, 1, 2, \dots, r\}$$

indicates a context window of r words on each side of the target word.

The training objective is to maximize the probability of predicting a word w_i given its context words w_j for $j \in N + i$:

$$\prod_{i \in C} \Pr(w_i \mid w_j : j \in N + i).$$

This objective captures the essence of the CBOW model, where the goal is to predict a target word based on its surrounding context.

Products are numerically unstable, let us convert it by taking the logarithm:

$$\sum_{i \in C} \ln \Pr \left(w_i \mid w_j : j \in N + i \right).$$

That is, we maximize the log-probability of the corpus.

A probability model is as follows: Given words $\{w_j : j \in N + i\}$, it takes their vector sum $v := \sum_{j \in N+i} v_{w_j}$, then take the dot-product-softmax with every other vector sum to obtain the probability:

$$\Pr \left(w \mid w_j : j \in N + i \right) := \frac{e^{v_w \cdot v}}{\sum_{w \in V} e^{v_w \cdot v}}.$$

The quantity to be maximized is then after simplifications:

$$\sum_{i \in C, j \in N+i} \left(v_{w_i} \cdot v_{w_j} - \ln \sum_{w \in V} e^{v_w \cdot v_{w_j}} \right).$$

The value on the left is calculated quickly, and the value on the right is slow, since it requires summing over the entire vocabulary set V for each word in the corpus. Additionally, to use gradient lifting that maximizes the log likelihood, one must calculate the gradient of the magnitude on the right, which is an intractable problem.

For skip-gram, the training objective is

$$\prod_{i \in C} \Pr \left(w_j : j \in N + i \mid w_i \right).$$

That is, we want to maximize the total probability for the corpus, as seen by a probability model that uses words to predict its word neighbors. We predict each word-neighbor independently, thus

$$\Pr \left(w_j : j \in N + i \mid w_i \right) = \prod_{j \in N+i} \Pr \left(w_j \mid w_i \right).$$

Products are numerically unstable, so one has to convert it by taking the logarithm:

$$\sum_{i\in C, j\in N+i} \ln \Pr(w_j \mid w_i).$$

The probability model is still the dot-product-softmax model, so the calculation proceeds as before.

$$\sum_{i\in C, j\in N+i} \left(v_{w_i} \cdot v_{w_j} - \ln \sum_{w\in V} e^{v_w \cdot v_{w_i}} \right).$$

To explain how the vector representations of words will be used in machine learning we give some basic definitions: Given two n-dimensional Euclidean vectors $a = (a_1, \ldots, a_n)$ and $b = (b_1, \ldots, b_n)$, their dot (or scalar) product is defined as

$$a \cdot b = \|a\| \|b\| \cos\theta,$$

where θ is the angle between vectors a and b.

The cosine similarity, $\cos(\theta)$, is represented using a dot product as

$$S_C(a, b) := \cos(\theta) = \frac{a \cdot b}{\|a\| \|b\|} = \frac{\sum_{i=1}^n a_i b_i}{\sqrt{\sum_{i=1}^n a_i^2} \cdot \sqrt{\sum_{i=1}^n b_i^2}}.$$

Note that the resulting similarity ranges from -1 meaning exactly opposite, to 1 meaning exactly the same, with 0 indicating orthogonality or decorrelation, but $\cos(\theta) \neq -1, 0, 1$ indicates intermediate similarity or dissimilarity.

For text matching, the attribute vectors a and b are usually the term frequency vectors of the documents. Cosine similarity can be seen as a method of normalizing document length during comparison. In the case of information retrieval, the cosine similarity of two documents will range from $0 \to 1$, since the term frequencies cannot be negative.

The process of word embedding learning in the Word2Vec objective function ensures that words appearing in similar contexts will have similar embeddings, as indicated by cosine similarity. This aligns with distributional hypothesis: linguistic items with similar distributions have similar meanings (see [239]).

1.3.4 *Statistical physics and Bayesian inference*

In this subsection, following [227], we provide an explanation of why the Gibbs measure is useful in machine learning. It may seem initially surprising that Physics has any bearing on minimization and probabilistic inference. The key lies in the Gibbs measure. From the perspectives of Statistics and Optimization, we consider two central problems:

(1) Minimization of a cost function.
(2) Sampling from a probability distribution.

In both contexts, the Gibbs measure proves remarkably useful. Let $E(x)$ denote a cost or energy function defined over $\mathbb{R}^d$. Statistical Mechanics associates to each configuration x a temperature-dependent Gibbs measure:

$$P(x) = \frac{1}{Z_N(\beta)} e^{-\beta E(x)},$$

where $\beta = 1/T$ is the inverse temperature, and the partition function $Z_N(\beta)$ is given by:

$$Z_N(\beta) = \int_{\mathbb{R}^d} e^{-\beta E(x)}\, dx.$$

This formulation proves very convenient. In particular, under some relatively mild assumptions, it can be shown that

$$\lim_{\beta \to \infty} \left(-\frac{\partial}{\partial \beta} \log Z(\beta) \right) = \min_x E(x),$$

which connects low-temperature limits to optimization.

Theorem 7. *Let $E : \mathbb{R}^d \to \mathbb{R}$ be an energy function and $Z(\beta) = \int_{\mathbb{R}^d} e^{-\beta E(x)} dx$ be the partition function. Assume:*

(1) E has a unique global minimum at $x^* \in \mathbb{R}^d$ with $E(x^*) = m$
(2) E is twice continuously differentiable in a neighborhood of x^*
(3) The Hessian matrix $H = \nabla^2 E(x^*)$ is positive definite
(4) $E(x) \to +\infty$ as $\|x\| \to \infty$ sufficiently fast to ensure convergence of $Z(\beta)$

Then, the following holds:

$$\lim_{\beta \to \infty} \left(-\frac{\partial}{\partial \beta} \log Z(\beta) \right) = \min_{x \in \mathbb{R}^d} E(x).$$

Proof. We begin by expressing the derivative in terms of the expected energy. Computing the derivative of $\log Z(\beta)$, we have

$$\frac{\partial}{\partial \beta} \log Z(\beta) = \frac{1}{Z(\beta)} \frac{\partial Z(\beta)}{\partial \beta}.$$

To evaluate the derivative of the partition function, we differentiate under the integral sign:

$$\frac{\partial Z(\beta)}{\partial \beta} = \frac{\partial}{\partial \beta} \int_{\mathbb{R}^d} e^{-\beta E(x)} dx = \int_{\mathbb{R}^d} (-E(x)) e^{-\beta E(x)} dx.$$

Therefore,

$$\frac{\partial}{\partial \beta} \log Z(\beta) = \frac{-\int_{\mathbb{R}^d} E(x) e^{-\beta E(x)} dx}{\int_{\mathbb{R}^d} e^{-\beta E(x)} dx} = -\langle E(x) \rangle_\beta,$$

where $\langle E(x) \rangle_\beta$ is the expected energy under the Gibbs distribution. Our goal now reduces to proving that $\lim_{\beta \to \infty} \langle E(x) \rangle_\beta = m$.

To analyze the asymptotic behavior, we perform a change of variables. Let $y = x - x^*$ and define $F(y) = E(x^* + y) - m$. Note that $F(0) = 0$ and $F(y) \geq 0$ for all y, with $F(y) > 0$ for $y \neq 0$. By the Taylor expansion around $y = 0$, we have

$$F(y) = \frac{1}{2} y^T H y + R(y),$$

where $R(y) = O(\|y\|^3)$.

Substituting this change of variables into the expected energy, we obtain

$$\langle E(x) \rangle_\beta = \frac{\int_{\mathbb{R}^d} (m + F(y)) e^{-\beta(m + F(y))} dy}{\int_{\mathbb{R}^d} e^{-\beta(m + F(y))} dy}.$$

Factoring out $e^{-\beta m}$ from both numerator and denominator yields

$$\langle E(x) \rangle_\beta = m + \frac{\int_{\mathbb{R}^d} F(y) e^{-\beta F(y)} dy}{\int_{\mathbb{R}^d} e^{-\beta F(y)} dy}.$$

We now apply Laplace's method to analyze the asymptotic behavior of both integrals as $\beta \to \infty$. For large β, both integrals are dominated by small neighborhoods of $y = 0$ where $F(y) \approx \frac{1}{2} y^T H y$.

For the denominator, we have

$$I_0(\beta) = \int_{\mathbb{R}^d} e^{-\beta F(y)} dy \approx \int_{\mathbb{R}^d} e^{-\frac{\beta}{2} y^T H y} dy = \left(\frac{2\pi}{\beta} \right)^{d/2} (\det H)^{-1/2},$$

where we used the standard Gaussian integral formula.

For the numerator, we need to evaluate

$$I_1(\beta) = \int_{\mathbb{R}^d} F(y) e^{-\beta F(y)} dy \approx \int_{\mathbb{R}^d} \frac{1}{2} y^T H y \cdot e^{-\frac{\beta}{2} y^T H y} dy.$$

To compute this integral, we use the identity

$$\frac{1}{2} y^T H y \cdot e^{-\frac{\beta}{2} y^T H y} = -\frac{1}{\beta} \frac{\partial}{\partial \beta} e^{-\frac{\beta}{2} y^T H y}.$$

This allows us to write

$$I_1(\beta) = -\frac{1}{\beta}\frac{\partial}{\partial\beta}\int_{\mathbb{R}^d} e^{-\frac{\beta}{2}y^T Hy}dy = -\frac{1}{\beta}\frac{\partial}{\partial\beta}\left[\left(\frac{2\pi}{\beta}\right)^{d/2}(\det H)^{-1/2}\right].$$

Computing the derivative, we find

$$\frac{\partial}{\partial\beta}\left(\frac{2\pi}{\beta}\right)^{d/2} = -\frac{d}{2\beta}\left(\frac{2\pi}{\beta}\right)^{d/2}.$$

Therefore,

$$I_1(\beta) = -\frac{1}{\beta}\cdot\left(-\frac{d}{2\beta}\right)\left(\frac{2\pi}{\beta}\right)^{d/2}(\det H)^{-1/2} = \frac{d}{2\beta^2}\left(\frac{2\pi}{\beta}\right)^{d/2}(\det H)^{-1/2}.$$

Taking the ratio of these integrals, we obtain

$$\frac{I_1(\beta)}{I_0(\beta)} = \frac{\frac{d}{2\beta^2}\left(\frac{2\pi}{\beta}\right)^{d/2}(\det H)^{-1/2}}{\left(\frac{2\pi}{\beta}\right)^{d/2}(\det H)^{-1/2}} = \frac{d}{2\beta^2}.$$

Consequently,

$$\langle E(x)\rangle_\beta = m + \frac{d}{2\beta^2} + O(\beta^{-3}).$$

Taking the limit as $\beta \to \infty$, we conclude that

$$\lim_{\beta\to\infty}\langle E(x)\rangle_\beta = m.$$

Finally, since $-\frac{\partial}{\partial\beta}\log Z(\beta) = \langle E(x)\rangle_\beta$, we have

$$\lim_{\beta\to\infty}\left(-\frac{\partial}{\partial\beta}\log Z(\beta)\right) = \lim_{\beta\to\infty}\langle E(x)\rangle_\beta = m = \min_{x\in\mathbb{R}^d} E(x).$$

□

The Gibbs framework is equally compelling in sampling problems, especially those common in Statistical Inference, Information Theory, and Machine Learning.

Consider the following illustrative scenario: suppose $X \sim \mathcal{N}(0,1)$ is an unknown latent variable, and $Y = X + Z$, with $Z \sim \mathcal{N}(0,1)$ representing Gaussian noise, independent of X. Given an observation $Y = y$, what is the distribution of X conditioned on Y?

By Bayes' theorem, the posterior distribution $P_{X|Y}(x|y)$ is:

$$P_{X|Y}(x|y) = \frac{P_{Y|X}(y|x)P_X(x)}{P_Y(y)},$$

where $P_Y(y)$ is the evidence or marginal likelihood. This can be recast in exponential form:

$$P_{X|Y}(x|y) = \frac{1}{Z} \exp(-E(x)),$$

where

- $E(x) = -\log P_{Y|X}(y|x)P_X(x)$,
- $Z = P_Y(y)$, the partition function.

This identification of the posterior with a Gibbs measure explains the growing popularity of physics-inspired terminology in Machine Learning—terms like energy-based models, free energy, and Boltzmann machines are now widespread ([3], [185], [233]).

In [227], the authors are particularly interested in computing the mean squared error across problem instances. To do so, one must analyze not only individual posteriors and partition functions but also ensemble averages, i.e., expectations over a distribution of models.

One important quantity in this context is the entropy:

$$S = -\sum_x P(x) \log P(x).$$

However, computing such averages is significantly harder than computing individual quantities. Fortunately, Statistical Physics provides tools precisely for this challenge. Spin glasses are systems in which the interaction strengths between particles are random variables. Their analysis has inspired numerous advances in Computer Science (see [54], [67], [227], [291] and references therein)

1.3.5 *Gibbs measure in machine learning problems on infinite graphs*

In this subsection, we provide an example where Gibbs measures on an infinite graph arise in the context of machine learning, specifically in Boltzmann machines and their extensions, including Restricted Boltzmann Machines (RBMs) and Deep Boltzmann Machines (DBMs).

We refer to [130], [154], [181], [185], and [332] for the theoretical foundations and practical applications of Boltzmann machines and their extensions in machine learning.

In a finite Boltzmann machine, one has a network of nodes (neurons) connected by weighted edges. If we imagine an infinite Boltzmann machine, it would have an infinite number of nodes and edges, creating an infinite graph.

The Gibbs measure is used to describe the probability distribution over the states of the nodes in the Boltzmann machine. The joint distribution of the visible and hidden nodes in a Boltzmann machine follows a Gibbs distribution, which is defined based on the energy function of the system.

The (Hamiltonian) energy function $E(\mathbf{v}, \mathbf{h})$ for a Boltzmann machine with visible nodes $\mathbf{v}$ and hidden nodes $\mathbf{h}$ is given by:

$$E(\mathbf{v}, \mathbf{h}) = -\sum_i b_i v_i - \sum_j c_j h_j - \sum_{i,j} v_i w_{ij} h_j,$$

where v_i and h_j are the states of the visible and hidden nodes, b_i and c_j are the biases of the visible and hidden nodes, and w_{ij} are the weights between visible node i and hidden node j.

Let this Hamiltonian be given on an infinite graph G. Take Λ a finite subgraph of the graph G. Denote by Ω the set of all configurations (states) of the visible and hidden nodes.

A probability measure μ on Ω is a Gibbs measure with respect to the Hamiltonian E if, for every finite subset $\Lambda \subset G$ and every configuration

$$\sigma := (\mathbf{v}, \mathbf{h})$$

outside Λ (denoted by σ_{Λ^c}), the conditional probability μ satisfies:

$$\mu(\sigma_\Lambda | \sigma_{\Lambda^c}) = \frac{1}{Z_\Lambda(\sigma_{\Lambda^c})} \exp\left(-\beta E_\Lambda(\sigma_\Lambda | \sigma_{\Lambda^c})\right),$$

where β is the inverse temperature (a positive constant), $E_\Lambda(\sigma_\Lambda|\sigma_{\Lambda^c})$ is the conditional Hamiltonian, which depends the visible and hidden nodes in Λ given the configuration outside Λ, and $Z_\Lambda(\sigma_{\Lambda^c})$ is the normalizing constant.

In practice, DBMs have a finite number of layers and nodes, but conceptually, they can be extended to consider an infinite number of hidden layers, making the graph lattice system infinite-dimensional. This theoretical construct helps in understanding the scalability and depth of neural networks.

For example, consider a DBM with multiple layers of hidden units. Each layer can be seen as a set of nodes in a lattice, and connections between nodes form the edges. If we extend this to an infinite number of layers, we have an infinite graph lattice system.

The energy function for a DBM extends the Boltzmann machine's energy function to multiple layers:

$$E(\mathbf{v}, \mathbf{h}^1, \mathbf{h}^2, \dots, \mathbf{h}^L) = -\sum_i b_i v_i - \sum_j c_j^1 h_j^1 - \sum_k c_k^2 h_k^2 - \dots - \sum_l c_l^L h_l^L$$
$$- \sum_{i,j} v_i w_{ij}^1 h_j^1 - \sum_{j,k} h_j^1 w_{jk}^2 h_k^2 - \dots$$

The Gibbs measure gives the joint distribution on finite set Λ over all layers:

$$\mu_\Lambda(\mathbf{v}, \mathbf{h}^1, \mathbf{h}^2, \ldots, \mathbf{h}^L) = \frac{1}{Z_\Lambda} \exp(-E_\Lambda(\mathbf{v}, \mathbf{h}^1, \mathbf{h}^2, \ldots, \mathbf{h}^L)).$$

Training DBMs involves adjusting the weights and biases to maximize the likelihood of the training data under the model's distribution.

1.3.6 *Gibbs measure in attention and transformers*

The definition of attention is due to [20], (see also [374], [376]) which we give here:

Definition 20 (Attention). *Let $K = (k_1, \ldots, k_N) \subset \mathbb{R}^{d_k}$ be a collection of keys, $V = (v_1, \ldots, v_N) \subset \mathbb{R}^{d_v}$ a collection of corresponding values, and $q \in \mathbb{R}^{d_q}$ a query. Also, let $a : \mathbb{R}^{d_q} \times \mathbb{R}^{d_k} \to \mathbb{R}$ be a measurable similarity function. Then* ***attention*** *is the mapping*

$$\mathit{Attention}(q, K, V) := \sum_{i=1}^{N} \mathit{softmatch}_a(q, K)_i \cdot v_i,$$

where Gibbs measure over the elements of K defined as

$$\mathit{softmatch}_a(q, K)_i := \frac{\exp(a(q, k_i))}{\sum_{j=1}^{N} \exp(a(q, k_j))}.$$

While $\text{Attention}(\cdot, K, V)$ is defined point-wise for a given query, it is almost always used to process a set of queries $Q = \{q_1, \ldots, q_M\} \subset \mathbb{R}^{d_q}$. Thus, one can write

$$\text{Attention}(Q, K, V) := \{\text{Attention}(q_i, K, V)\}_{i=1}^{M}.$$

Also, while $|K| = |V| = N$, in general M does not have to equal N. When $K = V = Q$, we call the following mapping **self-attention**:

$$Q \to \text{SelfAttention}(Q) := \text{Attention}(Q, Q, Q).$$

Transformers as Dynamical Systems on the Unit Sphere. Consider the unit sphere $\mathbb{S}^{d-1}$. A Transformer is modeled as a flow map on $(\mathbb{S}^{d-1})^n$. The input is a sequence of vectors of length n, namely,

$(x_i(0))_{i\in[n]} \in (\mathbb{S}^{d-1})^n$ serves as the initial condition, evolved through the dynamics:

$$\dot{x}_i(t) = P_{x_i(t)}\left(\frac{1}{Z_{\beta,i}(t)}\sum_{j=1}^{n} e^{\beta\langle Q(t)x_i(t),K(t)x_j(t)\rangle}V(t)x_j(t)\right) \tag{1.24}$$

for all $i \in [n]$ and $t \geq 0$. Here, $P_x y = y - \langle x, y\rangle x$ denotes the projection of $y \in \mathbb{S}^{d-1}$ onto the tangent space $T_x\mathbb{S}^{d-1}$.

The partition function $Z_{\beta,i}(t) > 0$ is defined as:

$$Z_{\beta,i}(t) = \sum_{k=1}^{n} e^{\beta\langle Q(t)x_i(t),K(t)x_k(t)\rangle}. \tag{1.25}$$

The matrices $Q(\cdot)$, $K(\cdot)$, and $V(\cdot)$ (Query, Key, Value) are learned parameters, and $\beta > 0$ is inverse temperature.

The *self-attention matrix*

$$A_{ij}(t) := \frac{e^{\langle Q(t)x_i(t),K(t)x_j(t)\rangle}}{Z_{\beta,i}(t)}, \quad (i,j) \in [n]^2, \tag{1.26}$$

which introduces nonlinear coupling. The matrix $A(t)$ is row-stochastic, with $A_{ij}(t)$ quantifying the attention particle i pays to particle j. Neighbor interactions are governed by $Q(t)$ and $K(t)$.

To simplify analysis (see for details [150]), one can focus on the case where Q, K, and V are constant (often identity matrices), yielding:

$$\dot{x}_i(t) = P_{x_i(t)}\left(\frac{1}{Z_{\beta,i}(t)}\sum_{j=1}^{n} e^{\beta\langle x_i(t),x_j(t)\rangle}x_j(t)\right) \tag{1.27}$$

Thus all above mentioned notation of Machine learning are either directly or formally based on Gibbs measures. A more detailed measure-theoretic interpretation of attention is given later in the book.

1.4 Gibbs measures in image models

This section is devoted to describe some image models of machine learning where Gibbs measures are applied (see, e.g. [41], [44], [98]- [102], [245] and the references therein).

1.4.1 *Image processing*

In digital imaging, a pixel (or picture element) is a physical point in a raster[8] image. The pixel is the smallest controllable element of a picture represented on the screen. Image processing involves the analysis and manipulation of a digitized image to improve its quality or extract useful information. It encompasses a wide range of techniques and methods aimed at enhancing the visual quality of images, removing noise, detecting features, segmenting objects, and much more.

In image processing, the energy function is designed to capture the characteristics of image. It often includes terms that account for pixel intensity, spatial relationships, and smoothness. For example:

$$H(x) = \sum_{\langle i,j \rangle} \phi(x_i, x_j),$$

where $\langle i, j \rangle$ are neighboring pixels and ϕ is a potential function that measures the interaction between pixel pairs.

Given an observed image y and a prior Gibbs measure $P(X)$, the goal is to find the most probable configuration x^* of the original image:

$$x^* := x^*(y) = \arg\max_x P(X = x | Y = y).$$

Simulated annealing is a common algorithm for finding the minimum energy configuration. It involves a stochastic process that gradually lowers the "temperature" of the system to converge to a global minimum.

Following [58] we give some examples.

Multiplicative noise.

Let X and Z be random fields on $S = \mathbb{Z}_m^2$ with phase space $\Lambda = \{-1, +1\}$.

Define the field $Y = XZ$ by $y(s) = x(s)z(s)$ for $s \in S$. The field Z is interpreted as multiplicative noise (or random flip field), flipping the phase at site s if $z(s) = -1$.

Since for any configuration $x \in \Omega = \{-1, 1\}^S$ we have that $x^2(s) \equiv 1$, for all $x, y, z \in \Omega$ we have $xy = z$ if and only if $y = xz$. Using this property we obtain

$$P(Y = y)P(X = x|Y = y) = P(X = x, Y = y) = P(X = x, ZX = y)$$

$$= P(X = x, Zx = y) = P(X = x, Z = yx).$$

[8] A raster is a dot matrix data structure that represents a generally rectangular grid of colored points in a display medium.

Denote

$$\pi(x|y) = P(X = x|Y = y).$$

If Z is independent of X, then

$$\pi(x|y) \propto P(X = x)P(Z = yx).$$

The random field X has the distribution $P(X = x) \propto e^{-H(x)}$.

To define the noise field Z, suppose $(Z(s), s \in S)$ is a family of i.i.d. random variables with

$$P(Z(s) = -1) = p, \quad \text{and} \quad P(Z(s) = +1) = q = 1 - p.$$

Then,

$$P(Z = z) = \prod_{s \in S} P(Z(s) = z(s)) \propto \exp\{\gamma \sum_{s \in S} z(s)\},$$

where $\gamma = \frac{1}{2} \ln \left(\frac{1-p}{p} \right)$.

The posterior distribution is

$$\pi(x|y) \propto \exp\{-H(x) + \gamma \sum_{s \in S} y(s)x(s)\}.$$

Here if $p > \frac{1}{2}$, then $\gamma > 0$.

Pixel-Edge model and image restoration. The restoration of degraded images is a branch of digital image processing, closely related to image segmentation and boundary detection. It has been extensively studied for its practical importance and theoretical interest.

In [148], an analogy is drawn between images and systems of statistical physics. Pixel gray levels, along with the presence and orientation of edges, are viewed as states of atoms or molecules in a lattice system. The analogous operation under the Gibbs measure yields the maximum a posteriori (MAP) estimate of the image, given the degraded observations. This results in a highly parallel "relaxation" algorithm for MAP estimation, with studied convergence properties.

Following [58] and [148], this subsection considers a stochastic model for the original image based on the Gibbs measure and a restoration algorithm based on stochastic relaxation and annealing. This approach computes the MAP estimate of the original image from the degraded image.

The essence of this restoration approach is a stochastic relaxation algorithm that generates a sequence of images converging, in an appropriate sense, to the MAP estimate. This sequence evolves through local changes in pixel gray levels and in the locations and orientations of boundary elements.

Here, the original image is represented as a pair $X = (F, L)$, where F is the matrix of observable pixel intensities and L is the matrix of unobservable edge elements. Thus, the usual gray levels are considered a marginal process. We refer to F as the intensity process and L as the line process.

The degradation model accounts for noise, blurring, and some nonlinearities, typical of most photochemical and photoelectric systems. Specifically, the degraded image G is of the form $G = \varphi(H(F)) + N$, where H is the blurring matrix, φ is a possibly nonlinear transformation, N is an independent noise field, and $+$ denotes any suitably invertible operation, such as addition or multiplication. It is shown that nonlinearities do not increase the computational burden.

Let $Z_m = \{(i,j) : 1 \leq i,j \leq m\}$ denote the $m \times m$ integer lattice. Then, $F = \{F_{i,j}\}_{(i,j)\in Z_m}$ represents the gray levels of the original, digitized image. Lowercase letters will denote the values assumed by these (random) variables; thus, for example, $\{f_{i,j}\}$ stands for $\{F_{i,j} = f_{i,j}, (i,j) \in Z_m\}$. We regard F as a sample realization of a random field, typically isotropic and homogeneous, with significant correlations extending well beyond nearest neighbors. The probability law of F is given by a Gibbs measure. Given a neighborhood system $\mathcal{N} = \{\mathcal{N}_{i,j} : (i,j) \in Z_m\}$, where $\mathcal{N}_{i,j} \subset Z_m$ denotes the neighbors of (i,j), an MRF over $(Z_m, \mathcal{N})$ is a stochastic process indexed by Z_m such that, for every (i,j) and every f,

$$P(F_{i,j} = f_{i,j} | F_{k,l} = f_{k,l}, (k,l) \neq (i,j))$$

$$= P(F_{i,j} = f_{i,j} | F_{k,l} = f_{k,l}, (k,l) \in \mathcal{N}_{i,j}).$$

By the Hammersley-Clifford theorem there exists real-valued "energy function" U defined over the states of the system such that

$$P(F = f) = \frac{1}{Z} \exp[-U(f)/T],$$

where Z is a normalizing constant and T is the temperature.

Here F is specified by this energy function:

$$U(f) = \sum_{c \in C} V_c(f),$$

where C is a collection of cliques, i.e., subsets of Z_m containing neighbors, and V_c is a potential function defined on the states of the clique c. The function U depends on the neighborhood system $\mathcal{N}$, and hence is of strictly local character.

Assume that the cliques are simply the pairs of neighbors:

$$C = \{\{(i,j),(k,l)\} : (k,l) \in \mathcal{N}_{i,j}\},$$

and take $V_c(f) = \beta|f_{i,j} - f_{k,l}|^2$, where $c = \{(i,j),(k,l)\}$, for some constant β. Then

$$U(f) = \beta \sum_{(i,j)} \sum_{(k,l) \in \mathcal{N}_{i,j}} |f_{i,j} - f_{k,l}|^2$$

and the field becomes the well-known Gaussian field.

In general, the posterior distribution (Gibbs measure) is given by

$$P(F, L|G) = \frac{1}{Z} \exp[-U(F, L, G)/T],$$

where the energy function $U(F, L, G)$ is given by

$$U(F, L, G) = U(F) + U(L) + U(G|F).$$

The term $U(G|F)$ accounts for the degradation process and is given by

$$U(G|F) = \sum_{i,j} |G_{i,j} - \varphi(H(F))_{i,j}|^2.$$

The relaxation algorithm generates a sequence of images that converges to the MAP estimate. The algorithm proceeds as follows:

(1) Initialize the temperature T to a high value.

(2) Repeat until convergence:

(a) For each pixel (i,j), update $F_{i,j}$ by sampling from the local conditional distribution

$$P(F_{i,j}|\{F_{k,l}\}, \{L_{k,l}\}, G).$$

(b) Lower the temperature T according to a predetermined schedule.

The annealing result (see [148, Section XII]) prescribes a schedule for lowering the temperature to ensure convergence to the global maximum of the posterior distribution. In practice, one can use this schedule as a guide and adjust it to balance convergence speed and solution quality.

1.4.2 *Image segmentation*

An image segmentation is[9] the process of partitioning a digital image into multiple segments (sets of pixels, also known as image objects).

The segmentation is needed to simplify and/or change the representation of an image into something that is more meaningful and easier to analyze. This is the process of assigning a label to every pixel in an image such that pixels with the same label share certain characteristics.

---

[9]https://en.wikipedia.org/wiki/Image_segmentation.

It is clear that more pixels typically provide more accurate representations of the original image. In color imaging systems, a color is typically represented by three or four component intensities such as red, green, and blue, or cyan, magenta, yellow, and black (which can be assigned to states of the Potts model).

In image analysis the problems involving incomplete data arise when some part of the data is missing or unobservable [74]. In this problems one wants to recover an original image which is hidden.

In [74] a Markov model-based image segmentation is studied, which involves hidden Markov random fields.

Let S be the set of pixels with a neighborhood system defined on it and $N(i)$ be the set of pixels considered as neighbors of pixel i.

The q-state (q-color) Potts model for the hidden data is defined as follows. Each pixel $i \in S$ takes one of q states, which can represent q different class assignments, represented as vectors $z_i \in \mathbb{R}^q$, such that one coordinate is 1, all others are equal to 0.

Then it is easy to see that the Kronecker delta has the representation $\delta_{z_i,z_j} = z_i^t z_j$.

The parameter β accounts for strength of the inter-pixel interaction.

The Gibbs measure of the configuration $\mathbf{z} = \{z_i, i \in S\}$ is defined by

$$\mu_\beta(\mathbf{z}) = \frac{1}{Z(\beta)} \exp\left(\frac{\beta}{2} \sum_{i \in S} \sum_{j \in N(i)} z_i^t z_j\right).$$

The parameter β is needed for control of spatial dependencies between pixels and leads to satisfying results in many cases (see [36], [74]).

This model has very nice symmetry (with respect to permutations of colors), it follows from this symmetry that the different colors in the image have the same probability (with respect to Gibbs measure of the Potts model). Therefore, to have distinct probabilities of the colors one can use the Potts model with external field which involves q-dimensional vector $v = (v_1, \dots, v_q)^t$ and the measure is

$$\mu_{\beta,v}(\mathbf{z}) = \frac{1}{Z(\beta, v)} \exp\left(\frac{\beta}{2} \sum_{i \in S} \sum_{j \in N(i)} z_i^t z_j + \sum_{i \in S} z_i^t v\right),$$

where v interpreted as weights for the different values of z_i.

In [74] the external fields are used to better adequacy especially when the color proportions are very unbalanced in the images to be recovered.

As usual the pixels set S in image analysis is $\mathbb{Z}^2$ with the second order neighborhood system: for each site, the neighbors are the eight sites surrounding it.

The image segmentation involves observed and unobserved data to be recovered.

In [74] the unobserved data is modeled as a discrete Markov random field, where a commonly used distribution for the random field is the q-color Potts model with a non-zero external field. Moreover, an illustration of the advantages of introducing the external field is given with numerical experiments.

1.4.3 *Image denoising*

Image denoising is the process of removing noise from an image. Noise can be introduced during image acquisition due to various factors such as poor lighting, high ISO settings in digital cameras, sensor heat, or transmission errors. The goal of image denoising is to reconstruct the original, clean image while preserving important details and features ([118], [160]).

Types of noise in images: Gaussian noise (known as white noise), it has a normal distribution of intensity values; Salt-and-pepper noise, it is random occurrences of black and white pixels; Poisson noise (known as shot noise), it occurs due to the discrete nature of electronic charge; Speckle noise, it is a multiplicative noise often found in medical and satellite images.

The mathematical foundation of image denoising revolves around the concept of removing noise from an observed image while preserving its underlying structure and features. This is typically modeled as an optimization problem where the goal is to find an estimate of the original image that minimizes some error criterion or maximizes some likelihood function.

Let $\mathbf{y}$ be the observed noisy image, $\mathbf{x}$ be the original clean image, and $\mathbf{n}$ be the noise. The relationship between these can be modeled as:

$$\mathbf{y} = \mathbf{x} + \mathbf{n}$$

Common noise models:

1. *Gaussian noise*: If the noise $\mathbf{n}$ is assumed to be Gaussian with zero mean and variance σ^2, then:

$$\mathbf{n} \sim \mathcal{N}(0, \sigma^2)$$

2. *Salt-and-pepper noise*: If the noise is randomly alters pixel values to the maximum or minimum intensity.

3. *Poisson noise*: If the noise follows a Poisson distribution, it is often due to photon counting in low-light conditions:

$$\mathbf{n} \sim \text{Poisson}(\mathbf{x})$$

Optimization Approaches:

1. A common approach to denoising is to minimize the squared difference between the noisy image and the estimated clean image:

$$\hat{\mathbf{x}} = \arg\min_{\mathbf{x}} \|\mathbf{y} - \mathbf{x}\|^2$$

This approach works well for Gaussian noise.

2. Total variation denoising aims to preserve edges while smoothing out noise:

$$\hat{\mathbf{x}} = \arg\min_{\mathbf{x}} \left(\|\mathbf{y} - \mathbf{x}\|^2 + \lambda \text{TV}(\mathbf{x})\right)$$

where $\text{TV}(\mathbf{x})$ is the total variation norm, often given by:

$$\text{TV}(\mathbf{x}) = \sum_{i,j} \sqrt{(\mathbf{x}_{i+1,j} - \mathbf{x}_{i,j})^2 + (\mathbf{x}_{i,j+1} - \mathbf{x}_{i,j})^2}$$

3. Non-local means denoising leverages the similarity of patches in the image:

$$\hat{\mathbf{x}}_i = \frac{1}{Z(i)} \sum_j \mathbf{y}_j e^{-\frac{\|\mathbf{y}_i - \mathbf{y}_j\|^2}{h^2}}$$

where $Z(i)$ is a normalizing constant and h is a filtering parameter.

4. Wavelet denoising involves transforming the image into the wavelet domain, thresholding the coefficients, and transforming back:

$$\hat{\mathbf{x}} = \mathcal{W}^{-1}(T(\mathcal{W}(\mathbf{y})))$$

where $\mathcal{W}$ and $\mathcal{W}^{-1}$ are the wavelet transform and its inverse, and T is a thresholding function.

In a probabilistic framework, denoising can be formulated as maximum a posteriori (MAP) estimation problem:

$$\hat{\mathbf{x}} = \arg\max_{\mathbf{x}} P(\mathbf{x}|\mathbf{y})$$

Using Bayes' theorem:

$$P(\mathbf{x}|\mathbf{y}) = \frac{P(\mathbf{y}|\mathbf{x})P(\mathbf{x})}{P(\mathbf{y})}$$

Assuming $\mathbf{n}$ is Gaussian, $P(\mathbf{y}|\mathbf{x})$ is Gaussian, and $P(\mathbf{x})$ is a prior distribution on the image.

Deep learning approaches:

1. Deep learning approaches like convolutional neural networks (CNNs) learn a mapping from noisy images to clean images using a large dataset of paired examples:

$$\hat{\mathbf{x}} = f(\mathbf{y}; \theta)$$

where f is a neural network parameterized by θ, and the parameters are learned by minimizing a loss function, typically the mean squared error:

$$\mathcal{L}(\theta) = \frac{1}{N} \sum_{i=1}^{N} \|\mathbf{y}_i - f(\mathbf{y}_i; \theta)\|^2$$

2. Generative adversarial networks (GANs) based denoising involves training a generator network to produce clean images from noisy ones, while a discriminator network tries to distinguish between real clean images and denoised images:

$$\mathcal{L}_{\text{GAN}} = \mathbb{E}[\log D(\mathbf{x})] + \mathbb{E}[\log(1 - D(G(\mathbf{y})))]$$

where D is the discriminator, G is the generator, and $\mathbf{y}$ is the noisy image.

These mathematical models and methods form the basis of various image denoising algorithms, each with its strengths and limitations depending on the type and level of noise present in the images.

1.4.4 *Image enhancement*

Image enhancement involves improving the visual quality of an image, making it more suitable for a specific application.

A fundamental operation in image processing, convolution is used to apply filters to an image. The convolution of an image I with a kernel K is given by:

$$(I * K)(x, y) = \sum_{i=-m}^{m} \sum_{j=-n}^{n} I(x - i, y - j) K(i, j)$$

Fourier transform converts an image from the spatial domain to the frequency domain, making it easier to analyze and enhance certain frequency components.

$$F(u, v) = \sum_{x=0}^{M-1} \sum_{y=0}^{N-1} I(x, y) e^{-j2\pi\left(\frac{ux}{M} + \frac{vy}{N}\right)}$$

Noise reduction techniques such as Gaussian smoothing use statistical properties of noise to reduce its impact on the image.

$$G(x, y) = \frac{1}{2\pi\sigma^2} e^{-\frac{x^2+y^2}{2\sigma^2}}$$

Bayesian methods use probabilistic models to infer the most likely enhanced image given the observed noisy image.

The choice of technique depends on the specific requirements and characteristics of the image and the application at hand.

1.4.5 *Texture synthesis*

Besag model: In this subsection we review a Besag model, which is used to describe the texture of various materials and involves the concepts of sites, neighborhood systems, and cliques in the context of image processing.

The set of sites is $S = \mathbb{Z}_m^2$, and the phase space is $\Lambda = \{0, 1, \ldots, L\}$. In the context of image processing, a site s is a pixel, and a phase $\lambda \in \Lambda$ is a shade of grey, or a color. The neighborhood system is

$$N_s = \{t \in S; t \neq s \text{ and } \|t - s\|^2 \leq d\},$$

where d is a fixed positive integer and where $\|t-s\|$ is the Euclidean distance between s and t. In this model, the only cliques participating in the energy function are singletons and pairs of mutual neighbors. The set of cliques appearing in the energy function is a disjoint sum of collections of cliques

$$C = \sum_{j=1}^{m(d)} C_j,$$

where C_1 is the collection of singletons, and all pairs $\{s, t\} \in C_j$, $2 \leq j \leq m(d)$, have the same distance $\|t - s\|$ and the same direction. The potential is given by

$$V_C(x) = \begin{cases} -\ln \binom{L}{x(s)} + \alpha_1 x(s) & \text{if } C = \{s\} \in C_1, \\ \alpha_j x(s) x(t) & \text{if } C = \{s, t\} \in C_j, \end{cases}$$

where $\alpha_j \in \mathbb{R}$. For any clique C not of type C_j, $V_C = 0$.

Define τ, depending on $x(N_s)$ as follows:

$$\tau = \tau(N_s) = \frac{e^{-(\alpha, b)}}{1 + e^{-(\alpha, b)}},$$

where (α, b) is the scalar product of

$$\alpha = (\alpha_1, \ldots, \alpha_{m(d)}), \quad b = (b_1, \ldots, b_{m(d)}),$$

here $b_1 = 1$, and for $j \in [2, m(d)]$,

$$b_j = b_j(x(N_s)) = x(r) + x(t),$$

and $\{r, s\}$ and $\{t, s\}$ are the two pairs in C_j containing s.

Note that τ is a function of (α, b). The parameter α_j controls the bond in the direction and at the distance that characterize C_j.

Proposition 3. *The local system has the form*

$$\pi^s(x) = \binom{L}{x(s)} \tau^{x(s)}(1-\tau)^{L-x(s)}.$$

Proof. Using general formula of a Gibbs measure, we can write the local characteristic at s as

$$\pi^s(x) = \frac{\exp\left[\ln\binom{L}{x(s)} - \alpha_1 x(s) - \left[\sum_{j=2}^{m(d)} \alpha_j \sum_{t:\{s,t\}\in C_j} x(t)\right] x(s)\right]}{\sum_{\lambda\in\Lambda} \exp\left[\ln\binom{L}{\lambda} - \alpha_1 \lambda - \left[\sum_{j=2}^{m(d)} \alpha_j \sum_{t:\{s,t\}\in C_j} x(t)\right] \lambda\right]}$$

The numerator of the fraction equals

$$\binom{L}{x(s)} e^{-(\alpha,b)x(s)},$$

and the denominator is

$$\sum_{\lambda\in\Lambda} \binom{L}{\lambda} e^{-(\alpha,b)\lambda} = \sum_{\ell=0}^{L} \binom{L}{\ell} (e^{-(\alpha,b)})^{\ell} = (1 + e^{-(\alpha,b)})^L.$$

This completes the proof. □

This proposition allows to call the model as the autobinomial model, because its local characteristics reduce to binomial distributions, where the probability of success is determined by the neighborhood configuration and the number of trials equal to the number of gray levels (see [4] and the references therein).

In practice, one can generalize this model by replacing $\binom{L}{x(s)}$ in the expression of the potential by a function of $x(s)$.

Acuna model: In [4], Acuna introduced a model for texture generation that integrates both local and global information from the image. The local information shapes the overall "aspect" of the picture, while the global information constrains the gray-level proportions. A method for parameter estimation was developed that is straightforward to implement and produces estimators with strong convergence properties. The estimated model parameters can be used to generate textures that are visually indistinguishable from the original samples.

Here we give some basic notations of Acuna model. As seen in the definition of the autobinomial model, the Gibbs measure depends on a parameter θ that ranges over a known parameter space Θ. This is a common occurrence in this type of application, leading us to consider a family $\mathcal{P} = \{P_\theta : \theta \in \Theta\}$ of parametric models. An important concept in this context is parameter identifiability.

For a family $\mathcal{P} = \{P_\theta : \theta \in \Theta\}$ of probability measures indexed by a parameter θ, where θ ranges over a known parameter space Θ, the parameter θ is said to be identifiable if and only if, for all $\theta_1, \theta_2 \in \Theta$ such that $\theta_1 \neq \theta_2$, we have $P_{\theta_1} \neq P_{\theta_2}$.

Since a finite Markov random field (Gibbs measure) is completely determined by its local characteristics, it is important to understand how the concept of identifiability applies to them.

Let

$$S = \{s = (i, j) : 1 \leq i, j \leq N\}.$$

$(S, \mathcal{G})$ is a graph. For $d > 0$, let

$$G^d_{(i,j)} = \{(k, l) \in S : 0 < (k - i)^2 + (l - j)^2 \leq d\},$$

$$\mathcal{G}_d = \{G^d_{(i,j)} : (i, j) \in S\}.$$

Furthermore, let $\mathcal{C}_1$ denote the class of cliques containing only one element, i.e., $\mathcal{C}_1 = S$, and $\mathcal{C}_j$, $j = 2, \ldots, m$, the class of pair cliques of type j, where m is the number of pair clique types associated with the graph $(S, \mathcal{G}_d)$.

Let $X = \{X_s, s \in S\}$ be a family of random variables indexed by S with a common state space $A = \{0, 1, \ldots, g - 1\}$, for some integer $g \geq 2$. Thus, the value of $X_s \in A$ for all $s \in S$. Let Ω denote the configuration space, $\Omega = \{x = (x_s)_{s \in S} : x_s \in A\}$.

For each configuration $x \in \Omega$ define

- the vector function $B(x) = (B_1(x), \ldots, B_m(x))$ as

$$B_1(x) = \sum_{s \in S} x_s, \quad B_j(x) = \sum_{\{s,t\} \in \mathcal{C}_j} x_s x_t, \quad j = 2, \ldots, m.$$

- the vector function $\mathbf{b}(x) = (b_1(x), \ldots, b_m(x))$ as

$$b_1(x) = 1, \quad b_j(x) = x_s + x_t, \quad \{s,t\} \in \mathcal{C}_j, \quad j = 2, \ldots, m.$$

The (parametric) energy function, defined in [4], is as follows

$$U(x; \theta, \sigma, \mu) = U(x; \theta) + \sigma N^2 \|M(x) - \mu\|^2,$$

where $U(x;\theta)$ is the energy function, $\sigma > 0$, $M(x) = (M_0(x), \ldots, M_{g-1}(x))$, $\mu = (\mu_0, \mu_1, \ldots, \mu_{g-1}) \in \mathbb{R}^g$, $\mu_i \geq 0$, $\sum_{i=0}^{g-1} \mu_i = 1$, and $M_i(x) = (1/N^2) \sum_{s \in S} I_i(x_s)$, with $I_i(\cdot)$ being the indicator function of the set $\{i\}$, and $\|\cdot\|$ denotes the usual Euclidean norm.

We shall see that σ can be chosen sufficiently large, but independently of the size of the lattice, in such a way that the distribution of gray levels will remain stable, close to the ideal values specified by μ, without inhibiting the local effect of the parameter θ.

For a configuration $x = (x_t)_{t \in S}$ denote by $x_{(s)}$ the configuration obtained from x after the pixel s is removed.

The local characteristics of the model are given by:

$$p_s(y \mid x_{(s)}; \Phi) = \frac{\binom{g-1}{x_s} \exp\left(y\theta \cdot \mathbf{b}(x_{(s)}) + \sigma R(y, x_{(s)}, \mu)\right)}{z(x_{(s)}; \Phi)},$$

for $y = 0, 1, \ldots, g-1$, where $\Phi = (\theta, \sigma, \mu)$,

$$R(u, x_{(s)}, \mu) = 2\mu_u - 2\tilde{M}_u(x_{(s)}),$$

where

$$\tilde{M}_u(x_{(s)}) = \frac{2}{N^2} \sum_{i \neq s} I_u(x_i), \quad u = 0, 1, \ldots, g-1,$$

and

$$z(x_{(s)}; \Phi) = \sum_{u=0}^{g-1} \binom{g-1}{u} \exp\left(u\theta \cdot \mathbf{b}(x_{(s)}) + \sigma R(u, x_{(s)}, \mu)\right).$$

The local characteristics are essentially independent of the image size. The conditional distribution of the brightness level at pixel s, given all other pixels in the image, incorporates both local and global interactions. The neighborhood of each pixel includes all other pixels in the square lattice. However, the local characteristics can be updated using only local information, as the global interaction intervenes only as proportions, not specific configurations.

One advantage of this model is that it offsets the biases of the autobinomial model more naturally by incorporating information about the image's gray-level histogram into the energy function. In contrast, to obtain interesting textures with the autobinomial model, the sample space must be artificially restricted through sampling. By specifying different values of μ for a fixed θ value, it is possible to obtain pictures of varying quality due to the relative importance of each color in the image.

Identifiability of the parameter: Let the family of measures $\mathcal{P}$ be indexed by $\theta = (\beta, \nu)$; that is,

$$\mathcal{P} = \{P_\theta : \theta \in \Theta\}$$

with

$$\Theta = \{(\beta, \nu) : \beta \in \mathbb{R}^m, \nu > 0\} \tag{1.28}$$

and assume that μ is known and

$$P_\theta(x) = \frac{\exp(-U(x,\theta))}{Z(\theta;\mu)}, x \in \Omega,$$

with

$$U(x,\theta) = -\sum_{s\in S} \ln \binom{g-1}{x_s} - \theta \cdot \gamma(x;\mu),$$

where $\gamma(x,\mu) = y = (B(x), -N^2\|M(x) - \mu\|^2)$.

The measure $P_\theta(\cdot)$ defined above induces a probability measure $\hat{P}_\theta(\cdot)$ on $\mathbb{R}^{m+1}$ as follows: for any $y = (y_1, \ldots, y_m, y_{m+1}) \in \mathbb{R}^{m+1}$, define

$$\hat{P}_\theta(y) = \sum_{x\in\mathcal{B}_y} P_\theta(x) \frac{\exp((\theta, y))}{Z(\theta;\mu)} \nu(\{y\}),$$

where

$$\mathcal{B}_y = \{x \in \Omega : B_k(x) = y_k, 1 \le k \le m, \text{ and } -N\|M(x) - \mu\|^2 = y_{m+1}\}$$

and

$$\nu(\{y\}) = \begin{cases} \sum_{x\in\mathcal{B}_y} \exp\left(\sum_{s\in S} \ln\binom{g-1}{x_s}\right) & \text{if } \mathcal{B}_y \ne \emptyset, \\ 0 & \text{if } \mathcal{B}_y = \emptyset, \end{cases}$$

$$Z(\theta;\mu) = \sum_{z\in\Omega} \exp\left(\sum_{s\in S} \ln\binom{g-1}{z_s} + \theta\cdot\gamma(z;\mu)\right) = \sum_{y\in\mathbb{R}^{m+1}} \exp((\theta, y))\nu(\{y\}).$$

Note that ν is a discrete measure with finite support and $\nu(y) \geq 1$ for all $y \in \text{supp}(\nu)$. Moreover, $\hat{P}_\theta(\cdot)$ is absolutely continuous with respect to ν and has density

$$\exp((\theta, y))/Z(\theta; \mu) = \exp(\theta \cdot y - \ln Z(\theta; \mu)).$$

Consequently, $\hat{P}_\theta(\cdot)$ is a member of a natural exponential family and y is the vector of sufficient statistics; $\Theta(\nu) = \mathbb{R}^{m+1}$, is open and convex.

One of main results in [4] is

Theorem 8. *For any integer m, the parameter θ is identifiable.*

Since $\hat{P}_\theta(\cdot)$ is a member of the exponential family and Θ is convex and locally compact, it is shown that the maximum likelihood estimator (MLE) of θ is unique. It converges with probability one to the true parameter value as the number of images in the sample increases, and it is asymptotically normal (see [4], [5] for proofs).

While the properties of MLE make them highly desirable, they have limited practical application because the solutions to the likelihood equations cannot be computed due to the intractability of partition function $Z(\theta; \mu)$. Now, following [4], we will present an alternative approach that addresses these computational challenges while still yielding consistent estimators.

Maximum bivariate pseudolikelihood estimation:

Let $\mathbf{X}^1, \ldots, \mathbf{X}^n$ be a random sample from the Gibbs measure $P_{\Phi_0}(\cdot)$, defined on Ω by

$$P_{\Phi_0}(\mathbf{x}) = \frac{1}{Z(\Phi_0)} \exp\left(-U(\mathbf{x}, \Phi_0)\right), \quad \mathbf{x} \in \Omega,$$

where

$$U(\mathbf{x}, \Phi_0) = -\sum_{s \in S} \ln \binom{g-1}{x_s} - \theta_0 \cdot B(\mathbf{x}) + \sigma_0 N^2 \left\| M(\mathbf{x}) - \mu_0 \right\|^2,$$

with $\Phi_0 = (\theta_0, \sigma_0, \mu_0)$ with $\theta_0 \in \mathbb{R}^m, \sigma_0 > 0$, and $\mu_0 \in \mathbb{R}^g$, such that $\sum_{l=0}^{g-1} \mu_{0,l} = 1, \mu_{0,l} \geq 0$.

Method of estimation makes use of the local structure of the model. Namely, one uses the bivariate conditional probabilities at pairs of sites belonging to the same clique type class.

Definition 21. *Let $\theta \in \mathbb{R}^m, \sigma > 0, \mu \in \mathbb{R}^g$. For a random sample $\mathbf{x}^1, \ldots, \mathbf{x}^n$ from $P_{\phi_0}(\cdot)$ and a fixed $j \in \{2, \ldots, m\}$, we define the bivariate* *pseudolikelihood function (bpsl)* $Q_n\left(\theta,\sigma,\mu;\mathbf{x}^1,\ldots,\mathbf{x}^n\right)$ *as*

$$Q_n\left(\theta,\sigma,\mu;\mathbf{x}^1,\ldots,\mathbf{x}^n\right) = \sum_{l=1}^{n}\sum_{\{s,t\}\in\mathcal{C}_j^0} \ln p\left(x_s^l, x_t^l, \mid \mathbf{x}_{(s,t)}^l, \theta,\sigma,\mu\right), \tag{1.29}$$

where $\mathbf{x}_{(s,t)}$ *denotes the configuration obtained from* $\mathbf{x}$ *after the sites* s *and* t *are removed, and* $\mathcal{C}_j^0$ *denotes the set of all cliques of type* j *such that each element in the clique has a complete neighborhood.*

In this definition, we restrict ourselves to cliques that are not too near the border of the lattice.

The bivariate conditional probabilities that appear in (1.29) are given by

$$p\left(x_s,x_t \mid \mathbf{x}_{(s,t)},\theta,\sigma,\mu\right) = \frac{\binom{g-1}{x_s}\binom{g-1}{x_t}\exp\left(\theta\cdot d\left(x_s,x_t,\mathbf{x}_{(s,t)}\right)+\sigma T\left(x_s,x_t,\mathbf{x}_{(s,t)},\mu\right)\right)}{\mathcal{Z}}, \tag{1.30}$$

where

$$\mathcal{Z} = \sum_{i,j=0}^{g-1}\binom{g-1}{i}\binom{g-1}{j}\exp\left(\theta\cdot d\left(i,j,\mathbf{x}_{(s,t)}\right)+\sigma T\left(i,j,\mathbf{x}_{(s,t)},\mu\right)\right)$$

and

$$d\left(i,j,\mathbf{x}_{(s,t)}\right) = \left(d_1\left(i,j,\mathbf{x}_{(s,t)}\right),\ldots,d_m\left(i,j,\mathbf{x}_{(x,t)}\right)\right)$$

with

$$d_1\left(i,j,\mathbf{x}_{(s,t)}\right) = i+j = ib_1\left(\mathbf{x}_{(s)}\right)+jb_1\left(\mathbf{x}_{(t)}\right),$$

$$d_k\left(i,j,\mathbf{x}_{(s,t)}\right) = ib_k\left(\mathbf{x}_{(s)}\right)+jb_k\left(\mathbf{x}_{(t)}\right)-ijI_{\mathcal{C}_k}(\{s,t\}), k=2,\ldots,m,$$

$$T\left(i,j,\mathbf{x}_{(s,t)},\mu\right) = 2\left\{\mu_i+\mu_j-\frac{\sum_{r\neq s,t}\left(I_i\left(x_r\right)+I_j\left(x_r\right)\right)+I_i(j)}{N^2}\right\}.$$

The parameter estimation method described in [97] utilizes available data to estimate the joint probability distribution of intensity values for small blocks of pixels. This method employs bivariate pseudolikelihood, which assumes a known theoretical distribution of gray levels that is incorporated into the algorithm as part of the maximization function. Here we use of bivariate pseudolikelihood because, in our efforts to sample from a constrained set of configurations, we are particularly interested in models with large values of σ. Generally speaking, the bivariate pseudolikelihood function exhibits less flatness compared to the univariate pseudolikelihood. In fact, the local characteristics of the restricted model (with $\sigma = \infty$) are degenerate, whereas the bivariate conditional distributions retain their distinctiveness.

Assume now that μ_0 is known, and the family of measures $\mathscr{P}$ is indexed by $\eta = (\theta, \sigma) \in \Theta$, with Θ as defined in (1.28). Then the bivariate pseudolikelihood function reduces to a function of θ and σ, and we define

$$\begin{aligned} q_n(\eta) &= Q_n\left(\theta, \sigma, \mu_0; \mathbf{x}^1, \ldots, \mathbf{x}^n\right) \\ &= \sum_{l=1}^{n} \sum_{\{s,t\} \in C_j^0} \left\{ \ln \binom{g-1}{x_s^l} + \ln \binom{g-1}{x_t^l} \right. \\ &\quad \left. + \eta \cdot \mathbf{w}\left(x_s^l, x_t^l, \mathbf{x}_{(s,t)}^l, \mu_0\right) - \ln \zeta\left(\mathbf{x}_{(s,t)}^l, \eta, \mu_0\right) \right\}, \end{aligned} \tag{1.31}$$

where $\mathbf{w}\left(i, j, \mathbf{x}_{(s,t)}, \mu\right) = \left(d\left(i, j, \mathbf{x}_{(s,t)}\right), T\left(i, j, \mathbf{x}_{(s,t)}, \mu\right)\right)$, and

$$\begin{aligned} &\zeta\left(\mathbf{x}_{(s,t)}, \eta, \mu_0\right) \\ &\quad = \sum_{i,j=0}^{g-1} \binom{g-1}{i} \binom{g-1}{j} \exp\left(\eta \cdot \mathbf{w}\left(i, j, \mathbf{x}_{(s,t)}, \mu_0\right)\right). \end{aligned}$$

The following lemmas are used to prove the existence and consistency of the maximum bivariate pseudolikelihood estimator (see [4] for detailed proofs).

Lemma 2. *The following assertions hold*

1. *For all n, the function $q_n(\cdot)$ defined in (1.31) is concave.*
2. *With probability one, for sufficiently large $n, q_n(\eta)$ is strictly concave.*

Since for sufficiently large $n, q_n(\eta)$ is a strictly concave smooth function, it is necessary to show only that the gradient $\nabla q_n(\eta)$ vanishes inside $\mathbb{R}^{m+1}$ to obtain a unique global maximum. This is equivalent to showing that $q_n(\eta)$ does not have directions of recession, meaning that, for any $\delta, \xi \in \mathbb{R}^{m+1}, \xi \neq 0$,

$$\lim_{\lambda \to \infty} q_n(\delta + \lambda \xi) = -\infty.$$

Lemma 3. *The following assertions hold*

a. *Let $\mu \in \mathbb{R}^g$. If for all $\xi \in \mathbb{R}^{m+1}$, $\xi \neq 0$, there exists an $l \in \{1, \ldots, n\}$ and an $\{s,t\} \in C_j^0$ such that*

$$\xi \cdot \mathbf{w}\left(x_s^l, x_t^l, \mathbf{x}_{(s,t)}^l, \mu\right) < \max_{i,j \in A} \xi \cdot \mathbf{w}\left(i, j, \mathbf{x}_{(s,t)}^l, \mu\right), \tag{1.32}$$

then $q_n(\eta)$ does not have directions of recession.

b. *With probability one, for a sufficiently large n, condition (1.32) is satisfied for any nonzero* $\xi \in \mathbb{R}^{m+1}$.

From these lemmas one obtains (see [4]):

Theorem 9. *With probability one, for all sufficiently large n, there exists a unique* $\hat{\eta}_n \in \mathbb{R}^{m+1}$ *that maximizes* $q_n(\cdot)$.

The value $\hat{\eta}_n$ is called the maximum bivariate pseudolikelihood estimator (MBPE).

Theorem 10. *[4] Let* $\mathbf{X}^1, \ldots, \mathbf{X}^n$ *be a random sample from* $P_{\Phi_0}(\cdot)$ *for* $\Phi_0 = (\eta_0, \mu_0)$ *with* $\eta_0 \in \Theta$. *Then the MBPE* $\hat{\eta}_n$ *of* η_0 *is strongly consistent, i.e.,*

$$P_{\Phi_0}(\hat{\eta}_n \to \eta_0) = 1.$$

1.4.6 *Signal reconstruction*

Piecewise constant signals and images represent a significant category of data, commonly found in various applications. Typical examples include bar code signals, logos, cartoons, Quick Response (QR) codes, and text images. These types of data are extensively utilized across general commercial sectors and the automotive industry [380].

The Potts model has found practical applications in signal reconstruction[10].

Consider noisy observation of a piecewise constant signal g in $\mathbb{R}^n$. To recover g from the noisy observation vector f in $\mathbb{R}^n$, one seeks a minimizer of the corresponding inverse problem, the L^p-Potts functional $P_\gamma(u)$ which is defined by

$$P_\gamma(u) = \gamma \|\nabla u\|_0 + \|u - f\|_p^p = \gamma \sum_i \left(1 - \delta_{u_i u_{i+1}}\right) + \sum_{i=1}^{n} |u_i - f_i|^p,$$

where the jump penalty $\|\nabla u\|_0$ forces piecewise constant solutions and the data term $\|u - f\|_p^p$ couples the minimizing candidate u to the data f. The parameter $\gamma > 0$ controls the trade-off between regularity and data fidelity. There are fast algorithms for the exact minimization of the L^1 and the L^2-Potts functional (see [138]). The algorithms are accompanied by a series of illustrative examples from molecular biology.

In [382], it is demonstrated that every sequence of minimizers for the discrete L^1-Potts functionals has an accumulation point, which corresponds

[10] https://en.wikipedia.org/wiki/Potts_model#cite_note-1

to a minimizer of the continuous model. Additionally, the paper presents an algorithm capable of computing an exact minimizer of the discrete L^1-Potts functional.

In image processing, the Potts functional is closely associated with the segmentation problem. However, solving this problem in two dimensions is NP-hard (see [56] for further details).

1.5 More applications in machine learning and related fields

Gibbs measures play an important role in various machine learning applications. Here, we provide a list of 15 examples (some of which will be discussed in detail in the upcoming chapters of this book):

(1) *Boltzmann machines* [181]: Gibbs measures are used to define the distribution of the states of Boltzmann machines, which are energy-based models used for learning probabilistic distributions.
(2) *Markov Random Fields (MRFs)* [213]: Gibbs measures are used to describe the joint distribution of MRFs, which model the dependencies between variables in a network.
(3) *Conditional Random Fields (CRFs)* [230]: Gibbs distributions are used in CRFs to model the conditional probability of the output given the input.
(4) *Ising Models* [148]: Inference and learning in Ising models, a type of MRF used in image processing and statistical physics, utilize Gibbs measures.
(5) *Gaussian Markov Random Fields (GMRFs)* [329]: GMRFs use Gibbs measures to define the joint distribution of Gaussian random variables with a Markov property.
(6) *Simulated Annealing* [215]: Gibbs measures guide the acceptance probability of new states in simulated annealing, an optimization technique.
(7) *Gibbs Sampling* [148]: A Markov Chain Monte Carlo (MCMC) method that generates samples from a multivariate probability distribution by iteratively sampling from the conditional distributions, often used in Bayesian inference.
(8) *Factor Graphs* [229]: Gibbs measures describe the probability distribution over the variables in factor graphs, which generalize MRFs and Bayesian networks.
(9) *Variational Inference* [46]: Gibbs measures are used to approximate complex posterior distributions in variational inference.

(10) *Monte Carlo Dropout* [141]: Gibbs measures underpin the probabilistic interpretation of dropout in neural networks for uncertainty estimation.
(11) *Probabilistic Graphical Models (PGMs)* [219]: Gibbs measures are fundamental in defining the joint distributions over the nodes in PGMs.
(12) *Energy-Based Models* [233]: Models like Hopfield networks and deep energy-based models use Gibbs measures to define their probability distributions.
(13) *Stereo Vision* [360]: Gibbs measures are used in stereo vision to model the disparity between stereo image pairs.
(14) *Generative Adversarial Networks (GANs)* [161]: Gibbs measures are used in some probabilistic GAN variants and other energy-based generative models to cast energies into probabilities via the Gibbs distribution [216, 401].
(15) *Nonparametric Bayesian Methods* [364]: Gibbs measures are used in nonparametric Bayesian methods like the Dirichlet Process to model infinite-dimensional distributions.

1.6 Conclusion

In this chapter, we have delved into the foundational aspects of Gibbs measures and their significance within the field of machine learning. We began by establishing a comprehensive understanding of measurable spaces and measures, which set the stage for exploring the more intricate concepts of graphs and configuration spaces. These preliminary topics were crucial in building the groundwork necessary to comprehend Gibbs measures effectively.

The introduction to Gibbs measures highlighted their mathematical formulation and properties, emphasizing their role in defining probability distributions over complex systems. We explored how Gibbs measures facilitate the modeling of dependencies and interactions within a system, which is particularly relevant in various machine learning contexts.

Gibbs measures were introduced in the context of image analysis, where they serve as priors in tasks such as segmentation, denoising, and enhancement. This section demonstrated how spatial interactions modeled through Gibbs distributions enable the capture of texture, structure, and other perceptual qualities.

In summary, this chapter provided a thorough introduction to Gibbs measures and their relevance in machine learning. By understanding these concepts, we gain valuable insights into how probabilistic modeling and sampling techniques can be leveraged to tackle complex problems in the field.

Recent Advances in Statistical Physics and Machine Learning

We refer to [227] for a comprehensive overview of the statistical physics approach to problems in computer science. This work emphasizes both heuristic and rigorous mathematical techniques, covering a wide range of topics including graph theory, constraint satisfaction problems, inference, machine learning, neural networks, and statistics.

In recent years (see [139] and the references therein), mean-field techniques originally developed in the context of complex systems, such as the TAP equations and the cavity approximation [272], along with their variants related to belief propagation, have found applications in machine learning. These methods have been applied to diverse problems, including: Compressed sensing [226], Community detection [93], among others.

In these applications, a Bayesian framework is often adopted, with the so-called teacher-student scenario introduced in [143]. This setup simplifies analysis in the Bayes-optimal case (see [397] for details), where the notion of Bayes optimality has been interpreted [207] in the context of error-correcting codes as equivalent to the Nishimori line [291] in statistical physics.

Recent references for further reading are [24], [63], [85], [92], [269], [298].

Chapter 2

Gibbs Sampling, Interacting Particle Systems and Markov Chain Monte Carlo Methods

The chapter starts with an in-depth treatment of Gibbs sampling methods, including their theoretical basis and practical applications.

This chapter explores the Markov Chain Monte Carlo (MCMC) methods and interacting particle systems (IPS), highlighting their applications in both statistical physics and machine learning. We begin by introducing and illustrating Gibbs sampling, an instance of MCMC methods, across various models, including the hard-core model, Potts model, disk packing, graph coloring, and neural networks. The chapter then presents the foundations of MCMC techniques, emphasizing their role in approximating complex distributions. Finally, we introduce IPS, outlining their mathematical structure, key examples, and contour methods, followed by a discussion of their theoretical and practical applications in modern machine learning.

2.1 Gibbs sampling

2.1.1 *Sampling procedure*

One of the key algorithms used in machine learning for sampling from complex probability distributions is Gibbs sampling. This algorithm iteratively samples from the conditional distributions of each component of the system, given the values of all other components. By repeatedly sampling from these conditional distributions, Gibbs sampling can approximate the target distribution defined by the Gibbs measure.

Gibbs Sampling is a Markov Chain Monte Carlo (MCMC) algorithm employed to sample from intricate, high-dimensional probability distributions. It proves especially valuable in Bayesian statistics for estimating the posterior distributions of model parameters when direct sampling is impractical [148, 156, 314]. The essence of Gibbs sampling lies in the concept that while sampling directly from a complex multivariate distribution may be difficult, sampling from the conditional distributions of individual variables given all others is often tractable. We aim to obtain k samples of $\mathbf{X} = (X_1, \ldots, X_n)$ from a joint distribution $P(X_1, \ldots, X_n)$. Let's denote the sample at iteration t as $\mathbf{X}^{(t)} = (X_1^{(t)}, \ldots, X_n^{(t)})$. The sampling procedure is as follows:

Algorithm 1 Gibbs Sampling

Initialize: Start with an initial value for each variable: $(X_1^{(0)}, X_2^{(0)}, \ldots, X_n^{(0)})$
for $t = 0, 1, 2, \ldots, k-1$ **do**
  Sample $X_1^{(t+1)} \sim P(X_1 | X_2^{(t)}, X_3^{(t)}, \ldots, X_n^{(t)})$
  Sample $X_2^{(t+1)} \sim P(X_2 | X_1^{(t+1)}, X_3^{(t)}, \ldots, X_n^{(t)})$
  $\vdots$
  Sample $X_n^{(t+1)} \sim P(X_n | X_1^{(t+1)}, X_2^{(t+1)}, \ldots, X_{n-1}^{(t+1)})$
end for
Return: Sequence of samples $\{\mathbf{X}^{(1)}, \mathbf{X}^{(2)}, \ldots, \mathbf{X}^{(k)}\}$

The sequence of samples will form a Markov chain.
Convergence:

- Burn-in Period: The initial samples may not represent the target distribution well, so they are discarded.
- Stationarity: After a sufficient number of iterations, the distribution of the samples should approximate the target distribution. This is when the chain has "converged".

Example 4. *Consider a simple case with two variables X and Y with a joint distribution $P(X, Y)$. Set initial values $X^{(0)}$ and $Y^{(0)}$ and do iterations: sample $X^{(t+1)}$ from $P(X|Y^{(t)})$, then sample $Y^{(t+1)}$ from $P(Y|X^{(t+1)})$. Perform this for many iterations, discarding the initial samples.*

This sampling method ensures the following:
- The samples approximate the joint distribution of all variables.
- The marginal distribution of any subset of variables can be approximated by considering only the samples of that subset.
- The expected value of any variable can be approximated by averaging over all the samples.

When performing the sampling:

- Initial values of the variables can be chosen randomly or by using an algorithm such as expectation–maximization.

- It's not necessary to determine an initial value for the first variable sampled.

- It's common to ignore some initial samples (burn-in period) and then consider only every nth sample. This helps because:

(1) the stationary distribution of the Markov chain is the desired joint distribution, but reaching it takes time;

(2) successive samples are correlated, forming a Markov chain. Sometimes, algorithms estimate the autocorrelation between samples to compute n, but in practice, it often involves some trial and error.

Simulated annealing is frequently used to mitigate the "random walk" behavior in the early stages of sampling, helping the process move quickly across the sample space.

Relation of conditional distribution and joint distribution:

The conditional distribution of one variable given all others is proportional to the joint distribution:

$$p(X_j \mid X_1, \ldots, X_{j-1}, X_{j+1}, \ldots, X_n) = \frac{p(X_1, \ldots, X_n)}{p(X_1, \ldots, X_{j-1}, X_{j+1}, \ldots, X_n)} \propto p(X_1, \ldots, X_n).$$

Here, "proportional to" means the denominator is not a function of X_j and thus is the same for all values of X_j; it forms part of the normalization constant for the distribution over X_j. In practice, determining the nature of the conditional distribution of a factor X_j involves factoring the joint distribution according to the individual conditional distributions defined by the graphical model over the variables, ignoring factors that are not functions of X_j (which, along with the denominator, form the normalization constant), and reinstating the normalization constant as needed. This means:

1. For discrete distributions, compute the probabilities of all possible values of X_j and sum them to find the normalization constant.

2. For continuous distributions of a known form, the normalization constant will also be known.

3. In other cases, the normalization constant can often be ignored, as most sampling methods do not require it.

2.1.2 *A theoretical justification*

To understand why Gibbs sampling converges to the correct distribution, we explore the underlying mathematical structure. In particular, we examine how Gibbs updates define a reversible Markov chain with the target distribution as its stationary distribution. Suppose that a sample X is taken from a distribution depending on a parameter vector $\theta \in \Theta$ of length d, with prior distribution $g(\theta_1, \ldots, \theta_d)$. It may be that d is very large and that numerical integration to find the marginal densities of the θ_i would be computationally expensive. Then an alternative method of calculating the marginal densities is to create a Markov chain on the space Θ by repeating these two steps:

(i) Pick a random index $1 \leq j \leq d$,

(ii) Pick a new value for θ_j according to $g(\theta_1, \ldots, \theta_{j-1}, \cdot, \theta_{j+1}, \ldots, \theta_d)$.

Proposition 4. *The steps (i) and (ii) define a reversible Markov chain with the desired invariant distribution g.*

Proof. Define $X \sim_j Y$ if $X_i = Y_i$ for all $i \neq j$ and let P_{XY} denote the probability of a jump from $X \in \Theta$ to $Y \in \Theta$. Then, the transition probabilities are

$$P_{XY} = \begin{cases} \frac{1}{d} \frac{g(Y)}{\sum_{Z \in \Theta: Z \sim_j X} g(Z)}, & X \sim_j Y, \\ 0, & \text{otherwise.} \end{cases}$$

So

$$g(X) P_{XY} = \frac{1}{d} \frac{g(X) g(Y)}{\sum_{Z \in \Theta: Z \sim_j X} g(Z)} = \frac{1}{d} \frac{g(Y) g(X)}{\sum_{Z \in \Theta: Z \sim_j Y} g(Z)} = g(Y) P_{YX}$$

since $X \sim_j Y$ is an equivalence relation. Thus the detailed balance equations are satisfied, implying the chain is reversible and has invariant distribution g. □

In practice, the index j is not chosen at random, and the chain cycles through the indexes in order. In general this gives a non-stationary Markov process, but each individual step will still be reversible, and the overall process will still have the desired stationary distribution (as long as the chain can access all states under the fixed ordering).

2.1.3 *Applications of Gibbs sampling*

Here we give some outline of items where a Gibbs sampling is useful. Detailed applications will be presented in the next chapters of the book.

Bayesian networks: Gibbs sampling is used to sample from the joint distribution of complex Bayesian networks.

Hierarchical models: In hierarchical Bayesian models, Gibbs sampling allows for efficient sampling of the posterior distributions of model parameters.

Latent Dirichlet allocation (LDA): In LDA, Gibbs sampling is used to infer the topic distributions in a set of documents.

Advantages and limitations:

Strengths: Efficient for high-dimensional distributions; Can handle complex models with many parameters; Provides approximate solutions where exact methods are infeasible.

Weaknesses: Can suffer from slow convergence; Requires careful tuning of parameters; May not perform well with highly correlated variables.

Advanced Topics:

Metropolis-Hastings within Gibbs: Combining Metropolis-Hastings updates with Gibbs sampling to handle variables with complex conditional distributions.

Hybrid Monte Carlo: A technique that combines Gibbs sampling with Hamiltonian dynamics to improve sampling efficiency.

Case Studies and Examples:

Simple linear regression: Gibbs sampling can be used to sample from the posterior distribution of regression coefficients.

Mixture models: In Gaussian mixture models, Gibbs sampling is used to infer the component memberships and parameters.

2.2 Gibbs sampler in models

2.2.1 *Hard core model*

In this subsection we give an application of the Gibbs sampler to the hard-core model (see [323, Chapter 7]), which is a model for independent sets on a graph. The Gibbs sampler operates as shown in Algorithm 2.

Algorithm 2 implements the Gibbs sampler for the hard-core model, where the parameter λ represents the fugacity or activity of the model. The algorithm's effectiveness lies in its ability to maintain the independent set property at each step while allowing the Markov chain to explore the configuration space efficiently. The key insight is that vertices can only be added to the independent set if doing so preserves the constraint that no two adjacent vertices are simultaneously included.

Algorithm 2 Gibbs Sampler for Hard-Core Model

Initialize: Start with an initial independent set I_0 of the graph G (i.e., $I_0 \subseteq V(G)$ such that no two vertices in I_0 are adjacent)
for $t = 0, 1, 2, \ldots$ **do**
  Select Vertex: Randomly select a vertex v from the graph G
  Update Configuration:
  if $v \in I_t$ **then**
    Remove v from the independent set with probability $\frac{1}{1+\lambda}$
    Set $I_{t+1} = I_t \setminus \{v\}$ with prob. $\frac{1}{1+\lambda}$, otherwise $I_{t+1} = I_t$
  else if $v \notin I_t$ **and** no neighbors of v are in I_t **then**
    Add v to the independent set with probability $\frac{\lambda}{1+\lambda}$
    Set $I_{t+1} = I_t \cup \{v\}$ with prob. $\frac{\lambda}{1+\lambda}$, otherwise $I_{t+1} = I_t$
  else
    Set $I_{t+1} = I_t$ (no change possible due to adjacency constraints)
  end if
end for
Output: After sufficient iterations for mixing, I_t represents a sample from the hard-core model's stationary distribution

The convergence properties of this sampler are particularly important in practice. After a sufficient burn-in period, the generated sequence of independent sets $\{I_t\}$ will approximate samples from the stationary distribution of the hard-core model. The mixing time—the number of iterations required for the chain to approach its stationary distribution—depends on the graph structure and the value of λ. For large λ, the model favors larger independent sets, while smaller λ values lead to sparser configurations.

The Gibbs sampler for the hard-core model proves particularly valuable for approximating the partition function and sampling from complex distributions where direct sampling is computationally intractable. This is especially relevant in statistical physics applications and combinatorial optimization problems involving independent sets.

2.2.1.1 *Transition probabilities*

The transition probabilities for this Gibbs sampler can be precisely characterized based on the vertex selection and update rules. Let I_t denote the current independent set and v be the randomly selected vertex. We distinguish between two primary cases:

Case 1: $v \in I_t$ (vertex is currently in the independent set)

Since removing a vertex from an independent set always preserves the independent set property, we have:

$$P(I_{t+1} = I_t \setminus \{v\} \mid I_t) = \frac{1}{1+\lambda} \tag{2.1}$$

$$P(I_{t+1} = I_t \mid I_t) = \frac{\lambda}{1+\lambda} \tag{2.2}$$

Case 2: $v \notin I_t$ (vertex is not currently in the independent set)

The transition depends critically on whether adding v would violate the independent set constraint:

- If any neighbor of v is in I_t, then v cannot be added without violating the independent set property:

$$P(I_{t+1} = I_t \mid I_t) = 1$$

- If no neighbors of v are in I_t, then v can potentially be added:

$$P(I_{t+1} = I_t \cup \{v\} \mid I_t) = \frac{\lambda}{1+\lambda} \tag{2.3}$$

$$P(I_{t+1} = I_t \mid I_t) = \frac{1}{1+\lambda} \tag{2.4}$$

These transition probabilities ensure that the Markov chain correctly explores the space of independent sets according to the hard-core model distribution with parameter λ. The probabilities are designed to satisfy detailed balance, guaranteeing that the stationary distribution corresponds to the desired Gibbs measure on independent sets.

We refer to [323, 351] and references therein for comprehensive treatments of Gibbs measures and hard-core models.

2.2.2 *Potts model*

The Potts model is a generalization of the Ising model used in statistical mechanics to describe interacting spins on a crystalline lattice. Each spin can take one of q different states, and the energy of a configuration is determined by the interactions between neighboring spins.

Consider a lattice Λ (e.g., $2D$ grid) where each site $i \in \Lambda$ has a spin σ_i that can take one of q states. The energy function (Hamiltonian) of the Potts model is given by:

$$H(\sigma) = -J \sum_{\langle i,j \rangle} \delta(\sigma_i, \sigma_j)$$

where J is the interaction strength, the sum is over neighboring pairs $\langle i, j\rangle$, and δ is the Kronecker delta, which is 1 if $\sigma_i = \sigma_j$ and 0 otherwise.

The probability of spin σ_i being in state k given the states of its neighbors is given by

$$P(\sigma_i = k \mid \sigma_{-i}) = \frac{\exp\left(\beta J \sum_{j\in N_i} \delta(k, \sigma_j)\right)}{\sum_{k'=1}^{q} \exp\left(\beta J \sum_{j\in N_i} \delta(k', \sigma_j)\right)},$$

where $\beta = \frac{1}{k_B T}$ (with k_B being the Boltzmann constant and T the temperature), N_i is the set of all nearest neighbors of i, and σ_{-i} represents the states of all spins except σ_i.

Algorithm 3 Gibbs Sampling for Potts Model

1: **Input:** Lattice Λ, number of states q, interaction strength J, temperature T
2: **Initialize:** Start with an initial configuration of spins $\sigma = \{\sigma_i\}_{i\in\Lambda}$
3: **while** not converged to equilibrium distribution **do**
4:   **for** each spin σ_i in the lattice Λ **do**
5:     Compute conditional probabilities for each possible state $k \in \{1, 2, \ldots, q\}$:
6:     $P(\sigma_i = k \mid \sigma_{-i}) = \frac{\exp\left(\beta J \sum_{j\in N_i} \delta(k,\sigma_j)\right)}{\sum_{k'=1}^{q} \exp\left(\beta J \sum_{j\in N_i} \delta(k',\sigma_j)\right)}$
7:     Sample a new state σ_i from the computed probability distribution
8:     Update the spin σ_i to the new state
9:   **end for**
10: **end while**
11: **Output:** Sample from the equilibrium distribution of the Potts model

Potts model appears in various fields including physics (for studying phase transitions), image processing (for image segmentation), and machine learning (for certain types of graphical models).

2.2.3 *Disk packing model*

The disk packing model (DPM) (see [341]) is a geometric model where disks (or circles) are packed into a container such that they do not overlap. This model finds applications in material science, biology, and computer graphics.

The model involves placing n disks with radii $r_1, \ldots, r_n$ in a bounded region $\Omega \subset \mathbb{R}^2$ (e.g., a unit square) without overlap. The state space is:

$$\mathcal{S} = \{\mathbf{X} = \{(x_1, y_1), \ldots, (x_n, y_n)\} \mid (x_i, y_i) \in \Omega,\ d((x_i, y_i), (x_j, y_j)) \\ \geq r_i + r_j\, \forall i \neq j\},$$

where d denotes the Euclidean distance. The target distribution is uniform over $\mathcal{S}$:

$$\pi(\mathbf{X}) \propto \mathbb{I}\{\mathbf{X} \in \mathcal{S}\},$$

where $\mathbb{I}$ is the indicator function for valid configurations.

The Gibbs sampler generates samples from $\pi(\mathbf{X})$ by iteratively updating each disk's position from its conditional distribution:

Algorithm 4 Gibbs Sampler for Disk Packing Model

1: **Initialize:** Start with $\mathbf{X} = \{(x_1, y_1), \ldots, (x_n, y_n)\} \in \mathcal{S}$
2: **for** each iteration $t = 1, 2, \ldots$ **do**
3:     **for** each disk $i = 1, 2, \ldots, n$ **do**
4:         **Define feasible region:** $\mathcal{F}_i = \{(x_i', y_i') \in \Omega \mid d((x_i', y_i'), (x_j, y_j)) \geq r_i + r_j\, \forall j \neq i\}$
5:         **Sample new position:** $(x_i', y_i') \sim \text{Uniform}(\mathcal{F}_i)$
6:         **Update configuration:** $\mathbf{X} \leftarrow \mathbf{X}'$, where $\mathbf{X}'$ differs from $\mathbf{X}$ only in disk i's position
7:     **end for**
8: **end for**
9: **Monitor convergence:** Use diagnostics such as trace plots or Gelman-Rubin statistic

2.2.3.1 *Conditional distribution and sampling strategy*

For each disk i, the conditional distribution given the positions of all other disks is:

$$P((x_i', y_i') \mid \mathbf{X}_{-i}) = \frac{1}{|\mathcal{F}_i|}\mathbb{I}\{(x_i', y_i') \in \mathcal{F}_i\},$$

where $|\mathcal{F}_i|$ denotes the area of the feasible region and $\mathbf{X}_{-i}$ represents the positions of all disks except disk i.

In practice, sampling from this conditional distribution can be implemented using rejection sampling:

(1) Sample (x_i', y_i') uniformly from Ω
(2) Check if $(x_i', y_i') \in \mathcal{F}_i$ (i.e., verify no-overlap and boundary constraints)
(3) If valid, accept the sample; otherwise, reject and repeat

For dense configurations where $|\mathcal{F}_i|$ is small, this approach may be inefficient. Alternative strategies include using spatial data structures (e.g., quadtrees) to optimize constraint checking or employing local perturbations around the current position.

2.2.3.2 *Transition probabilities*

For a single update of disk i, the transition probability from configuration $\mathbf{X}$ to configuration $\mathbf{X}'$ (differing only in disk i's position) is:

$$P(\mathbf{X} \to \mathbf{X}') = P((x_i', y_i') \mid \mathbf{X}_{-i}) = \frac{1}{|\mathcal{F}_i|}\mathbb{I}\{(x_i', y_i') \in \mathcal{F}_i\}.$$

For a complete cycle where all n disks are updated sequentially, the transition probability is:

$$P(\mathbf{X}^{(0)} \to \mathbf{X}^{(n)}) = \prod_{i=1}^{n} P((x_i^{(i)}, y_i^{(i)}) \mid \mathbf{X}_{-i}^{(i-1)}),$$

where $\mathbf{X}^{(k)}$ denotes the configuration after updating the first k disks, and $\mathbf{X}_{-i}^{(i-1)}$ represents the positions of all disks except disk i after updating disks $1, \ldots, i-1$.

2.2.3.3 *Practical considerations*

Convergence: The Gibbs sampler may exhibit slow convergence in dense packing scenarios due to limited feasible regions for individual disks. Convergence diagnostics such as trace plots, autocorrelation analysis, or the Gelman-Rubin statistic should be employed to assess chain mixing.
Initialization: Starting configurations should satisfy all constraints. Random sequential placement or physics-based relaxation methods can generate valid initial states.
Efficiency: For problems with many disks or tight packing constraints, hybrid approaches combining Gibbs sampling with other MCMC methods (e.g., parallel tempering) may improve sampling efficiency.

The Gibbs sampler for the disk packing model is particularly valuable for exploring the space of valid configurations and estimating properties of random packings, such as packing density distributions or structural correlations.

2.2.4 *Neural networks*

Neural networks provide a natural application domain for Gibbs sampling, particularly in models where neurons can exist in multiple discrete states and interactions follow probabilistic rules.

2.2.4.1 *Binary neural networks with Gibbs sampling*

Consider a neural network on a lattice $S = \mathbb{Z}^2$ where each site s represents a neuron with binary states $\Phi = \{0, 1\}$. A neuron is excited if $x(s) = 1$ and inhibited if $x(s) = 0$.

For neighboring neurons s and t, the synaptic connection has strength w_{st}, which is excitatory if $w_{st} > 0$ and inhibitory otherwise. The energy function (Hamiltonian) of this neural network is:

$$H(x) = \sum_{s\in S} \sum_{t\in \mathcal{N}_s} w_{ts} x(t) x(s) - \sum_{s\in S} h_s x(s),$$

where h_s is the threshold of neuron s and $\mathcal{N}_s$ denotes the neighborhood of site s.

The Gibbs sampler updates each neuron's state by sampling from the local conditional distribution:

$$\pi_T(x(s) \mid x(\mathcal{N}_s)) = \frac{\exp\left(-\frac{1}{T} H_s(x(s) \mid x(\mathcal{N}_s))\right)}{\sum_{y\in\{0,1\}} \exp\left(-\frac{1}{T} H_s(y \mid x(\mathcal{N}_s))\right)},$$

where $T > 0$ is the temperature parameter and $H_s(x(s) \mid x(\mathcal{N}_s))$ is the local energy contribution.

For the binary case, the probability of choosing state 0 is:

$$p_0 = \frac{1}{1 + \exp\left(\frac{\alpha}{T}\right)},$$

where $\alpha = \sum_{t\in\mathcal{N}_s} w_{ts} x(t) - h_s$ represents the net input to neuron s, and state 1 is chosen with probability $p_1 = 1 - p_0$.

2.2.4.2 *Multi-state Potts neural networks*

The framework extends naturally to multi-state neurons using the q-state Potts-glass model [255], [390]. Each neuron can exist in one of q states, modeled as a Potts spin $\sigma_i \in \{1, 2, \ldots, q\}$.

Research has shown [390] that for q-state Potts-glass neural networks with $q \geq 3$, there exists a novel dynamical phase where spurious memories are completely suppressed, preventing the network from being trapped in meta-stable states. This property makes multi-state Potts neural networks particularly advantageous for practical applications.

The storage capacity of such networks scales as $Nq(q-1)$ [205], where N is the network size and q is the number of states per neuron. This quadratic scaling in the number of states provides significantly higher memory capacity compared to binary networks.

Gibbs sampling in neural networks offers several key advantages:

Associative Memory: The probabilistic dynamics allow for robust pattern retrieval even from corrupted inputs, as demonstrated in Hopfield-type networks with Potts neurons [255].

Learning Rules: Gibbs sampling can be combined with various learning rules, including Hebbian learning and pseudo-inverse rules [251], to store patterns without interference between memories.

Phase Transition Studies: Neural networks trained on Potts model configurations can effectively identify phase transitions, matching the accuracy of traditional Monte Carlo methods [362].

Semantic Memory Modeling: Potts-based neural networks provide insights into human semantic memory, with storage capacity scaling quadratically with the number of possible feature values [225].

The temperature parameter T in Gibbs sampling provides a natural mechanism for annealing, allowing the network to explore the solution space initially and then converge to low-energy configurations as the temperature is reduced.

2.2.5 *Other applications of Gibbs sampling*

We shall list a few other example applications of Gibbs sampling in common tasks of statistical inference and machine learning.

Example 5. *Sampling from a Bivariate Normal Distribution [146, 158, 315]:*

Consider sampling from a bivariate normal distribution with mean $\boldsymbol{\mu} = [\mu_1, \mu_2]^T$ and covariance matrix $\boldsymbol{\Sigma} = \begin{pmatrix} \sigma_{11} & \sigma_{12} \\ \sigma_{12} & \sigma_{22} \end{pmatrix}$.

The conditional distributions are:

$$X_1 | X_2 \sim \mathcal{N}\left(\mu_1 + \frac{\sigma_{12}}{\sigma_{22}}(X_2 - \mu_2), \sigma_{11} - \frac{\sigma_{12}^2}{\sigma_{22}}\right) \tag{2.5}$$

$$X_2 | X_1 \sim \mathcal{N}\left(\mu_2 + \frac{\sigma_{12}}{\sigma_{11}}(X_1 - \mu_1), \sigma_{22} - \frac{\sigma_{12}^2}{\sigma_{11}}\right) \tag{2.6}$$

The Gibbs sampling algorithm:

(1) Initialize $(X_1^{(0)}, X_2^{(0)})$
(2) For $t = 1, 2, \ldots, T$:
   - Sample $X_1^{(t)} \sim p(X_1|X_2^{(t-1)})$
   - Sample $X_2^{(t)} \sim p(X_2|X_1^{(t)})$

Example 6. *Topic Modeling with Latent Dirichlet Allocation [47, 165, 177]:*

LDA models documents as mixtures of topics, where each topic is a distribution over words. For document d with words $\mathbf{w}_d$ and topic assignments $\mathbf{z}_d$, Gibbs sampling iteratively samples topic assignments.

The conditional probability for assigning word w_{dn} to topic k is:

$$P(z_{dn} = k|\mathbf{z}_{-dn}, \mathbf{w}) \propto \frac{n_{dk}^{(-dn)} + \alpha}{n_d^{(-dn)} + K\alpha} \cdot \frac{n_{kw}^{(-dn)} + \beta}{n_k^{(-dn)} + V\beta}$$

where $n_{dk}^{(-dn)}$ is the count of words in document d assigned to topic k (excluding current word), $n_{kw}^{(-dn)}$ is the count of word w assigned to topic k (excluding current assignment), K is the number of topics, V is vocabulary size, and α, β are Dirichlet priors.

The algorithm alternates between sampling topic assignments and updating count statistics until convergence.

Example 7. *Bayesian Inference in Gaussian Mixture Models ([41], [278]):*

In Bayesian GMMs, both cluster assignments $\mathbf{z}$ and parameters $\boldsymbol{\theta} = \{\boldsymbol{\mu}_k, \boldsymbol{\Sigma}_k, \pi_k\}_{k=1}^K$ are treated as random variables with appropriate priors.

Gibbs sampling alternates between:

(1) **Sample cluster assignments**: For each data point $\mathbf{x}_i$:

$$P(z_i = k|\mathbf{x}_i, \boldsymbol{\theta}) \propto \pi_k \mathcal{N}(\mathbf{x}_i; \boldsymbol{\mu}_k, \boldsymbol{\Sigma}_k)$$

(2) **Sample parameters**: Given assignments, sample from posterior distributions:

$$\boldsymbol{\mu}_k|\mathbf{X}, \mathbf{z} \sim \mathcal{N}(\hat{\boldsymbol{\mu}}_k, \hat{\boldsymbol{\Sigma}}_k) \tag{2.7}$$

$$\boldsymbol{\Sigma}_k^{-1}|\mathbf{X}, \mathbf{z} \sim \text{Wishart}(\hat{\nu}_k, \hat{\mathbf{W}}_k^{-1}) \tag{2.8}$$

$$\boldsymbol{\pi}|\mathbf{z} \sim \text{Dirichlet}(\hat{\boldsymbol{\alpha}}) \tag{2.9}$$

where $\hat{\boldsymbol{\mu}}_k, \hat{\boldsymbol{\Sigma}}_k, \hat{\nu}_k, \hat{\mathbf{W}}_k, \hat{\boldsymbol{\alpha}}$ are updated hyperparameters.

This approach naturally incorporates uncertainty in parameter estimates and handles model selection through the posterior distribution.

Example 8. *State Inference in Hidden Markov Models ([306], [403]):*

For HMMs with state sequence $\mathbf{s} = (s_1, \ldots, s_T)$ and observations $\mathbf{o} = (o_1, \ldots, o_T)$, Gibbs sampling samples states sequentially from their conditional distributions.

The conditional probability for state s_t given all other states and observations is:

$$P(s_t = j | s_{-t}, \mathbf{o}) \propto P(s_{t-1} \to j) \cdot P(s_t = j \to s_{t+1}) \cdot P(o_t | s_t = j)$$

where $P(s_{t-1} \to j)$ and $P(s_t = j \to s_{t+1})$ are transition probabilities, and $P(o_t | s_t = j)$ is the emission probability.

For Bayesian HMMs with unknown parameters, Gibbs sampling alternates between sampling state sequences and sampling transition/emission parameters from their posterior distributions, enabling joint inference over both states and parameters.

Below is a list of 9 more applications of Gibbs sampling for the attention of an interested readers.

(1) Restricted Boltzmann machines (RBMs) [183]: RBMs use Gibbs sampling for training and inference by drawing samples from the model's distribution.
(2) Deep Belief networks (DBNs) [184]: DBNs, which are stacks of RBMs, rely on Gibbs sampling for training each layer.
(3) Latent Dirichlet Allocation (LDA) [165]: Gibbs sampling is used for inferring the topic distribution in LDA, a topic modeling technique.
(4) Bayesian Networks [219]: Gibbs sampling is applied to perform inference in Bayesian networks when exact inference is intractable.
(5) Image Denoising [321]: Gibbs sampling is used in image denoising techniques to model the pixel intensities.
(6) Markov Chain Monte Carlo (MCMC) Methods [156]: Gibbs sampling is a key MCMC method used in various applications for drawing samples from complex distributions.
(7) Sequence Modeling [281]: Gibbs sampling is applied in sequence modeling tasks to generate samples from the posterior distribution over sequences.
(8) Collaborative Filtering [330]: Gibbs sampling is used in collaborative filtering techniques for recommendation systems to model user preferences.

(9) Hierarchical Bayesian Models [146]: Gibbs sampling is used to perform inference in hierarchical Bayesian models, which model data with multiple levels of abstraction.

2.3 Markov chain Monte Carlo methods

Markov chain Monte Carlo (MCMC) methods encompass a class of algorithms used to sample from probability distributions by constructing a Markov chain that converges to the target distribution. These include Gibbs sampling as discussed in the previous section as a special instance.

MCMC methods, Gibbs sampling, Metropolis-Hastings algorithm, and Hamiltonian Monte Carlo, are widely used in Bayesian inference, statistical physics, and machine learning. These methods provide a flexible framework for exploring complex distributions and estimating quantities of interest.

Following [58], let us give basic notations of MCMC. Then, we shall give several models where MCMC is used.

To evaluate the expectation $E[\phi(Z)]$ of a k-dimensional random vector Z with a probability density $f(x)$, where $\phi : \mathbb{R}^k \to \mathbb{R}$ and $E[|\phi(Z)|] < \infty$, the formula

$$E[\phi(Z)] = \int_{\mathbb{R}^k} \phi(x) f(x) dx$$

can be used if one is able to compute the integral analytically.

In case when the computation is difficult, the first approximation way is numerical integration. A second alternative is to generate a sequence of i.i.d random vectors $\{Z_n\}_{n \geq 1}$ with the same distribution as Z, and apply the strong law of large numbers:

$$E[\phi(Z)] = \lim_{n \to \infty} \frac{1}{n} \sum_{i=1}^{n} \phi(Z_i)$$

to obtain an estimate based on a finite sample $(Z_1, \ldots, Z_n)$, namely $\frac{1}{n} \sum_{i=1}^{n} \phi(Z_i)$. The choice of the sample size n can be made via the theory of confidence intervals in order to attain a given accuracy (see [58] and corresponding references therein). This method, known as Monte Carlo, is competitive with numerical integration for large dimension k.

In some situations, the probability density $f(x)$ is known only up to a normalizing factor, that is, $f(x) = K\tilde{f}(x)$, and when, of course, the integral of $\tilde{f}(x)$ that gives the normalizing factor is difficult to compute. For example, in physics, the partition function of a Gibbs measure is usually uncomputable in closed form, due to the size of the state space in the corresponding models.

2.3.1 *Inverse distribution method*

Recall that replicas of a random variable Z refer to multiple independent copies of Z. These copies are denoted as $Z_1, Z_2, \ldots, Z_n$, where each Z_i is distributed identically to Z, but all are mutually independent.

Replicas are used in various statistical and probabilistic analyses, especially in fields like statistical physics, machine learning, and Monte Carlo methods, to study the properties of random variables by examining their independent instances.

To generate replicas of Z, the method involves transforming a sequence of i.i.d. random variables from a known distribution into samples from the desired distribution of Z. This can be achieved by using the cumulative distribution function (CDF) of Z.

Algorithm 5 Inverse Distribution Method

1: Compute the CDF of Z, denoted as $F_Z(x)$, which gives the probability that $Z \leq x$.
2: Generate a random variable U from a uniform distribution on $[0, 1]$.
3: Find the value x such that $F_Z(x) = U$, i.e., compute $x = F_Z^{-1}(U)$.
4: Repeat steps 2 and 3 to generate as many replicas of Z as needed.

The process described in Algorithm 5 uses a sequence of independent and identically distributed (i.i.d.) random variables $\{U_n\}_{n\geq 1}$, uniformly distributed on $[0, 1]$, to generate a sequence $\{Z_n\}_{n\geq 1}$ via

$$Z_n = F_Z^{-1}(U_n),$$

where $F_Z^{-1}(u)$ is the generalized inverse distribution function[1] of the CDF $u = F_Z(x) = \int_{-\infty}^{x} f(y)\,dy$, and $f(x)$ is the probability density function of Z. The applicability of this method is limited by the difficulty of obtaining $F_Z^{-1}(u)$ analytically, and it requires complete knowledge of the probability density function $f(x)$.

Example 9. *The inverse function of* $F(x) = 1 - e^{-\theta x}$ *is* $F^{-1}(y) = -\frac{1}{\theta}\ln(1-y)$. *Hence, if* U *is uniformly distributed on* $[0, 1]$, *then* $\frac{1}{\theta}\ln(1-U)$ *is an exponential random variable with mean* θ^{-1}. *So is*

$$Z = \frac{1}{\theta}\ln(U),$$

since U *and* $1 - U$ *are identically distributed.*

[1]The generalized inverse distribution function is given by $F_Z^{-1}(u) = \inf\{x \in \mathbb{R} : F_Z(x) \geq u\}$.

2.3.2 *Acceptance-rejection method*

Another method, called *acceptance-rejection* (also known as *rejection sampling*), brings us nearer to an efficient solution. The acceptance-rejection method introduced the fundamental concept of probabilistic acceptance of proposals, which later became a key component of the Metropolis-Hastings algorithm and other MCMC methods.

Assume one can generate a sequence of i.i.d. random vectors $\{Y_n\}_{n\geq 1}$ with the probability density $g(x)$ satisfying

$$\frac{f(x)}{g(x)} \leq c, \text{for all } \ x,$$

where c is a finite constant.

Let $\{U_n\}_{n\geq 1}$ be a sequence of i.i.d. random variables uniformly distributed on $[0,1]$. If we define τ to be the first index $n \geq 1$ for which

$$U_n \leq \frac{f(Y_n)}{cg(Y_n)},$$

and let $Z = Y_\tau$, then Z admits the probability density function $f(x)$ and $E[\tau] = c$.

MCMC is explained as follows. One constructs an irreducible aperiodic HMC $\{X_n\}_{n\geq 0}$ with state space E with the stationary distribution π. Since E is finite, the chain is ergodic, and for any initial distribution μ and all $i \in E$,

$$\lim_{n\to\infty} P_\mu(X_n = i) = \pi(i) \tag{2.10}$$

and

$$\lim_{N\to\infty} \frac{1}{N}\sum_{n=1}^{N} \phi(X_n) = E_\pi[\phi(X)]. \tag{2.11}$$

It is obvious that, when n is large, we can consider that X_n has a distribution close to π. It is interesting to know how accurately X_n imitates an E-valued random variable Z with distribution π. That is one seeks to obtain estimates of the form

$$\|\mu\mathbf{P}^n - \pi\| \leq A\alpha^n,$$

where $\alpha < 1$.

The Perron-Frobenius theorem gives a theoretical answer, in terms of the second-largest eigenvalue modulus (SLEM). Let us first recall the Perron-Frobenius (P-F) theorem (see e.g. [335]):

Theorem 11. (Perron-Frobenius) *Let* $\mathbf{P}$ *be an* $n \times n$ *irreducible non-negative matrix. The following assertions hold:*

1. Spectral radius: *The matrix* $\mathbf{P}$ *has a unique largest positive eigenvalue, called the P-F root, denoted by* λ_{max}[2].
2. Simplicity: *The P-F root* λ_{max} *is a simple eigenvalue (i.e., its algebraic multiplicity is 1). Moreover, any other eigenvalue* λ *of* $\mathbf{P}$ *satisfies* $|\lambda| < \lambda_{max}$.
3. P-F eigenvector: *There exists a corresponding positive eigenvector* v *(i.e., all entries of* v *are positive) such that* $\mathbf{P}v = \lambda_{max}v$. *This vector is unique up to multiplication by a positive scalar.*

In particular, for a non-negative irreducible matrix that is also a stochastic matrix (i.e., all row sums are 1), the Perron-Frobenius theorem implies:
- The P-F root $\lambda_{\max} = 1$. This eigenvalue corresponds to the steady-state distribution in Markov chains.
- The corresponding positive eigenvector can be normalized to form a probability distribution, which is the stationary distribution of the Markov chain.

To apply Theorem 11, it is necessary to find good upper bounds of the SLEM, since the eigenstructure of a transition matrix is often very difficult to find exactly. One has mainly to use reversible chains, this is the case in MCMC, presented as follows.

First, the fundamental issue lies in designing the MCMC algorithm. We need to identify an ergodic transition matrix $\mathbf{P} = (p_{ij})_{i,j \in E}$ whose stationary distribution matches the target distribution. There exist countless transition matrices meeting this criterion, and within them, there are infinitely many that define a reversible chain, that is, such that

$$\pi(i)p_{ij} = \pi(j)p_{ji}. \tag{2.12}$$

We seek solutions of the form

$$p_{ij} = q_{ij}\alpha_{ij} \tag{2.13}$$

for $j \neq i$, where $Q = (q_{ij})_{i,j \in E}$ is an arbitrary irreducible transition matrix on E, called the *candidate-generating matrix.* When the present state is i, the next tentative state j is chosen with probability q_{ij}. When $j \neq i$, this new state is accepted with probability α_{ij}. Otherwise, the next state is the same state i. Thus, the resulting probability of moving from i to j when

[2] This eigenvalue is also called the spectral radius of $\mathbf{P}$.

$i \neq j$ is given by (2.13). It remains to select the *acceptance probabilities* α_{ij}. Take

$$\alpha_{ij} = \frac{s_{ij}}{1 + t_{ij}}, \tag{2.14}$$

where $\Sigma = (s_{ij})_{i,j \in E}$ is a *symmetric matrix* and

$$t_{ij} = \frac{\pi(i) q_{ij}}{\pi(j) q_{ji}}. \tag{2.15}$$

Here Σ must be selected in such a way that the constraint $\alpha_{ij} \in [0, 1]$ is respected. It is easy to see that the reversibility condition (2.12) is satisfied, and therefore π is the stationary distribution, by the detailed balance test.

Example 10. Metropolis algorithm:

In order to satisfy the constraint $\alpha_{ij} \in [0, 1]$, one must have

$$s_{ij} \leq 1 + \min(t_{ij}, t_{ji}).$$

The acceptance probability for the Metropolis algorithm is given by:

$$\alpha_{ij} = \min\left(1, \frac{\pi(j) q_{ji}}{\pi(i) q_{ij}}\right),$$

where $\pi(i)$ is the target distribution and q_{ij} is the proposal distribution from state i to state j.

A special case occurs when the candidate-generation mechanism is purely random ($q_{ij} = q_{ji}$), leading to:

$$\alpha_{ij} = \min\left(1, \frac{\pi(j)}{\pi(i)}\right).$$

Example 11. Barker's algorithm*:*

For Barker's algorithm, with $s_{ij} = 1$, the acceptance probability is:

$$\alpha_{ij} = \frac{\pi(j) q_{ji}}{\pi(j) q_{ji} + \pi(i) q_{ij}}.$$

In the special case of a purely random selection of the candidate ($q_{ij} = q_{ji}$), this simplifies to:

$$\alpha_{ij} = \frac{\pi(j)}{\pi(i) + \pi(j)}.$$

Example 12. Independent sampling*:*

For independent sampling, with $q_{ij} = \pi(j)$ and $\alpha_{ij} = 1$ for all $i, j \in E$, we have:

$$p_{ij} = \pi(j) \text{ for all } i, j \in E.$$

Observe that

- Hastings algorithms require knowledge of the target distribution π only up to a normalizing constant, as they depend on the ratios $\pi(j)/\pi(i)$.

- This statement holds true only if the candidate-generating matrix Q is known. This is not necessarily the case in independent sampling.

Example 13. The Traveling Salesman Problem (TSP). *TSP is a classic optimization problem in computer science and operations research. It involves finding the shortest possible route for a salesman to visit a given set of cities exactly once and return to the original city. TSP is known for its computational complexity and is classified as an NP-hard problem, meaning that there is no known algorithm that can solve all instances of the problem efficiently (i.e., in polynomial time).*

Formulation of TPS: Given a set of cities, $C = \{c_1, c_2, \ldots, c_n\}$ *and a distance matrix* $D = (d_{ij})$*, where* d_{ij} *represents the distance between city* c_i *and city* c_j*. The problem is to find a permutation* π *of the cities that minimizes the total travel distance, i.e., minimize*

$$d(\pi) := \sum_{i=1}^{n-1} d_{\pi(i)\pi(i+1)} + d_{\pi(n)\pi(1)}, \tag{2.16}$$

where $\pi(n)$ *is the last city in the permutation, and* $\pi(1)$ *is the starting city.*

For solving the TSP a Metropolis chain in the context of a simulated annealing algorithm is used, which is a sequence of solutions (or states) generated by the algorithm as it attempts to find a near-optimal solution. The Metropolis criterion is used to decide whether to accept a new solution or not, based on the difference in the "cost" (or total distance in the case of TSP) between the current solution and the proposed solution, as well as the current "temperature" of the system. Here's a detailed description of how it works:

1. *- Start with an initial tour (solution)* S_0 *of the cities.*
   - Calculate the cost $C(S_0)$*, which is the total distance of the tour.*
   - Set an initial temperature T_0*.*
2. *For each iteration, do the following:*
   - Create a new tour S' *by making a small change to the current tour* S*. Common methods include:*
   (i) Select two cities at random and swap their positions.
   (ii) Select a subsequence of the tour and reverse it.
   (iii) Remove a city from the tour and reinsert it at a different position.
   - Calculate the cost $C(S')$ *of the new tour* S'*.*

- If $C(S')$ is less than $C(S)$ (i.e., the new solution is better), accept S' as the new current solution.
- If $C(S')$ is greater than $C(S)$ (i.e., the new solution is worse), accept S' with a probability P given by the Metropolis criterion:

$$P = \exp\left(\frac{C(S) - C(S')}{T}\right)$$

where T is the current temperature. This allows for occasional acceptance of worse solutions to escape local minima.
- Decrease the temperature T according to a cooling schedule. Common cooling schedules include:
Linear: $T_{new} = T - \Delta T$
Geometric: $T_{new} = \alpha T$, where α is a constant factor less than 1.
Exponential: $T_{new} = T_0 \exp(-kt)$, where k is a constant and t is the iteration number.

3. *Termination:*
- The algorithm terminates when the temperature T reaches a predefined minimum value or after a certain number of iterations.
- The best solution found during the process is returned as the approximate solution to the TSP.

The algorithm starts with a random tour and iteratively improves it using the Metropolis criterion while gradually cooling the temperature. The best tour found during the process is returned as the solution.

The Complete graph: In the last example the given connectivity, distance between the n cities is represented by a weighted, non directed, connected graph $G(V, E)$ with n vertices whose edges and edge weights represent the given distances between the cities. For simplicity, let us assume that the underlying graph is complete, meaning $d_{ij} > 0$ for all $i, j \in V$. The state space is $S = S_n$, (where S_n is the permutation group of n distinguishable objects). Two permutations in S are neighbors, denoted by $\sim$, if one permutation arises from the other by reversing a substring of the tour. More precisely, if $\pi = (\pi_1, \pi_2, \ldots, \pi_n)$, then for $1 \leq i < j \leq n$, the corresponding substring is $(\pi_i, \pi_{i+1}, \ldots, \pi_j)$. Reversing the order of the substring $(\pi_i, \pi_{i+1}, \ldots, \pi_j)$ in π yields the new permutation $\pi' = (\pi_1, \ldots, \pi_{i-1}, \pi_j, \pi_{j-1}, \ldots, \pi_i, \pi_{j+1}, \ldots, \pi_n)$.

The algorithm will only allow one-step transitions between neighbors. The target distribution for the Metropolis chain is the Gibbs measure with energy function $d(\pi)$ defined in (2.16). At each step, the proposal chain T

chooses two indices i, j uniformly at random from $\{1, 2, \ldots, n\}$ with $i < j$. Given that the current state is π, the proposal chain proposes the neighboring permutation π' constructed as described above. In other words, the proposal chain T is a simple random walk on a graph with vertex set $S = S_n$ (each vertex is identified with a permutation in S_n) and for which each vertex has $\binom{n}{2}$ neighbors.

For a fixed temperature T, we get the following transition probabilities P for the Metropolis chain:

$$P(\pi \to \pi') = \begin{cases} \frac{2}{n(n-1)} \min\left(1, e^{(d(\pi)-d(\pi'))/T}\right) & \text{if } \pi \sim \pi' \\ 0 & \text{if } \pi \not\sim \pi' \end{cases}$$

Note that for this algorithm:

$$P(\pi \to \pi) = 1 - \sum_{\pi' \sim \pi} P(\pi \to \pi').$$

Additionally, we have:

$$d(\pi') - d(\pi) = d_{i-1,i} + d_{j,j+1} - d_{i-1,j} - d_{i,j+1}.$$

which shows that the computation of $d(\pi') - d(\pi)$ does not depend on n, resulting in a low computational cost for this step of the algorithm.

The last step for implementing the algorithm is to decide on a cooling schedule $T_1, T_2, \ldots$ with corresponding iterations $N_1, N_2, \ldots$. Finding a suitable cooling schedule for simulated annealing often requires some experimentation.

Example 14. The knapsack problem: *The knapsack problem is a classic combinatorial optimization problem where the goal is to select a subset of items with given weights and values to maximize the total value without exceeding a given weight capacity. Solving the knapsack problem can be challenging, especially for large instances, which is why approximate methods like MCMC using Gibbs measures are useful.*

The problem is described as follows: Consider n items labeled $1, 2, \ldots, n$. Each item i (for $i = 1, \ldots, n$) has a weight w_i and a value v_i. We want to select a subset of these items to put into a knapsack that has an upper weight limit of W. The goal is to find an optimal selection of items that maximizes the total value of the items in the knapsack.

We represent any selection $A \subseteq \{1, 2, \ldots, n\}$ of items with a binary vector $\mathbf{x} = (x_1, \ldots, x_n) \in \{0, 1\}^n$, where $x_i = 1$ if and only if item i is included in A. The state space S is defined as:

$$S = \left\{ \mathbf{x} \in \{0,1\}^n : \sum_{i=1}^{n} w_i x_i \le W \right\}$$

The value function f on S is defined by:

$$f(\mathbf{x}) = \sum_{i=1}^{n} v_i x_i.$$

The optimization problem is to find:

$$\max_{\mathbf{x} \in S} f(\mathbf{x}).$$

This problem has been widely studied in computer science and is known to be NP-complete as a decision problem. No efficient algorithm is known (or is likely to exist) for finding an exact solution to the knapsack problem for large n.

One uses MCMC approach to solve an optimization problem of this kind is to simulate a Metropolis chain on S that converges to a stationary distribution, which places high probability on states with extreme values of f. A standard distribution used in this approach is the Gibbs measure:

$$\mu_T(s) = Z_T^{-1} e^{-f(s)/T} \quad \textit{for } s \in S,$$

where $T > 0$ temperature, $Z_T = \sum_{s \in S} e^{-f(s)/T}$ is the normalizing constant.

The energy function f is the function arises from the given optimization problem. For a fixed temperature T, the Gibbs measure places higher probability on states with relatively small values of f than on states with relatively large values of f. If the goal is to maximize f, then one would use $-f$ as the energy function instead.

Note that for a high temperature T, the measure μ_T is almost uniform on S. For lower temperature, the μ_T concentrates near states that minimize f.

In practice, MCMC and Gibbs measures provide a powerful framework for tackling combinatorial optimization problems like the knapsack problem, offering a way to approximate solutions when exact methods are infeasible.

2.4 Interacting particle systems

Interacting particle systems (IPS) are mathematical models used to study systems composed of a large number of components (particles) that interact with each other. These models are particularly useful in statistical mechanics, probability theory, and related fields to understand the collective behavior emerging from local interactions among particles.

2.4.1 *Basic definitions and the main problem of IPS*

Consider the integer lattice $\mathbb{Z}^d$, $d \in \mathbb{N}$, where each site $x \in \mathbb{Z}^d$ can be occupied by a particle. The state of the system at time t is described by a configuration $\eta(t) = \{\eta_x(t)\}_{x\in\mathbb{Z}^d}$, where $\eta_x(t)$ represents the state of the site x at time t. For example, in a simple case, $\eta_x(t)$ could be 0 or 1, indicating whether the site x is empty or occupied by a particle.

The dynamics of the system are typically described by a continuous-time Markov process. The generator $\mathcal{L}$ of the Markov process defines the evolution of the system. For a function f on the state space (which is a function of the configuration η), the generator $\mathcal{L}$ is given by:

$$\mathcal{L}f(\eta) = \sum_{x\in\mathbb{Z}^d} c_x(\eta)\left(f(\eta^x) - f(\eta)\right),$$

where η^x denotes the configuration obtained from η by flipping the state at site x, and $c_x(\eta)$ is the rate at which this flip occurs.

The evolution of the system is typically studied via the master equation, which describes the time evolution of the probability distribution over the state space.

The process $\eta(t)$ is Markovian, meaning that the future evolution of the system depends only on its current state, not on its past history.

The main problem in the study of IPS is the long-term behavior of the system. For many systems, there exists a stationary distribution μ, which is invariant under the dynamics (i.e., $\mathcal{L}^*\mu = 0$, where $\mathcal{L}^*$ is the adjoint of the generator $\mathcal{L}$). Ergodicity ensures that the system, regardless of its initial state, converges to this stationary distribution.

Techniques such as coupling (constructing two processes on a common probability space) and duality (relating the IPS to another process that is easier to analyze) are often used to study these systems. The book [249] presents a comprehensive introduction to the theory of IPS. It covers basic models like the Voter model, contact process, and exclusion process, along with advanced topics. A more introductory text that also touches on percolation theory is the book [110]. It's a good starting point for those new to the field.

2.4.2 *Examples of IPS*

The Voter model: The Voter model used in statistical physics, sociophysics, and network theory to study the dynamics of opinion formation and consensus in a population. It was introduced in the 1970s and has since become a fundamental model for understanding how individual decisions or states can lead to collective behavior over time (see [72], [84] and the references therein).

The model consists of a set of agents (or nodes in a network), each of which can be in one of two states. These states often represent different opinions, political choices, or cultural traits (e.g., 0 or 1).

Interaction rule of Voter model is as follows: At each time step, a randomly chosen agent adopts the state of one of its randomly selected neighbors. This rule mimics social influence, where individuals are influenced by their peers. If we denote the state of agent i as s_i, the probability that agent i changes its state at time t depends on the state of its neighbors. Specifically, if agent i has k_i neighbors, and n_i of them are in state 1, then the probability of s_i switching to state 1 (if it was previously 0) is n_i/k_i.

The evolution of the probability distribution of the system can be described by a master equation, which tracks the probability of finding the system in a particular configuration over time.

The model is neutral, meaning that there is no inherent preference for either state. The dynamics are purely driven by interactions between agents.

The system has absorbing states, where all agents eventually reach the same state (either all 0s or all 1s). Once the system reaches an absorbing state, it remains there forever.

In the Voter model on a regular lattice, domains of the same state tend to grow over time, leading to a coarsening process where larger and larger regions of the system become homogeneous.

The behavior of the Voter model can differ depending on the structure of the network (e.g., regular lattices, random graphs, or complex networks). In mean-field models, where every agent interacts with every other agent, the dynamics are different from those on finite or structured networks.

The Voter model has been applied to various problems, including the spread of opinions in social networks, language competition, and cultural dissemination. It serves as a simple yet powerful framework for studying consensus formation and polarization.

The basic Voter model can be extended in various ways, such as introducing bias, multi-state versions, or more complex rules for state updating. These extensions allow the model to capture more realistic social dynamics.

The Contact process: The Contact Process is a stochastic model used in mathematical biology, particularly in the study of the spread of diseases, population dynamics, and spatially structured populations. The process is a type of IPS, which is a general framework for modeling complex systems with many components interacting locally [57], [111], [249].

The Contact Process can be defined on a graph, typically a lattice $\mathbb{Z}^d$, but it can be generalized to any countable graph. Each site of the graph can be in one of two states: "infected" or "healthy" (or equivalently, "occupied" or "vacant"). The state of the process at time t is given by η_t, where $\eta_t(x) = 1$ if site x is infected at time t, and $\eta_t(x) = 0$ if it is healthy.

The evolution of the process is governed by two types of transitions:

1. Infection: An infected site x can infect a neighboring site y with rate λ, where $\lambda > 0$ is the infection rate. The neighbors of x are defined by the edges of the graph (e.g., nearest neighbors on $\mathbb{Z}^d$). Mathematically, the infection event can be written as:

$$\eta_t(x) = 1 \quad \text{and} \quad \eta_t(y) = 0 \quad \text{transition to} \quad \eta_{t+dt}(y) = 1 \quad \text{with rate} \quad \lambda.$$

2. Recovery: An infected site x can recover (become healthy) with rate equal to 1. This is usually modeled as a Poisson process, meaning each infected site has an independent exponential clock with rate 1, and when the clock rings, the site becomes healthy:

$$\eta_t(x) = 1 \quad \text{transition to} \quad \eta_{t+dt}(x) = 0 \quad \text{with rate} \quad 1.$$

The dynamics of the contact process can be described by a continuous-time Markov process. The generator $\mathcal{L}$ of the process, which gives the infinitesimal evolution of the system, can be written as:

$$\mathcal{L}f(\eta) = \sum_{x \in \mathbb{Z}^d} \left[\lambda \eta(x) \sum_{y \sim x} \left(f(\eta^{x,y}) - f(\eta) \right) + (1 - \eta(x)) \left(f(\eta^x) - f(\eta) \right) \right],$$

where

- f is a function on the configuration space $\{0, 1\}^{\mathbb{Z}^d}$.

- $y \sim x$ denotes that y is a neighbor of x.

- $\eta^{x,y}$ is the configuration where y is infected in configuration η, i.e., $\eta^{x,y}(z) = \eta(z)$ for all $z \neq y$ and $\eta^{x,y}(y) = 1$.

- η^x is the configuration where x is recovered, i.e., $\eta^x(z) = \eta(z)$ for all $z \neq x$ and $\eta^x(x) = 0$.

Phase transition. Many IPSs exhibit phase transitions, where the qualitative behavior of the system changes dramatically as some parameter (e.g., particle density, interaction strength) crosses a critical value. For example, the contact process has a phase transition between a regime where particles survive indefinitely and one where they die out. There exists a critical infection rate λ_c such that:

- If $\lambda < \lambda_c$, the infection dies out almost surely, meaning that for any finite initial configuration, the system eventually reaches the all-healthy state (extinction).
- If $\lambda > \lambda_c$, there is a positive probability that the infection survives indefinitely (persistence).

The value of λ_c depends on the underlying graph or lattice. For example, on $\mathbb{Z}^d$, the critical value λ_c is non-trivial and depends on the dimension d.

Long-term behavior. For $\lambda > \lambda_c$, the contact process exhibits a non-trivial stationary distribution, where the fraction of infected sites is non-zero. The stationary distribution ν_λ satisfies the detailed balance condition for the generator $\mathcal{L}$.

The Contact Process is a rich mathematical model that captures the essence of local interactions leading to global phenomena, such as the spread of diseases or the persistence/extinction of species.

The Ising model (Glauber dynamics): Glauber dynamics is a type of stochastic process used in statistical mechanics, particularly in the study of spin systems like the Ising model. It describes the time evolution of a system of interacting spins (or particles) on a lattice, where each spin can flip randomly according to certain probabilistic rules. The dynamics are designed to drive the system toward thermal equilibrium at a given temperature (see [103], [112], [166], [249], [264]).

Consider a lattice $\Lambda \subset \mathbb{Z}^d$, where each site $i \in \Lambda$ has a spin σ_i that can take values in $\{-1, +1\}$. The state of the entire system is given by the configuration $\sigma = (\sigma_i)_{i\in\Lambda}$. Recall the Ising model Hamiltonian of a configuration σ given by:

$$H(\sigma) = -J \sum_{\langle i,j \rangle} \sigma_i \sigma_j - h \sum_{i\in\Lambda} \sigma_i,$$

where

- $J > 0$ is the interaction strength between neighboring spins.
- $\langle i, j \rangle$ denotes pairs of neighboring sites.
- h is an external magnetic field.
- σ_i and σ_j are the spins at sites i and j.

Glauber dynamics are defined by transition rates that specify how likely a spin at site i is to flip, depending on the configuration of its neighbors and the temperature T. The transition rate $c_i(\sigma)$ for the spin at site i to flip (i.e., $\sigma_i \to -\sigma_i$) is often given by the following rule, based on detailed balance:

$$c_i(\sigma) = \frac{1}{1 + \exp(\beta \Delta H_i(\sigma))},$$

where

- $\beta = \frac{1}{k_B T}$ is the inverse temperature, with k_B being the Boltzmann constant.

- $\Delta H_i(\sigma) = H(\sigma^i) - H(\sigma)$ is the change in energy if the spin at site i is flipped, where σ^i denotes the configuration obtained by flipping the spin σ_i in configuration σ.

Alternatively, some versions of Glauber dynamics use the Metropolis-Hastings rule:

$$c_i(\sigma) = \min\left(1, \exp(-\beta \Delta H_i(\sigma))\right).$$

The evolution of the probability distribution $P_t(\sigma)$ of the system's configuration over time can be described by the master equation:

$$\frac{dP_t(\sigma)}{dt} = \sum_{\sigma'} \left[P_t(\sigma') c_{\sigma' \to \sigma} - P_t(\sigma) c_{\sigma \to \sigma'}\right],$$

where $c_{\sigma \to \sigma'}$ is the transition rate from configuration σ to σ'. For Glauber dynamics, $c_{\sigma \to \sigma'}$ involves flipping a single spin.

The stationary distribution $\pi(\sigma)$ for Glauber dynamics corresponds to the Gibbs measure, which is the equilibrium distribution of the Ising model:

$$\pi(\sigma) = \frac{1}{Z} \exp\left(-\beta H(\sigma)\right),$$

where Z is the partition function, ensuring normalization:

$$Z = \sum_{\sigma} \exp\left(-\beta H(\sigma)\right).$$

Glauber dynamics satisfies detailed balance with respect to the Gibbs measure:

$$\pi(\sigma) c_i(\sigma) = \pi(\sigma^i) c_i(\sigma^i).$$

Convergence to equilibrium: A key question in the study of Glauber dynamics is the rate of convergence to the equilibrium (Gibbs) measure. This is often quantified by the mixing time, which is the time it takes for the system's distribution to become close to the stationary distribution. The mixing time depends on factors such as the size of the system, the temperature, and the structure of the lattice.

Glauber dynamics is often studied in the context of phase transitions, where the system undergoes a qualitative change in behavior at a critical temperature T_c. For the Ising model, this corresponds to a transition between ordered (ferromagnetic) and disordered (paramagnetic) phases. The dynamics near the critical temperature are particularly slow, a phenomenon known as critical slowing down.

Thus Glauber dynamics helps to understand how macroscopic properties emerge from microscopic interactions.

Other models: IPSs are widely used to model various complex phenomena across different fields

Epidemiology: IPSs are employed to model the spread of diseases by representing individuals as particles that can be in different states, such as susceptible, infected, or recovered. The interactions between these particles, such as infection spread from one individual to another, are governed by local rules (see [110] for a detailed exposition of IPSs, including their application in epidemiology).

Ecology: In ecology, IPSs can model population dynamics, where the particles represent individuals or species. The local interactions, such as competition, predation, or reproduction, determine the changes in the population over time (see [370] for the application of IPS in ecological modeling).

Sociology: IPSs are used to model opinion dynamics in social networks, with the above discussed Voter model being a classic example (see [84]).

Physics: The Ising model, which we have already discussed many times, is a prominent example of an IPS used in statistical mechanics. It models ferromagnetism by representing magnetic spins as particles that interact with their neighbors, leading to phenomena such as phase transitions (see [136] for detailed properties of Ising model).

2.4.3 *Contour method*

Contour arguments, while traditionally rooted in statistical mechanics and variational calculus, have intriguing applications in machine learning. They provide a framework for understanding and optimizing complex systems, particularly in models where energy landscapes, stability, and phase transitions play a crucial role.

2.4.3.1 *Contours for Potts model*

Assume we have a very big set $\mathcal{D}$ of data. In machine learning by classification $\mathcal{D}$, one would like to learn the boundary separating the instances of one class from the instances of all other classes. In statistical mechanical language, the classification can be interpreted as introduction of contours, each consisting (inside) some fixed information of the data set. Then investigation of $\mathcal{D}$ can be treated by the properties (i.e. the length of boundary) of the contours. This contour method first was introduced by Peierls [301] in his study of the Ising model. The original argument benefited from the particular symmetries of the Ising model. The adaptation of the method to the treatment of non-symmetric models is not trivial, and was developed by Pirogov and Sinai [343], [396]. In [323], [325], [326] the contour method on Cayley tree was developed.

Here we illustrate the contour method for the IPS described by q-state Potts model, where $q \geq 2$ (see [343], [326]).

This method identifies interfaces (or contours) between regions of different phases (Gibbs measures) and assigns weights to these interfaces to analyze the system's behavior, particularly in the low-temperature regime. The q-state Potts model is given by the Hamiltonian:

$$H(\sigma) = -J \sum_{\langle i,j \rangle} \delta_{\sigma_i, \sigma_j},$$

where

- $\sigma_i \in \{1, 2, \dots, q\}$ represents the spin at site i,
- $J > 0$ (parameter, interpreted as a ferromagnetic interaction),
- $\langle i, j \rangle$ denotes nearest-neighbor pairs,
- $\delta_{\sigma_i, \sigma_j}$ is the Kronecker delta, which equals 1 if $\sigma_i = \sigma_j$ and 0 otherwise.

The system is defined on a lattice (e.g., $\mathbb{Z}^d$ or Cayley tree) with boundary conditions specified.

In the low-temperature regime ($\beta = \frac{1}{T}$ is large), spins align predominantly in one of the q possible states, forming ordered regions (phases). Interfaces between these regions are rare but crucial for understanding the system's behavior.

Contours are the boundaries separating regions of different spin values $1, 2, \dots, q$. For a configuration σ, its boundary (denoted by $\partial\sigma$) is defined as the set of edges where neighboring spins differ:

$$\partial\sigma = \{\langle i, j \rangle : \sigma_i \neq \sigma_j\}.$$

The boundary is divided to several contours. See Figure 2.1 for an example of configuration and its contours. Since the Potts model has only nearest-neighbor interactions, the γ_1 (red) and γ_2 (black) have interactions. Other contours are independent. For convenience of the contour argument union of all interacting contours can be considered as one contour. For example, in

Figure 2.1 one can take $\gamma_1 \cup \gamma_2$ as one contour. Then new formed contours do not interact. Such consideration allows us to use a *contour-removal* operation (see [325] for details). This operation is similar to the one in ordinary Peierls argument [301]: Given a family of contours defining a configuration σ the family obtained by omitting one of them is also the family of contours of a (different) configuration. There is an algorithm of the contour-removal operation to obtain a new configuration as follows. Take the configuration σ and change all the spins in the interior of γ (which must be removed) to value 1. This makes γ disappear, but leaves intact the other contours.

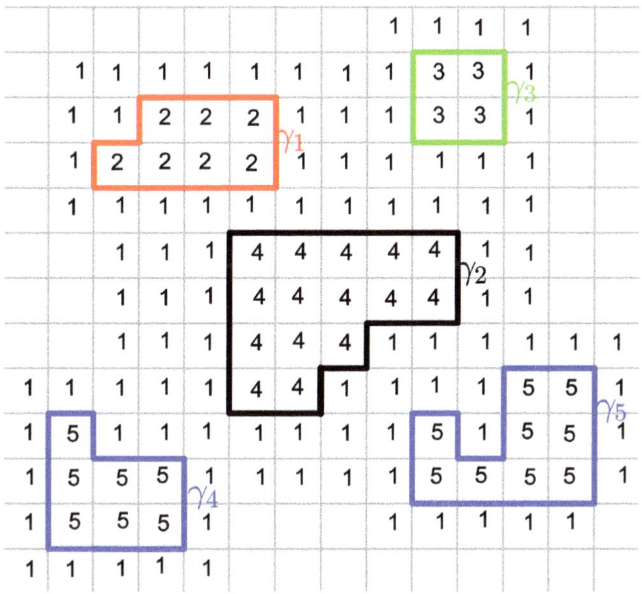

Fig. 2.1: An example of configuration of 5-state Potts model. Its boundary $\partial\sigma = \cup_{i=1}^{5}\gamma_i$. Each γ_i is a contour.

It is crucial (in particular for machine learning problems) to understand that if we know all the contours that make up the boundary of a configuration with q spin values, we can assign color "i" to each contour that inside has all spins equal to i. This establishes a one-to-one correspondence between each configuration and a set of contours, each associated with a specific color. In machine learning this helps to represent a big data $\mathcal{D}$ by the contours of separating data set to different classes.

This representation of configuration by the set of contours is also useful to rewrite Hamiltonian $H(\sigma)$ depending on the lengths of contours. For example, in case of Potts model on the Cayley tree, the Hamiltonian of a configuration on the ball V_n of the tree is given by (see [326]):

$$H_{V_n}(\sigma) = 1 - |V_{n+1}| + |\partial\sigma|.$$

Consider V_n and take the configuration $\varphi_1(x) \equiv 1$ as a boundary configuration outside V_n. Let μ^1 be the Gibbs measure corresponding to the boundary configuration. Fix a contour γ and denote

$$p_1(\gamma) = \mu^1\{\sigma : \gamma \subset \partial(\sigma(V_n))\}.$$

Then the following Peierls inequality holds (see [326, Lemma 8]):

$$p_i(\gamma) \leq \exp\{-\beta|\gamma|\}.$$

This inequality indicates that, in the system, a large contour (with significant length) occurs with a low probability. Furthermore, the inequality plays a crucial role in proving the following result (see [326, Theorem 11]):

Theorem 12. *For all sufficiently large β there are at least q Gibbs measures for the ferromagnetic Potts model on Cayley tree of order two.*

Note that q distinct measures mentioned in Theorem 12 are obtained by q distinct boundary configurations $\varphi_i(x) \equiv i$, $i = 1, \ldots, q$.

The above mentioned argument works for general models too, i.e., each contour is associated with:

- *Geometry*: The shape and size of the contour, often described using percolation theory or graph-theoretic constructs.

- *Weight*: The Boltzmann weight associated with the contour is proportional to its energy cost, typically growing with the contour's length (or surface area in higher dimensions).

- *Energy of contours*: The energy of a contour (say γ) is proportional to the number of edges in the boundary:

$$E_\gamma \propto J|\gamma|,$$

where $|\gamma|$ is the total length (or area) of the interface.

The partition function Z can be expressed as a sum over configurations or equivalently as a sum over contours:

$$Z = \sum_{\sigma} e^{-\beta H(\sigma)} \approx \sum_{\partial\sigma} e^{-\beta \sum_{\gamma\in\partial\sigma} E_\gamma}.$$

At low temperatures, the sum is dominated by configurations with minimal total contour energy.

Key results from the contour method are

- *Existence of multiple Gibbs measures*: The presence of low-energy contours corresponds to the stability of multiple ordered phases.
- *Phase coexistence*: Interfaces represent regions of phase coexistence.
- *Rigorous bounds*: By estimating the weights of contours, one can rigorously bound the probability of large deviations, ensuring that large contours are exponentially suppressed.

Here's an exploration of contours relevance in machine learning:

Contour arguments help analyze how energy levels define regions of high or low probability. They explain sudden changes in the learned distribution, akin to transitions in statistical mechanics [199].

By studying the contours of the loss surface, one can understand the behavior of optimization algorithms ([162], [212]).

Contour arguments can aid in interpreting how models generalize:

- Flat vs. sharp minima: In neural networks, contours of the loss landscape reveal whether the model has reached a flat minimum (associated with better generalization) or a sharp one (see [189]).

- Complexity penalties: Regularizers like weight decay can reshape the contours of the optimization surface, promoting smoother solutions ([222]).

In variational approaches to learning (see [310]), contour methods can:

- Examine how well the variational family captures the true posterior by comparing contour levels of the approximating and target distributions.

- Study how the support of distributions is constrained within certain regions of the parameter space.

For unsupervised learning tasks like clustering or image segmentation:

- Contours are used to define boundaries between clusters or segments [338].

- They reveal features like connected components and percolation phenomena [131], particularly in graph-based clustering.

Contour arguments can illuminate the dynamics of stochastic processes ([262]):

- In optimization, contour levels track the flow of parameters toward equilibrium.

- Understanding the progression of probability mass in generative models.

Contour arguments play a role in understanding adversarial vulnerabilities [361]:

- Contours of the decision function expose regions susceptible to adversarial perturbations.

- They help characterize the robustness of models to input noise or attacks [261].

2.4.3.2 *Multiple classes*

The contour argument is closely related to Learning Multiple Classes ([7, page 56], [386]). A multi-class classification problem involves a training set: $\{(x_i, y_i)\}_{i=1}^N$, where $x_i \in \mathbb{R}^n$ are the feature vectors and $y_i \in \{1, \dots, K\}$ are class labels.

Define the hypothesis space $\mathcal{H}$, as set of functions $f : \mathbb{R}^n \to \mathbb{R}^K$, where each $f(x)_k$ indicates the score or probability of x belonging to class k.

The learning algorithm selects a hypothesis f^* that minimizes a loss function $\mathcal{L}(f)$, typically formulated as:

$$f^* = \arg\min_{f \in \mathcal{H}} \mathbb{E}_{(x,y)\sim\mathcal{D}}[\mathcal{L}(f(x), y)],$$

where $\mathcal{D}$ is the data distribution.

Examples of multi-class classification models are

Linear models.

1. Softmax regression (or Generalized logistic regression) [41]:

Define $p(y = k|x)$ using a linear score and the softmax function:

$$p(y = k|x) = \frac{\exp(w_k^\top x)}{\sum_{j=1}^K \exp(w_j^\top x)}.$$

Cross-entropy loss function is

$$\mathcal{L} = -\frac{1}{N}\sum_{i=1}^N \log(p(y_i|x_i)).$$

The training of multi-class classification models, particularly those based on deep learning, often relies on gradient-based optimization techniques. These methods aim to minimize the loss function by iteratively adjusting the model parameters. For the details of gradient descent and its popular variants like Stochastic Gradient Descent (SGD) and Adam see [52], [53], [214].

2. Multi-class support vector machines (SVMs).

Extend binary SVMs using a "max-margin" principle for multi-class data. Define

$$\min_{w_k,\xi} \frac{1}{2}\sum_{k=1}^K ||w_k||^2 + C\sum_{i=1}^N \sum_{k\neq y_i} \xi_{i,k},$$

subject to:

$$w_{y_i}^\top x_i - w_k^\top x_i \geq 1 - \xi_{i,k}, \quad \forall k \neq y_i, \quad \xi_{i,k} \geq 0.$$

Probabilistic models.

1. *Naive Bayes classifier*: Assumes feature independence

$$p(y|x) \propto p(y) \prod_{j=1}^{n} p(x_j|y).$$

This is simple, computationally efficient, and works well for text classification. But fails when features are highly correlated.

2. *Gaussian processes*: Models a distribution over functions for multi-class classification using covariance functions. Employs Laplace approximation or variational methods due to non-Gaussian likelihoods.

Neural networks. Neural networks are powerful tools for multi-class problems, especially in cases with high-dimensional data (e.g., images, text):

- Input layer corresponds to features.
- Output layer has K neurons (one for each class), often with a softmax activation.
- Hidden layers extract features, potentially hierarchically.
- Cross-entropy is the standard loss for multi-class classification.

Training is the backpropagation with optimization algorithms like SGD, Adam, or RMSProp.

Regularization by techniques such as dropout, L_2-regularization, and batch normalization prevent overfitting.

Advanced algorithms. 1-vs-All (OvA) Strategy:

- Train K binary classifiers $f_1, f_2, \ldots, f_K$.
- For a new point x, predict the class with the highest confidence:

$$y = \arg\max_k f_k(x).$$

1-vs-1 (OvO) Strategy:

- Train $\binom{K}{2}$ binary classifiers, each distinguishing between a pair of classes (i, j).
- Aggregate predictions using: Majority voting and Weighted scores based on classifier confidence.

Hinge Loss (for multi-class SVM) function:

$$\mathcal{L} = \sum_{i=1}^{N} \sum_{k \neq y_i} \max(0, 1 - f(x_i, y_i) + f(x_i, k)).$$

2.5 IPS in machine learning

In the context of machine learning, interacting particle systems can be used to model dependencies between variables in a probabilistic model and simulate the evolution of the system over time.

2.5.1 *Theoretical foundations of IPS in machine learning*

In machine learning (ML), Interacting Particle Systems (IPS) and Gibbs measures underpin algorithms for sampling, optimization, and probabilistic modeling. This subsection explores the synergy between IPS and Gibbs measures, presenting key theorems on their convergence, phase transitions, and generalization guarantees. We include proofs of central results and discuss applications in energy-based models, Markov chain Monte Carlo (MCMC), and PAC-Bayesian learning. Our goal is to bridge IPS dynamics and Gibbs measures, highlighting their role in modern ML challenges.

Recall that an IPS is defined by:
- A state space $\mathcal{X}$ (e.g., $\{-1,1\}^d$).
- A set of particles $\{X_t^i\}_{i=1}^N$.
- Interaction rules (e.g., Glauber dynamics, exclusion processes).

Example 15. *The Glauber dynamics for Ising models updates particles via:*

$$\mathbb{P}(X_t^i = x \mid X_t^{-i}) \propto \exp\left(-\beta H(x, X_t^{-i})\right),$$

where H is an energy function and β is inverse temperature.

A Gibbs measure μ_β on $\mathcal{X}$ is defined as:

$$\mu_\beta(x) = \frac{1}{Z(\beta)} \exp\left(-\beta H(x)\right),$$

where, as always, $Z(\beta) = \sum_{x\in\mathcal{X}} \exp(-\beta H(x))$ is the partition function.

The following theorem gives convergence of Glauber dynamics to Gibbs measure:

Theorem 13. *Let $\{X_t\}$ be an IPS following Glauber dynamics with energy H. If the dynamics are irreducible and aperiodic, then:*

$$\lim_{t\to\infty} \mathbb{P}(X_t = x) = \mu_\beta(x) \quad \forall x \in \mathcal{X}.$$

Proof. The sketch of proof (see [249] and [243] for details): Verify detailed balance $\mu_\beta(x)\mathbb{P}(x \to y) = \mu_\beta(y)\mathbb{P}(y \to x)$. Show ergodicity: Irreducibility and aperiodicity ensure a unique stationary distribution. For convergence apply the Perron-Frobenius theorem for finite Markov chains. □

This theorem is applied to MCMC sampling (e.g., Gibbs sampling) in Bayesian inference.

Consider the Curie-Weiss Gibbs measure

$$\mu_\beta(x) \propto \exp\left(\frac{\beta}{N}\sum_{i<j} x_i x_j\right). \tag{2.17}$$

Theorem 14. *For measure (2.17), there exists a critical $\beta_c = 1$ such that:*

1. *For $\beta < \beta_c$, μ_β is unimodal.*
2. *For $\beta > \beta_c$, μ_β bifurcates into two modes.*

Proof. The sketch (see [54], [119, Chapter 2], [136, Sec.3.2]): define mean magnetization: $m = \frac{1}{N}\sum_{i=1}^{N} x_i$. Free energy minimization: Show m solves $m = \tanh(\beta m)$. Bifurcation analysis: Solve for β where $\frac{d}{dm}\tanh(\beta m) = 1$. □

An implication of this theorem is that sampling becomes exponentially slow near β_c, affecting ML models like Restricted Boltzmann Machines.

Now we describe a bound on the generalization error of a hypothesis selected from a distribution based on a prior. It involves several concepts from statistical learning theory, particularly from the perspective of Bayesian learning. Let us give some necessary notations:

1. $\mathcal{D}$: A data distribution. This refers to the true distribution of the data used for learning.

2. $\mathcal{H}$: A hypothesis class. This is the set of all possible hypotheses or models that could be used to predict the outcomes from the data.

3. π: A prior distribution over the hypothesis class $\mathcal{H}$. This distribution encodes any prior knowledge or assumptions about the hypotheses before observing the data.

4. δ: A confidence parameter, where $\delta \in (0,1)$. This parameter specifies the probability that the bound holds, i.e., with probability $1-\delta$, the bound on the generalization error will hold.

5. h: An element of the hypothesis class $\mathcal{H}$, representing a specific hypothesis.

6. $\mathbb{E}_{h\sim\rho}[L_\mathcal{D}(h)]$: The expected generalization error of the hypothesis h, where the expectation is taken over the Gibbs posterior ρ. The generalization error is defined as $L_\mathcal{D}(h)$, which is the loss of the hypothesis h with respect to the data distribution $\mathcal{D}$.

7. $\mathbb{E}_{h\sim\rho}[L_S(h)]$: The expected empirical loss of the hypothesis h, where the expectation is taken over the Gibbs posterior ρ, and $L_S(h)$ is the loss of hypothesis h on the sample S (the finite data used for training).

8. N: The sample size. This is the number of data points in the sample S.

9. $\mathrm{KL}(\rho\|\pi)$: The Kullback-Leibler divergence between the Gibbs posterior ρ and the prior π. This measures how much the posterior distribution differs from the prior distribution.

10. $\rho(h) \propto \pi(h)\exp(-\beta L_S(h))$: The Gibbs posterior distribution ρ. The posterior is proportional to the product of the prior $\pi(h)$ and the exponential of the negative empirical loss scaled by a factor β. Here, β is a parameter that controls the influence of the empirical loss on the posterior.

The following theorem provides a Bayesian perspective on generalization, bounding the difference between the generalization error and the empirical error using the Gibbs posterior and the Kullback-Leibler divergence:

Theorem 15. *Let $\mathcal{D}$ be a data distribution and $\mathcal{H}$ a hypothesis class. For any prior π over $\mathcal{H}$ and $\delta \in (0,1)$, with probability $1-\delta$:*

$$\mathbb{E}_{h\sim\rho}[L_{\mathcal{D}}(h)] \leq \mathbb{E}_{h\sim\rho}[L_S(h)] + \sqrt{\frac{KL(\rho\|\pi) + \log\frac{1}{\delta}}{2N}},$$

where ρ is a Gibbs posterior $\rho(h) \propto \pi(h)\exp(-\beta L_S(h))$.

Proof. Sketch (see [73], [169], [267]): Change of measure inequality by using Donsker-Varadhan's variational formula. Apply Hoeffding's inequality to the empirical risk. □

Application: PAC-Bayes learning with Gibbs posteriors [169]. PAC-Bayesian learning algorithms combine the flexibility of prior distributions on models with the performance guarantees of PAC algorithms [267]. PAC-Bayesian stochastic model selection can provide performance guarantees that surpass those of deterministic PAC-Bayesian model selection. These guarantees naturally apply to continuous concept classes and introduce a straightforward notion of an optimal posterior distribution for stochastically selecting a model. While the optimal posterior is a Gibbs measure, it is demonstrated that, under mild assumptions, simpler posterior distributions perform almost as well. An open question remains as to whether stronger guarantees can be provided for model averaging as opposed to stochastic model selection.

Theorem 16. *(Mean-field approximation guarantee) Let μ_β be a Gibbs measure and $\mathcal{Q}$ be the set of product distributions. The mean-field approximation $q^* = \arg\min_{q\in\mathcal{Q}} KL(q\|\mu_\beta)$ satisfies:*

$$KL(q^*\|\mu_\beta) \leq \inf_{q\in\mathcal{Q}} KL(q\|\mu_\beta) + \epsilon,$$

under suitable convexity conditions on H.

Proof. Sketch (see [41, 378]): Variational Principle: Maximize $\mathbb{E}_q[\log \mu_\beta] - \text{Ent}(q)$. Fixed-Point Iteration: Show convergence of iterative updates (e.g., coordinate ascent). □

This theorem applied to variational inference in latent variable models.

2.5.2 *Applications of IPS in machine learning*

Interacting particle systems (IPS) are mathematical models that describe the collective behavior of a set of particles that interact with each other according to certain rules. These systems have applications in various fields, including statistical mechanics, computer science, and machine learning. In machine learning, IPS models are used to analyze and understand various phenomena such as optimization, learning dynamics, and the behavior of algorithms. Below are some key applications of interacting particle systems in machine learning, along with references.

1. Particle-based optimization methods. Interacting particle systems are used to model optimization processes in machine learning. These processes can be used for tasks like parameter estimation or finding the global minimum in high-dimensional spaces.

Application: Swarm intelligence, such as particle Swarm optimization (PSO), models particles that update their positions based on interactions with other particles and their own experiences. This can be used in optimization tasks like hyperparameter tuning in deep learning models or feature selection [211].

2. Monte Carlo methods (Markov chain Monte Carlo). Interacting particle systems are closely related to Monte Carlo simulations, where particles represent random samples that interact according to certain transition rules. This approach is widely used in statistical learning, particularly in probabilistic graphical models and Bayesian inference.

Application: MCMC algorithms (like Metropolis-Hastings) are based on the idea of simulating the interactions of particles to sample from complex probability distributions. These methods are used in Bayesian deep learning, where models like variational autoencoders (VAEs) require efficient sampling of latent variables [41, 156].

3. Deep learning and neural networks. Interacting particle systems can be applied to the study of neural networks and their training dynamics, especially in models like reservoir computing or spiking neural networks (SNNs). These models often rely on the interaction between "neurons" or "particles" to simulate collective behavior and learning.

Application: In spiking neural networks (SNNs), individual neurons can be modeled as interacting particles, where each neuron has a "state" that evolves over time based on interactions with neighboring neurons. SNNs are used in various areas of machine learning, including robotics and real-time signal processing [162, 259].

4. Mean-Field approximation and large-scale models. Interacting particle systems are sometimes used to approximate the behavior of large-scale machine learning models, where the interactions between particles represent the dependencies between components of the system.

Application: Mean-field methods are used in machine learning for approximating the dynamics of complex systems. For example, in variational inference, each particle (sample) interacts with others to update estimates of posterior distributions. This approach can be applied in large-scale Bayesian networks or deep learning models [378].

5. Graphical models and belief propagation. Interacting particle systems are useful for understanding the behavior of graphical models, particularly in terms of message passing algorithms like belief propagation. These algorithms are used in machine learning to infer dependencies and make predictions in probabilistic graphical models.

Application: In a probabilistic graphical model (such as a Markov random field or a Bayesian network), each node (representing a random variable) can be seen as a particle interacting with other nodes to exchange information about the joint distribution. This method is used in tasks like image segmentation, computer vision, and recommendation systems [300, 394].

6. Reinforcement learning and multi-agent systems. Multi-agent systems, where agents (or particles) interact with one another and their environment, have become increasingly popular in reinforcement learning (RL). IPS models can be applied to multi-agent systems where agents evolve based on interactions with their environment and each other.

Application: Particle-based approaches can be used to model the collective behavior of agents in decentralized systems, such as in multi-agent RL.

In this setting, each "particle" (agent) may influence the behavior of others, leading to cooperative or competitive dynamics [295, 353].

7. Data assimilation and Kalman filters.

Data assimilation is used to combine predictions from a model with observed data, and Kalman Filters can be seen as a form of interacting particle system where each particle represents a state estimate that evolves according to the model dynamics and new observations.

Application: Kalman filters and particle filters are used in machine learning for state estimation, particularly in time series forecasting, robotics (for localization and mapping), and signal processing [19].

8. Statistical mechanics and learning dynamics. IPS models are used to understand the dynamics of learning systems, particularly in the context of energy-based models and systems exhibiting phase transitions. These models are useful for understanding convergence behavior and optimization in complex systems.

Application: In energy-based models, such as restricted Boltzmann machines (RBMs), interacting particles (neurons) interact to find the optimal configuration that minimizes energy, which relates to finding good model parameters [185, 319].

These applications demonstrate how interacting particle systems provide powerful tools for modeling complex, dynamic systems, which are central to many problems in machine learning. Each reference can provide further details on the theoretical foundations and practical applications of these approaches.

2.6 Conclusion

This chapter mainly devoted to Gibbs sampling, we detailed the sampling procedure and illustrated its importance as a method for obtaining samples from complex, high-dimensional probability distributions. The discussion included various applications of Gibbs sampling, showcasing its versatility and efficacy in different scenarios.

We examined the applications of Gibbs measures and Gibbs sampling in machine learning. This section underscored the practical relevance of these concepts, illustrating their utility in tasks such as parameter estimation, Bayesian inference, and probabilistic graphical models.

This chapter has provided a comprehensive overview of Gibbs sampling and its foundational role in probabilistic modeling, statistical physics, and machine learning. Beginning with applications of the Gibbs sampler in classical models such as the hard-core and Potts models, we demonstrated how this powerful algorithm enables simulation of high-dimensional distributions that are otherwise intractable. The use of Gibbs sampling in graph coloring and disk packing illustrates its flexibility in combinatorial and spatial problems.

A key highlight was the connection between Gibbs sampling and Markov chain Monte Carlo (MCMC) methods, which underpin modern Bayesian inference. We examined the construction and convergence of Markov chains, emphasizing their relevance to statistical inference, especially in machine learning contexts like neural networks and data augmentation.

Finally, we introduced interacting particle systems (IPS), a class of stochastic processes that model the dynamics of large systems of interacting components. We explored their mathematical foundations, applications in statistical mechanics, and recent extensions into machine learning theory and algorithms. IPS methods are emerging as robust tools for distributed inference, message passing, and optimization in complex models.

Altogether, the chapter underscores the deep interconnections between statistical mechanics, computational statistics, and machine learning. Through Gibbs sampling, MCMC, and IPS, we gain powerful techniques for modeling and solving high-dimensional, structured, and probabilistic systems.

For further reading covering Gibbs sampling, MCMC, interacting particle systems, and their applications in machine learning and image analysis see [12], [23], [37], [227], and [265].

Chapter 3

Gibbs Measures for Parameter Estimation in Machine Learning Models

This chapter examines the role of Gibbs measures in parameter estimation problems, bridging concepts from statistical physics and statistical inference. It begins with a general overview of parameter estimation models and introduces how Gibbs measures can be used to model uncertainty in such settings. Several illustrative examples are presented, including the Ising model, the Solid-on-Solid model, and a DNA interaction model. The chapter also discusses variational methods for parameter estimation and contrasts parametric and non-parametric approaches, highlighting how Gibbs measures can be applied in both frameworks.

3.1 What is parameter estimation model?

In the field of machine learning, parameter estimation is a crucial step in building accurate and reliable models. One popular approach to parameter estimation is the use of Gibbs measures, which are a type of probability distribution that captures the uncertainty in the parameters of a model.

Gibbs measures are particularly useful in cases where the parameters of a model are interdependent and cannot be estimated independently. In such cases, Gibbs measures provide a way to sample from the joint distribution of the parameters, taking into account their dependencies.

The key idea behind Gibbs measures is to iteratively sample from conditional distributions of each parameter given the values of the other parameters. This process continues until convergence is reached, at which point the samples can be used to estimate the parameters of the model.

One advantage of using Gibbs measures for parameter estimation is that they can handle complex, high-dimensional models with many interdependent parameters. This makes them particularly well-suited for tasks such as image recognition, natural language processing, and reinforcement learning.

3.1.1 *Introduction to parameter estimation*

Parameter estimation involves determining the values of parameters in a mathematical model that best fit observed data. It plays a critical role in various fields such as physics, biology, machine learning, and economics. The mathematical foundation of parameter estimation models is built on optimization theory [55], probability theory [41, 42] and statistics [67, 210].

Mathematical Framework: The fundamental model in parameter estimation is defined as a mathematical representation $f(x, \theta)$ of a system, where x represents the input or independent variables and θ denotes the parameters to be estimated. This model output is commonly referred to as the predicted or modeled output. Given a dataset of observed data points $\{(x_i, y_i)\}_{i=1}^n$, the primary objective is to estimate the parameter vector θ such that the model $f(x, \theta)$ accurately approximates the true underlying relationship between the input variables x and the output y.

Objective Functions: The core of parameter estimation lies in defining an appropriate objective function that quantifies the discrepancy between observed data and model predictions. The choice of objective function depends on the nature of the problem and the underlying assumptions about the data distribution.

For regression problems, the Mean Squared Error (MSE) serves as a widely adopted objective function, expressed mathematically as:

$$J(\theta) = \frac{1}{n} \sum_{i=1}^{n} \left(y_i - f(x_i, \theta)\right)^2$$

In probabilistic modeling contexts, the likelihood function provides a natural framework for parameter estimation. The likelihood function is defined as:

$$\mathcal{L}(\theta) = \prod_{i=1}^{n} P(y_i \mid x_i, \theta)$$

For computational convenience, it is common to work with the negative log-likelihood, which transforms the product into a sum and converts the maximization problem into a minimization one:

$$-\log \mathcal{L}(\theta) = -\sum_{i=1}^{n} \log P(y_i \mid x_i, \theta)$$

Estimation Methods: Parameter estimation encompasses several distinct methodological approaches, each with its own theoretical foundations and practical applications.

Least Squares Estimation (LSE) operates under the fundamental assumption that the errors $\epsilon_i = y_i - f(x_i, \theta)$ are independently and identically distributed following a normal distribution. This method seeks to minimize the sum of squared residuals by solving the optimization problem:

$$\hat{\theta} = \arg\min_{\theta} \sum_{i=1}^{n} \left(y_i - f(x_i, \theta)\right)^2.$$

The least squares approach provides optimal estimates when the error assumptions are satisfied and offers computational advantages due to its quadratic nature.

Maximum Likelihood Estimation (MLE) takes a probabilistic approach by seeking parameter values that maximize the likelihood of observing the given data. The MLE estimator is obtained by solving:

$$\hat{\theta} = \arg\max_{\theta} \prod_{i=1}^{n} P(y_i \mid x_i, \theta).$$

Equivalently, this can be formulated as a minimization problem using the negative log-likelihood:

$$\hat{\theta} = \arg\min_{\theta} -\sum_{i=1}^{n} \log P(y_i \mid x_i, \theta).$$

Maximum likelihood estimation provides asymptotically efficient estimators and forms the theoretical foundation for many statistical inference procedures.

Bayesian Estimation distinguishes itself by incorporating prior knowledge about the parameters through a prior distribution. This approach applies Bayes' theorem to combine prior beliefs with observed data, yielding a posterior distribution over the parameters:

$$P(\theta \mid \text{data}) \propto P(\text{data} \mid \theta) P(\theta).$$

The posterior distribution $P(\theta \mid \text{data})$ encapsulates all available information about the parameters, providing not only point estimates but also uncertainty quantification through the full distributional characterization.

Optimization Algorithms: The practical implementation of parameter estimation typically requires solving complex optimization problems. Several algorithmic approaches have been developed to address these computational challenges effectively.

Gradient descent represents one of the most fundamental optimization techniques, employing an iterative update rule based on the negative gradient of the objective function:

$$\theta_{k+1} = \theta_k - \eta \nabla J(\theta_k),$$

where η denotes the learning rate, a crucial hyperparameter that controls the step size in the optimization process. The method's simplicity and broad applicability make it particularly valuable for large-scale problems.

Newton's method offers a more sophisticated approach by incorporating second-order derivative information through the Hessian matrix. The Newton update rule is given by:

$$\theta_{k+1} = \theta_k - [H_J(\theta_k)]^{-1} \nabla J(\theta_k),$$

where H_J represents the Hessian matrix of the objective function J. While Newton's method typically exhibits faster convergence rates compared to gradient descent, it requires computing and inverting the Hessian matrix, which can be computationally expensive for high-dimensional parameter spaces.

3.1.2 *Models with latent variables*

Statistical and machine learning models frequently incorporate unobserved or hidden variables that, while not directly observable, play a crucial role in capturing the underlying structure of the data. These latent variables influence the observed data patterns and require specialized estimation techniques to handle their inherent uncertainty.

The probabilistic framework governing models with latent variables establishes the relationship between observed data X, latent variables Z, and model parameters θ through the joint probability distribution:

$$P(X, Z \mid \theta) = P(X \mid Z, \theta) P(Z \mid \theta)$$

The primary objective in latent variable modeling is to simultaneously estimate the parameters θ and infer the latent variables Z from the observed data alone.

Below we present some examples of Latent Variable Models.

Example 16 (Gaussian Mixture Models). *Gaussian Mixture Models (GMMs) represent one of the most fundamental applications of latent variable modeling. These models assume that the observed data originates from a mixture of Gaussian distributions, where the latent variable Z indicates the component assignment for each data point. The model specification is:*

$$P(X) = \sum_{k=1}^{K} \pi_k \mathcal{N}(x \mid \mu_k, \Sigma_k),$$

where π_k denotes the mixture weights and $\mathcal{N}(x \mid \mu_k, \Sigma_k)$ represents the Gaussian density function for the k-th component.

Example 17 (Hidden Markov Models). *Hidden Markov Models (HMMs) extend the latent variable framework to sequential data by modeling time series with hidden states. The latent variable represents the sequence of hidden states Z, characterized by transition probabilities $P(Z_t \mid Z_{t-1})$ that govern state evolution and emission probabilities $P(X_t \mid Z_t)$ that relate hidden states to observed outputs.*

Example 18 (Latent Dirichlet Allocation). *Latent Dirichlet Allocation (LDA) models provide a sophisticated approach to document modeling by representing documents as mixtures of topics. The latent variables correspond to topic assignments Z for individual words, with document-topic distributions $\theta_d \sim \text{Dirichlet}(\alpha)$ and topic-word distributions $\phi_k \sim \text{Dirichlet}(\beta)$ governing the generative process.*

Example 19 (Variational Autoencoders). *Variational Autoencoders (VAEs) represent a modern approach to learning probabilistic latent representations of high-dimensional data. The latent variable Z corresponds to the encoded representation in a lower-dimensional latent space, with the model specified as:*

$$P(X) = \int P(X \mid Z) P(Z) \, dZ.$$

Parameter Estimation for Latent Variable Models: The presence of latent variables necessitates specialized estimation techniques that can handle the inherent uncertainty in the unobserved components.

The Expectation-Maximization (EM) algorithm provides an iterative framework specifically designed to handle latent variables in parameter estimation. The algorithm alternates between two fundamental steps that collectively maximize the likelihood of the observed data. The expectation step computes the expected value of the log-likelihood function with respect to the conditional distribution of the latent variables given the current parameter estimates:

$$Q(\theta \mid \theta^{(t)}) = \mathbb{E}_{Z \sim P(Z|X,\theta^{(t)})}[\log P(X, Z \mid \theta)].$$

The maximization step then updates the parameter estimates by maximizing the Q function computed in the expectation step. This iterative process continues until convergence, guaranteeing monotonic improvement in the likelihood function.

Variational inference offers an alternative approach that approximates the intractable posterior distribution of latent variables $P(Z \mid X)$ using a simpler, tractable distribution $q(Z)$. The method optimizes the Evidence Lower BOund (ELBO), which provides a lower bound on the marginal log-likelihood:

$$\log P(X) \geq \mathbb{E}q(Z)[\log P(X, Z)] - \mathbb{E}q(Z)[\log q(Z)].$$

This variational approach transforms the challenging inference problem into an optimization problem over the variational parameters defining $q(Z)$.

Markov Chain Monte Carlo (MCMC) methods provide a sampling-based approach to parameter estimation in latent variable models. These techniques generate samples from the posterior distribution $P(Z \mid X)$ to estimate both the latent variables and model parameters through Monte Carlo integration.

3.1.3 *Statistical properties of estimators*

Understanding the statistical properties of parameter estimators is crucial for assessing their reliability and performance in practical applications.

3.1.3.1 *Bias and variance*

The bias of an estimator quantifies the systematic error introduced by the estimation procedure, measuring the difference between the expected value of the estimator and the true parameter value:

$$\text{Bias}(\hat{\theta}) = \mathbb{E}[\hat{\theta}] - \theta_{\text{true}}.$$

The variance characterizes the variability of the estimator across different samples, indicating the precision of the estimation procedure:

$$\text{Var}(\hat{\theta}) = \mathbb{E}[(\hat{\theta} - \mathbb{E}[\hat{\theta}])^2].$$

These two quantities capture the fundamental trade-off in estimation theory, where reducing bias may increase variance and vice versa, leading to the well-known bias-variance decomposition of the mean squared error.

3.1.3.2 *Consistency*

Consistency represents a fundamental asymptotic property that ensures the estimator converges to the true parameter value as the sample size increases without bound.

Definition 22 (Consistent estimator). *An estimator $\hat{\theta}$ is consistent for a parameter θ_{true} if it satisfies:*

$$\lim_{n\to\infty} P(|\hat{\theta}_n - \theta_{true}| > \epsilon) = 0, \quad \forall \epsilon > 0$$

This condition can be equivalently expressed in terms of convergence in probability:

$$\hat{\theta}_n \xrightarrow{P} \theta_{\text{true}} \quad \text{as} \quad n \to \infty$$

There exist different types of consistencies for estimators. *Weak consistency* requires convergence in probability, as defined above:

$$\lim_{n\to\infty} P(|\hat{\theta}_n - \theta_{\text{true}}| > \epsilon) = 0.$$

Strong consistency imposes the more stringent requirement of almost sure convergence:

$$P\left(\lim_{n\to\infty} \hat{\theta}_n = \theta_{\text{true}}\right) = 1.$$

Strong consistency implies weak consistency, but the converse is not necessarily true.

To ensure consistency, an estimator typically must satisfy several fundamental conditions. *Asymptotic unbiasedness* requires that the bias vanishes as the sample size increases:

$$\mathbb{E}[\hat{\theta}_n] \to \theta_{\text{true}} \quad \text{as} \quad n \to \infty.$$

Additionally, the estimator must exhibit vanishing variance:

$$\text{Var}(\hat{\theta}_n) \to 0 \quad \text{as} \quad n \to \infty$$

The Cramér-Rao lower bound establishes a fundamental limit on the variance of any unbiased estimator:

$$\text{Var}(\hat{\theta}) \geq \frac{1}{I(\theta)},$$

where $I(\theta)$ represents the Fisher information.

The consistency of many estimators is directly connected to the Law of Large Numbers (LLN), which establishes that sample averages converge to their corresponding population means as the sample size approaches infinity.

Below we present some examples of consistent estimators.

Example 20. *The sample mean provides a fundamental example of a consistent estimator. For a sample $X_1, X_2, \ldots, X_n$ drawn from a distribution with mean μ, the sample mean:*

$$\hat{\mu}_n = \frac{1}{n}\sum_{i=1}^{n} X_i$$

is consistent for μ due to the Law of Large Numbers.

Example 21. *The sample variance, when properly normalized, also exhibits consistency. For a distribution with variance σ^2, the sample variance:*

$$\hat{\sigma}_n^2 = \frac{1}{n-1}\sum_{i=1}^{n}(X_i - \hat{\mu})^2$$

provides a consistent estimator for σ^2.

Maximum Likelihood Estimators (MLEs) also possess consistency properties under appropriate regularity conditions, making them particularly valuable in practical applications where theoretical guarantees are important.

Consistent estimators often exhibit *asymptotic normality*, meaning their sampling distribution approaches a normal distribution as the sample size increases. This property is typically expressed as:

$$\sqrt{n}(\hat{\theta}_n - \theta_{\text{true}}) \xrightarrow{d} \mathcal{N}(0, \sigma^2).$$

This asymptotic normality enables the construction of confidence intervals and hypothesis tests for finite samples, even when the exact sampling distribution is unknown.

3.1.3.3 *Bias versus consistency*

Not all estimators possess the desirable property of consistency. The relationship between bias and consistency in statistical estimation reveals important theoretical insights about estimator behavior. It is worth highlighting that these properties are independent of each other, meaning an estimator can exhibit various combinations of bias and consistency.

Unbiased but not Consistent: An estimator can be unbiased yet fail to achieve consistency. Consider an independent and identically distributed (i.i.d.) sample $\{x_1, \ldots, x_n\}$ where we define the estimator $T_n(X) = x_n$ for estimating the population mean $\mathbb{E}[X]$. In this case, the sampling distribution of T_n remains identical to the underlying distribution for any sample size n, as it considers only the last observation while ignoring all others.

Since $\mathbb{E}[T_n(X)] = \mathbb{E}[X]$ for any n, the estimator is unbiased. However, it does not converge to any specific value as the sample size increases, thus failing the consistency requirement.

Conversely, if a sequence of estimators is both unbiased and converges to some value, then it must be consistent, as unbiased estimators that converge necessarily converge to the correct parameter value.

Biased but Consistent: Alternatively, an estimator can exhibit bias while maintaining consistency. For instance, consider estimating the mean using the modified estimator:

$$\hat{\mu}_n = \frac{1}{n}\sum_{i=1}^{n} x_i + \frac{1}{n}$$

This estimator is clearly biased since $\mathbb{E}[\hat{\mu}_n] = \mu + \frac{1}{n} \neq \mu$. However, as $n \to \infty$, the estimator approaches the true value μ, demonstrating consistency despite the presence of bias.

Important practical examples include the sample variance and sample standard deviation. When computed without Bessel's correction (using the sample size n instead of the degrees of freedom $n-1$), both estimators are negatively biased but remain consistent. The uncorrected sample variance converges to the true population variance, while the uncorrected sample standard deviation converges to the true population standard deviation.

With Bessel's correction applied, the corrected sample variance becomes unbiased, while the corrected sample standard deviation remains biased (though less so than the uncorrected version). Importantly, both corrected estimators maintain consistency, as the correction factor approaches unity as the sample size grows.

A Pathological Example: Consider another illustrative example where T_n represents a sequence of estimators for parameter θ:

$$\Pr(T_n) = \begin{cases} 1 - 1/n, & \text{if } T_n = \theta \\ 1/n, & \text{if } T_n = n\delta + \theta \end{cases}$$

In this case, we observe that $T_n \xrightarrow{P} \theta$, indicating convergence in probability to the true parameter. However, the expected value is $\mathbb{E}[T_n] = \theta + \delta$, showing that the bias equals δ and does not converge to zero as n increases. This example demonstrates that an estimator can be consistent while maintaining a constant bias that never diminishes.

3.2 Gibbs measures in parameter estimation

Gibbs measures provide a fundamental probabilistic framework for parameter estimation in systems where the structure and dependencies between variables are of paramount importance. These measures find extensive application in structured prediction problems, graphical models, and systems inspired by statistical mechanics, where the interaction patterns between components carry significant information about the underlying phenomena.

The essence of a Gibbs measure lies in its ability to assign probabilities to configurations based on an energy function that encodes the system's structure. For a configuration x (or (x, y) in supervised learning contexts), the Gibbs measure is defined as:

$$p(x \mid \theta) = \frac{\exp(-H(x;\theta))}{Z(\theta)},$$

where $H(x;\theta)$ represents the energy function (equivalently, the negative log-likelihood) that depends on both the configuration x and the parameter vector θ. The partition function $Z(\theta)$ serves as a normalization constant that ensures the probability distribution integrates to unity:

$$Z(\theta) = \sum_x \exp(-H(x;\theta)).$$

The energy function $H(x;\theta)$ embodies the structural assumptions of the model, capturing dependencies between variables and encoding prior knowledge about the system's behavior. The parameters θ are typically estimated through maximum likelihood estimation, which involves maximizing the likelihood of observed data $\{x_1, \ldots, x_N\}$, or equivalently, minimizing the negative log-likelihood:

$$L(\theta) = -\sum_{n=1}^{N} \log p(x_n \mid \theta).$$

Substituting the Gibbs measure formulation yields:

$$L(\theta) = \sum_{n=1}^{N} H(x_n;\theta) + N \log Z(\theta).$$

The computational challenge inherent in this approach stems from the partition function $Z(\theta)$, which requires summation over all possible configurations x. For systems with large state spaces or complex interaction patterns, this summation becomes computationally intractable, necessitating the development of approximation methods.

In supervised learning contexts, Gibbs measures extend naturally to conditional distributions over labels y given inputs x:

$$p(y \mid x, \theta) = \frac{\exp(-H(x, y; \theta))}{Z(x, \theta)},$$

where the energy function $H(x, y; \theta)$ now incorporates both input and output variables, and the partition function is conditioned on the input.

The appeal of Gibbs measures for parameter estimation lies in their ability to naturally incorporate structural dependencies in data while providing a principled connection between statistical mechanics and machine learning. Furthermore, they offer a rigorous probabilistic framework for reasoning about uncertainty and variability in parameter estimates, making them particularly valuable in applications where understanding model confidence is crucial.

3.2.1 *Parameter estimation in the Ising model*

The Ising model serves as a canonical example of parameter estimation using Gibbs measures. Originally developed to model ferromagnetism in statistical physics, it has found widespread application in machine learning for modeling binary variables with pairwise interactions. The model considers binary variables (spins) $y_i \in \{-1, 1\}$ that can represent diverse phenomena: atomic states in physical systems, pixel intensities in image analysis, or interaction patterns in biological networks.

The parameter estimation problem in the Ising model involves inferring the interaction parameters θ_{ij} (representing pairwise couplings between variables) and external field parameters h_i (representing individual biases) from observed configurations $\mathbf{y}$.

The Gibbs measure for an Ising model configuration $\mathbf{y} = (y_1, y_2, \ldots, y_n)$ is expressed as:

$$P(\mathbf{y} \mid \boldsymbol{\theta}) = \frac{1}{Z(\boldsymbol{\theta})} \exp\left(\sum_{(i,j) \in E} \theta_{ij} y_i y_j + \sum_{i \in V} h_i y_i \right),$$

where $\boldsymbol{\theta} = \{\theta_{ij}\}$ denotes the collection of pairwise interaction parameters, $\{h_i\}$ represents the external field parameters, E is the edge set defining the graph structure, V is the vertex set, and $Z(\boldsymbol{\theta})$ is the partition function:

$$Z(\boldsymbol{\theta}) = \sum_{\mathbf{y}} \exp\left(\sum_{(i,j) \in E} \theta_{ij} y_i y_j + \sum_{i \in V} h_i y_i \right).$$

The fundamental challenge in Ising model parameter estimation arises from the computational intractability of the partition function, which requires summation over all 2^n possible spin configurations.

3.2.1.1 *Maximum likelihood estimation*

For a dataset $\mathcal{D} = \{\mathbf{y}^{(1)}, \ldots, \mathbf{y}^{(M)}\}$ of observed configurations, the likelihood function is:

$$L(\boldsymbol{\theta}) = \prod_{m=1}^{M} P(\mathbf{y}^{(m)} \mid \boldsymbol{\theta}).$$

The corresponding log-likelihood takes the form:

$$\ell(\boldsymbol{\theta}) = \sum_{m=1}^{M} \left(\sum_{(i,j)\in E} \theta_{ij} y_i^{(m)} y_j^{(m)} + \sum_{i\in V} h_i y_i^{(m)} \right) - M \log Z(\boldsymbol{\theta}).$$

The maximum likelihood estimator is obtained by maximizing $\ell(\boldsymbol{\theta})$. To derive the gradient, we differentiate the log-likelihood with respect to θ_{ij}:

$$\frac{\partial \ell(\boldsymbol{\theta})}{\partial \theta_{ij}} = \frac{\partial}{\partial \theta_{ij}} \left[\sum_{m=1}^{M} \sum_{(k,l)\in E} \theta_{kl} y_k^{(m)} y_l^{(m)} - M \log Z(\boldsymbol{\theta}) \right].$$

The first term yields the empirical correlation:

$$\frac{\partial}{\partial \theta_{ij}} \sum_{m=1}^{M} \sum_{(k,l)\in E} \theta_{kl} y_k^{(m)} y_l^{(m)} = \sum_{m=1}^{M} y_i^{(m)} y_j^{(m)} = M \langle y_i y_j \rangle_{\text{data}}.$$

For the second term, we use the identity $\frac{\partial \log Z}{\partial \theta_{ij}} = \frac{1}{Z} \frac{\partial Z}{\partial \theta_{ij}}$ to obtain:

$$\frac{\partial \log Z(\boldsymbol{\theta})}{\partial \theta_{ij}} = \frac{1}{Z(\boldsymbol{\theta})} \sum_{\mathbf{y}} y_i y_j \exp\left(\sum_{(k,l)\in E} \theta_{kl} y_k y_l + \sum_{k\in V} h_k y_k \right) = \langle y_i y_j \rangle_{\text{model}}.$$

Therefore, we have

$$\frac{\partial \ell(\boldsymbol{\theta})}{\partial \theta_{ij}} = M(\langle y_i y_j \rangle_{\text{data}} - \langle y_i y_j \rangle_{\text{model}}),$$

where $\langle y_i y_j \rangle_{\text{data}} = \frac{1}{M} \sum_{m=1}^{M} y_i^{(m)} y_j^{(m)}$ is the empirical correlation observed in the data, and $\langle y_i y_j \rangle_{\text{model}} = \sum_{\mathbf{y}} y_i y_j P(\mathbf{y} \mid \boldsymbol{\theta})$ is the model-predicted correlation under the current parameter estimates.

The computational bottleneck lies in evaluating $\langle y_i y_j \rangle_{\text{model}}$, which requires computing expectations under the Gibbs distribution. Computing this expectation requires summing over all 2^n possible configurations $\mathbf{y}$, making it computationally intractable for systems of even modest size. The partition function $Z(\boldsymbol{\theta})$ must also be evaluated for each configuration, compounding the computational burden.

3.2.1.2 *Pseudo-likelihood estimation*

The pseudo-likelihood approach [36] circumvents the partition function by decomposing the joint likelihood into a product of conditional probabilities:

$$P(\mathbf{y} \mid \boldsymbol{\theta}) \approx \prod_{i=1}^{n} P(y_i \mid \mathbf{y}_{\setminus i}, \boldsymbol{\theta}),$$

where $\mathbf{y}_{\setminus i}$ denotes all spins except y_i. The conditional probability for each spin is:

$$P(y_i \mid \mathbf{y}_{\setminus i}, \boldsymbol{\theta}) = \frac{\exp\left(y_i \left(\sum_{j \in N(i)} \theta_{ij} y_j + h_i\right)\right)}{2\cosh\left(\sum_{j \in N(i)} \theta_{ij} y_j + h_i\right)},$$

where $N(i)$ denotes the neighborhood of node i.

The pseudo-likelihood function becomes:

$$\ell_{\mathrm{PL}}(\boldsymbol{\theta}) = \sum_{m=1}^{M} \sum_{i=1}^{n} \log P\left(y_i^{(m)} \mid \mathbf{y}_{\setminus i}^{(m)}, \boldsymbol{\theta}\right).$$

The computational efficiency of pseudo-likelihood stems from the fact that each conditional probability has a tractable normalization constant. Unlike the global partition function $Z(\boldsymbol{\theta})$ which requires summing over all 2^n configurations, each conditional probability $P(y_i \mid \mathbf{y}_{\setminus i}, \boldsymbol{\theta})$ only involves normalizing over the 2 possible values of y_i. The denominator $2\cosh(\cdot)$ represents the sum of exponentials for $y_i = +1$ and $y_i = -1$, requiring only $O(|N(i)|)$ operations where $|N(i)|$ is the neighborhood size. This reduces the computational complexity from $O(2^n)$ per gradient step to $O(n \cdot d)$ where d is the average graph degree.

Nevertheless, the "no free lunch" principle applies: This computational gain comes at the cost of approximation accuracy. The pseudo-likelihood provides only an approximation to the true likelihood and may underestimate parameter uncertainties, particularly when the graph structure is densely connected and long-range correlations are strong.

3.2.1.3 *Numerical simulations*

To systematically evaluate the performance of pseudo-likelihood estimation for Ising models, we conducted numerical experiments using synthetic data generated from two-dimensional Ising models with periodic boundary conditions.

Data Generation Protocol
We generated synthetic datasets using Gibbs sampling from 2D Ising models defined by their Hamiltonian. The sampling procedure follows the standard Gibbs algorithm, where each spin is updated according to its conditional probability:

$$P(y_i = +1 \mid \mathbf{y}_{\setminus i}) = \sigma\left(2\left(\theta \sum_{j \in N(i)} y_j + h\right)\right) \tag{3.1}$$

where $\sigma(x) = (1 + e^{-x})^{-1}$ is the sigmoid function. To ensure equilibrium sampling, we employed a burn-in period of 1000 iterations before collecting the desired number of configurations.

Simulation parameters

We tested for four distinct scenarios:

(1) **Baseline performance assessment**: We established reference performance metrics using a 10×10 lattice with coupling strength $\theta = 0.5$ and external field $h = 0.2$, generating 2000 equilibrium configurations.
(2) **Sample size dependence**: To characterize the convergence properties, we varied the number of samples $M \in \{100, 200, 500, 1000, 2000\}$ while maintaining fixed system parameters.
(3) **Coupling strength analysis**: We investigated performance across different coupling regimes by varying $\theta \in [0.1, 0.8]$, with particular attention to the critical coupling $\theta_c \approx 0.44$ for the 2D Ising model.
(4) **System size scaling**: We examined finite-size effects and computational scaling using lattice sizes $L \in \{6, 8, 10, 12, 15\}$, corresponding to system sizes $n \in \{36, 64, 100, 144, 225\}$.

Parameter estimation was performed using the L-BFGS-B algorithm with bounds: $\theta \in [0.01, 2.0]$ and $h \in [-1.0, 1.0]$. The optimization was initialized at $(\theta_0, h_0) = (0.1, 0.0)$ and aimed to minimize the negative pseudo-log-likelihood.

3.2.1.4 *Results and analysis*

Baseline performance
Our baseline experiments on the 10×10 system with 2000 samples demonstrated good parameter recovery. The true parameters $(\theta, h) = (0.500, 0.200)$ were estimated as $(\hat{\theta}, \hat{h}) = (0.493, 0.217)$, yielding absolute errors of $|\hat{\theta} - \theta| = 0.0066$ and $|\hat{h} - h| = 0.0175$. The computation completed in 262.8 seconds, establishing the feasibility of the method for moderate-sized systems.

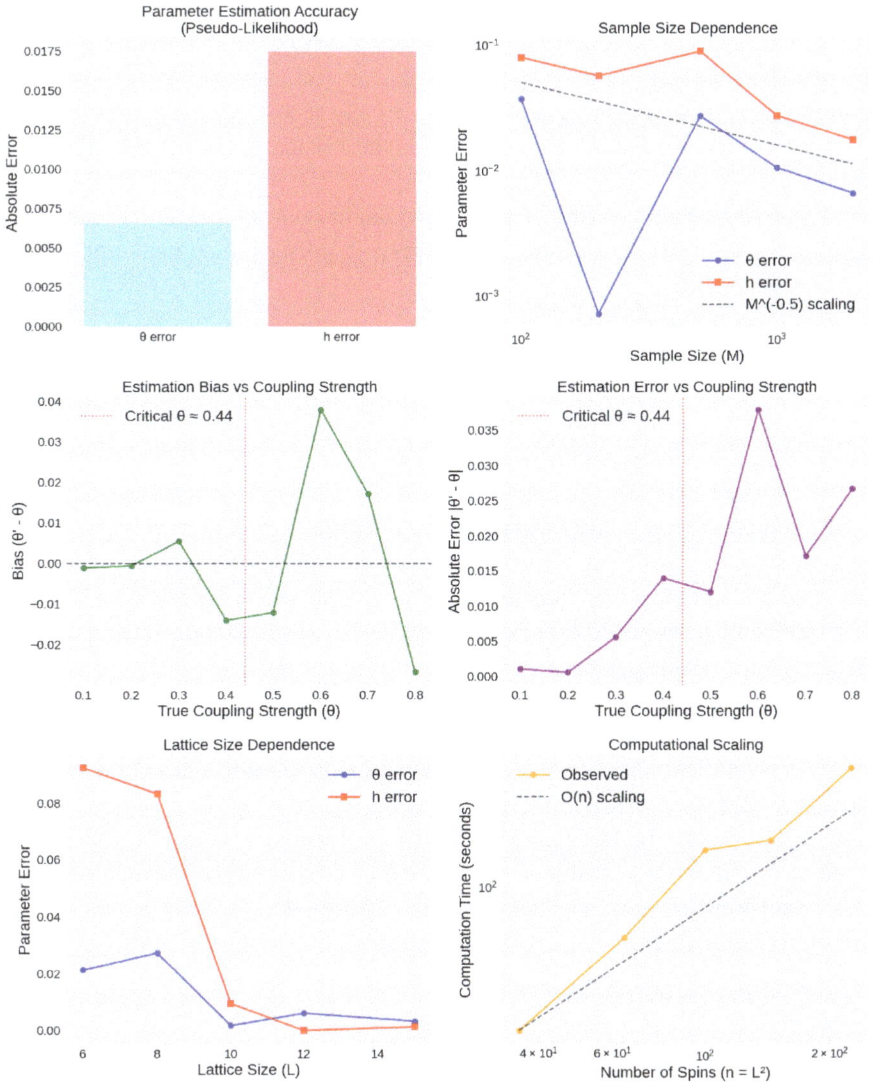

Fig. 3.1: Pseudo-likelihood estimation results showing parameter accuracy, sample size scaling, critical point behavior, and computational scaling.

Sample size scaling analysis

The relationship between estimation accuracy and sample size follows expected statistical scaling laws. Figure 3.1 (top right) illustrates the error convergence, revealing that both parameter errors decrease approximately as $M^{-0.5}$, consistent with the central limit theorem.

Critical point behavior

The analysis of coupling strength dependence reveals interesting behavior near the critical point. Figure 3.1 (second row) shows both the estimation bias and absolute error as functions of the true coupling strength. The vertical dashed line at $\theta_c \approx 0.44$ marks the critical point of the 2D Ising model.

The results demonstrate that pseudo-likelihood estimation is most reliable in the subcritical regime, with degraded performance near and above the critical point.

Computational scaling

The computational complexity analysis confirms approximately linear scaling with system size. Figure 3.1 (bottom right) shows the computation time versus the number of spins $n = L^2$. The observed scaling follows the theoretical $O(n)$ behavior reasonably well.

Finite-size effects

Our lattice size analysis reveals that estimation accuracy varies with system size, as shown in Figure 3.1 (bottom left). The coupling parameter θ shows improved estimation accuracy with larger systems, while the external field parameter h demonstrates particularly strong improvement, with errors decreasing from 0.0924 ($L = 6$) to 0.0013 ($L = 15$). This improvement reflects the increased statistical information available in larger systems and the reduced relative importance of boundary effects.

The numerical simulations demonstrate that pseudo-likelihood estimation provides a practical and computationally efficient framework for parameter inference in Ising models, with performance characteristics that make it particularly suitable for moderate to large-scale applications where exact methods become computationally intractable.

3.2.1.5 *Monte Carlo methods*

One could alleviate the problem and avoid merely optimizing an approximation to the true distribution by referring back to the sampling techniques introduced in earlier chapters. Specifically, Monte Carlo methods offer a principled approach to handling the intractable partition function by generating samples from the target distribution. Let us briefly mention two instances of MCMC methods tailored for the task at hand.

Gibbs sampling

Gibbs sampling generates samples $\mathbf{y} \sim P(\mathbf{y} \mid \boldsymbol{\theta})$ through iterative conditional sampling. At each step, one spin is updated according to its conditional distribution:

$$P(y_i \mid \mathbf{y}_{\setminus i}, \boldsymbol{\theta}) = \frac{\exp\left(y_i \left(\sum_{j \in N(i)} \theta_{ij} y_j + h_i\right)\right)}{2\cosh\left(\sum_{j \in N(i)} \theta_{ij} y_j + h_i\right)}.$$

The generated samples enable Monte Carlo estimation of $\langle y_i y_j \rangle_{\text{model}}$, facilitating gradient-based optimization of the *true* likelihood function.

Metropolis-Hastings algorithm

The Metropolis-Hastings algorithm generates samples using a proposal distribution and an acceptance criterion that ensures convergence to the target distribution $P(\mathbf{y} \mid \boldsymbol{\theta})$. This approach offers greater flexibility in the choice of proposal distributions but may require careful tuning for efficient sampling.

3.2.1.6 *Variational inference*

Variational inference approximates the intractable distribution $P(\mathbf{y} \mid \boldsymbol{\theta})$ with a simpler distribution $Q(\mathbf{y})$ by minimizing the Kullback-Leibler divergence:

$$\text{KL}(Q \parallel P) = \sum_{\mathbf{y}} Q(\mathbf{y}) \log \frac{Q(\mathbf{y})}{P(\mathbf{y} \mid \boldsymbol{\theta})}.$$

The parameters $\boldsymbol{\theta}$ are then estimated using the variational approximation to the log-likelihood. This approach trades exactness for computational tractability and can provide reasonable approximations when the variational family is well-chosen.

3.2.2 *Parameter estimation in the solid-on-solid model*

The solid-on-solid (SOS) model represents a fundamental framework in statistical physics for describing surface growth phenomena and roughness characterization. This classical lattice model captures the essential physics of surface fluctuations by treating the height $h_i \in \mathbb{Z}$ at each lattice site i as a discrete variable, thereby enabling a tractable yet meaningful representation of complex surface dynamics (see [178] and references therein).

The energetics of the SOS model are encoded in its Hamiltonian, which emphasizes the role of local height variations in determining the system's configuration energy. The model Hamiltonian takes the form:

$$H(\mathbf{h}) = \sum_{\langle i,j \rangle} V(h_i - h_j),$$

where $\mathbf{h} = \{h_i\}$ denotes the collection of height variables across all lattice sites, and the summation extends over all nearest-neighbor pairs $\langle i, j \rangle$. The interaction potential $V(h_i - h_j)$ typically assumes a quadratic form that penalizes large height differences between adjacent sites:

$$V(h_i - h_j) = \frac{\theta}{2}(h_i - h_j)^2,$$

where $\theta > 0$ represents the coupling strength parameter that controls the degree of surface smoothness.

Within the framework of statistical mechanics, the equilibrium properties of the SOS model are governed by the Gibbs measure, which assigns probabilities to different height configurations according to their energetic favorability. The probability distribution over height configurations is given by:

$$P(\mathbf{h} \mid \theta) = \frac{1}{Z(\theta)} \exp\left(-H(\mathbf{h})\right),$$

where the partition function $Z(\theta)$ ensures proper normalization:

$$Z(\theta) = \sum_{\mathbf{h}} \exp\left(- \sum_{\langle i,j \rangle} \frac{\theta}{2}(h_i - h_j)^2 \right).$$

The central challenge in parameter estimation for the SOS model lies in inferring the coupling strength θ from observed height configurations $\mathcal{D} = \{\mathbf{h}^{(1)}, \ldots, \mathbf{h}^{(M)}\}$. This inference problem requires careful consideration of both the empirical statistics of the observed data and the theoretical predictions of the model.

The empirical characterization of the data begins with computing the mean squared height difference between neighboring sites across all observed configurations:

$$\langle (h_i - h_j)^2 \rangle_{\text{data}} = \frac{1}{M} \sum_{m=1}^{M} \frac{1}{|E|} \sum_{\langle i,j \rangle} \left(h_i^{(m)} - h_j^{(m)} \right)^2,$$

where $|E|$ denotes the total number of edges in the lattice structure. This quantity provides a direct measure of the surface roughness present in the observed data.

The parameter estimation proceeds through maximum likelihood estimation, which requires evaluating the gradient of the Hamiltonian with respect to the coupling parameter as in the previous section. The derivative of the Hamiltonian takes the simple form:

$$\frac{\partial H(\mathbf{h})}{\partial \theta} = \frac{1}{2} \sum_{\langle i,j \rangle} (h_i - h_j)^2,$$

reflecting the linear dependence of the energy on the coupling strength for fixed height differences.

The log-likelihood function for the observed data incorporates both the direct energy contributions and the normalization constant:

$$\ell(\theta) = -\sum_{m=1}^{M} \sum_{\langle i,j \rangle} \frac{\theta}{2} \left(h_i^{(m)} - h_j^{(m)} \right)^2 - M \log Z(\theta).$$

Taking the gradient of the log-likelihood with respect to the parameter reveals the fundamental structure of the estimation problem:

$$\frac{\partial \ell(\theta)}{\partial \theta} = -\frac{1}{2} \langle (h_i - h_j)^2 \rangle_{\text{data}} + \frac{1}{2} \langle (h_i - h_j)^2 \rangle_{\text{model}}.$$

This gradient expression embodies the core principle of maximum likelihood estimation: the optimal parameter value balances the empirical statistics of the observed data against the theoretical predictions of the model. The iterative update rule for the maximum likelihood estimator follows:

$$\theta^{(t+1)} = \theta^{(t)} + \eta \frac{\partial \ell(\theta^{(t)})}{\partial \theta},$$

where $\eta > 0$ represents the learning rate that controls the step size of the optimization procedure (e.g. gradient ascent).

As in the previous section, the primary computational challenge in implementing this estimation scheme lies in evaluating the model expectation $\langle (h_i - h_j)^2 \rangle_{\text{model}}$, which requires knowledge of the partition function $Z(\theta)$. This quantity is generally intractable for realistic system sizes, necessitating the use of approximation methods.

Analogously, Monte Carlo sampling provides one avenue for addressing this computational bottleneck. By employing algorithms such as Metropolis-Hastings or Gibbs sampling to generate height configurations $\mathbf{h} \sim P(\mathbf{h} \mid \theta)$, one can estimate the model expectation through sample averaging:

$$\langle (h_i - h_j)^2 \rangle_{\text{model}} \approx \frac{1}{K} \sum_{k=1}^{K} \frac{1}{|E|} \sum_{\langle i,j \rangle} (h_i^{(k)} - h_j^{(k)})^2,$$

where K denotes the number of sampled configurations.

3.2.3 *Parameter estimation for DNA models*

DNA models that leverage Gibbs measures focus on capturing the interactions and structural properties of nucleotide sequences [324], [328].

These models find widespread application in computational biology for tasks such as motif finding, sequence alignment, and predicting secondary structures. The Gibbs measure framework provides a principled approach for modeling DNA sequences by treating them as realizations of a random field with specific energy functions that encode biological constraints and dependencies.

Consider a DNA sequence represented as $\mathbf{x} = \{x_1, x_2, \dots, x_N\}$, where each position x_i takes one of the four nucleotide values from the alphabet $\{A, T, G, C\}$. The core of the Gibbs measure approach lies in defining an appropriate energy function that captures the essential features of DNA sequence organization.

The DNA model is described by a Hamiltonian that includes pairwise interactions and external influences:

$$H(\mathbf{x}; \theta, h) = -\sum_{i=1}^{N-1} \theta_{x_i, x_{i+1}} - \sum_{i=1}^{N} h_{x_i},$$

where $\theta_{x_i, x_{i+1}}$ represents the interaction strength between neighboring bases x_i and x_{i+1}, and h_{x_i} denotes the external field or bias for base x_i. The first term captures pairwise interactions between adjacent nucleotides, quantifying phenomena such as CpG suppression, while the second term represents intrinsic nucleotide preferences independent of sequence context.

The probability of a DNA sequence under the Gibbs measure is given by:

$$P(\mathbf{x} \mid \theta, h) = \frac{1}{Z(\theta, h)} \exp\left(-H(\mathbf{x}; \theta, h)\right),$$

where the partition function ensures proper normalization:

$$Z(\theta, h) = \sum_{\mathbf{x}} \exp\left(-H(\mathbf{x}; \theta, h)\right).$$

The central challenge in parameter estimation for DNA models lies in inferring two sets of parameters from observed sequences $\mathcal{D} = \{\mathbf{x}^{(1)}, \mathbf{x}^{(2)}, \dots, \mathbf{x}^{(M)}\}$:

(1) $\theta_{x,x'}$: A 4×4 interaction matrix representing pairwise nucleotide interactions
(2) h_x: A vector of length 4 representing external field parameters for base preferences

The empirical characterization of the data involves computing the frequencies of dinucleotides and individual nucleotides across all observed sequences. These empirical statistics provide the foundation for parameter estimation through maximum likelihood methods.

Following the general framework for Gibbs measure parameter estimation, the log-likelihood function for the observed sequences takes the form:

$$\ell(\theta, h) = -\sum_{m=1}^{M}\left[\sum_{i=1}^{N-1}\theta_{x_i^{(m)},x_{i+1}^{(m)}} + \sum_{i=1}^{N} h_{x_i^{(m)}}\right] - M\log Z(\theta, h).$$

The gradients with respect to the parameters reveal the fundamental structure of the estimation problem. For interaction parameters $\theta_{x,x'}$:

$$\frac{\partial \ell(\theta,h)}{\partial \theta_{x,x'}} = \langle \delta(x_i = x, x_{i+1} = x')\rangle_{\text{data}} - \langle \delta(x_i = x, x_{i+1} = x')\rangle_{\text{model}},$$

where $\delta(\cdot)$ is the indicator function. Similarly, for external field parameters h_x:

$$\frac{\partial \ell(\theta,h)}{\partial h_x} = \langle \delta(x_i = x)\rangle_{\text{data}} - \langle \delta(x_i = x)\rangle_{\text{model}}.$$

These gradient expressions embody the core principle of maximum likelihood estimation: the optimal parameter values balance the empirical statistics of the observed data against the theoretical predictions of the model. The iterative update rules follow:

$$\theta_{x,x'}^{(t+1)} = \theta_{x,x'}^{(t)} + \eta\frac{\partial \ell(\theta,h)}{\partial \theta_{x,x'}}, \quad h_x^{(t+1)} = h_x^{(t)} + \eta\frac{\partial \ell(\theta,h)}{\partial h_x},$$

where $\eta > 0$ represents the learning rate that controls the step size of the optimization procedure.

As in other Gibbs measure models, the primary computational challenge lies in evaluating the model expectations $\langle\cdot\rangle_{\text{model}}$, which require knowledge of the partition function $Z(\theta, h)$. For sequences of realistic length, exact computation becomes intractable due to the exponential number of possible sequences (4^N configurations).

Monte Carlo sampling provides a practical approach to address this computational bottleneck. By employing algorithms such as Metropolis-Hastings or Gibbs sampling to generate sequences from $P(\mathbf{x} \mid \theta, h)$, one can estimate the model expectations through sample averaging. Alternative approximation methods include mean field approximations, which replace the complex joint distribution with a product of independent marginal distributions, and variational methods that optimize a lower bound on the log-likelihood.

3.3 Variational approaches to parameter estimation

Variational approaches to parameter estimation constitute a powerful framework rooted in optimization theory and statistical inference. These methods have become indispensable tools in machine learning, statistics, and Bayesian inference, particularly for approximating complex posterior distributions and performing efficient parameter estimation in high-dimensional settings.

3.3.1 *Bayesian inference framework*

In Bayesian parameter estimation, the fundamental goal is to estimate the posterior distribution $p(\theta|\mathcal{D})$, where $\mathcal{D}$ represents the observed data, $p(\mathcal{D}|\theta)$ is the likelihood function, and $p(\theta)$ is the prior distribution over parameters θ.

By Bayes' theorem, we obtain:

$$p(\theta|\mathcal{D}) = \frac{p(\mathcal{D}|\theta)p(\theta)}{p(\mathcal{D})}, \tag{3.2}$$

where the evidence (or marginal likelihood) is given by:

$$p(\mathcal{D}) = \int p(\mathcal{D}|\theta)p(\theta)\, d\theta. \tag{3.3}$$

The central challenge in Bayesian inference arises from the fact that the evidence integral in Equation (3.3) is often intractable for high-dimensional or complex models. This intractability necessitates the development of approximation methods, among which variational approaches stand out as particularly effective.

3.3.2 *Variational approximation*

Variational approaches circumvent the intractability of the posterior by approximating $p(\theta|\mathcal{D})$ using a simpler, tractable distribution $q(\theta;\phi)$ parameterized by variational parameters ϕ. The optimal approximation is found by solving the optimization problem:

$$q^*(\theta;\phi) = \arg\min_{q\in\mathcal{Q}} \mathrm{KL}(q(\theta;\phi)\|p(\theta|\mathcal{D})), \tag{3.4}$$

where KL denotes the Kullback-Leibler (KL) divergence and $\mathcal{Q}$ represents the family of candidate distributions.

Since the KL divergence involves the intractable posterior, we reformulate this optimization problem using the evidence lower bound (ELBO).

The KL divergence minimization in Equation (3.4) is equivalent to maximizing the ELBO, defined as:

$$\mathcal{L}(\phi) = \mathbb{E}_{q(\theta;\phi)}[\log p(\mathcal{D}|\theta)] - \mathrm{KL}(q(\theta;\phi)\|p(\theta)). \tag{3.5}$$

The ELBO provides a tractable lower bound on the log evidence:

$$\log p(\mathcal{D}) \geq \mathcal{L}(\phi). \tag{3.6}$$

Maximizing $\mathcal{L}(\phi)$ with respect to ϕ yields the optimal variational approximation:

$$q^*(\theta;\phi^*) \approx p(\theta|\mathcal{D}). \tag{3.7}$$

3.3.3 *ELBO optimization and parameter estimation*

The practical implementation of variational inference requires solving the optimization problem in Equation (3.4), which is achieved by maximizing the ELBO with respect to the variational parameters ϕ. This optimization process simultaneously serves two purposes: finding the best variational approximation and enabling subsequent parameter estimation.

3.3.3.1 *Gradient-based optimization*

The optimization of the ELBO is typically achieved using gradient-based methods. The gradient of the ELBO with respect to the variational parameters is:

$$\nabla_\phi \mathcal{L}(\phi) = \nabla_\phi \mathbb{E}_{q(\theta;\phi)}[\log p(\mathcal{D},\theta) - \log q(\theta;\phi)], \tag{3.8}$$

where $p(\mathcal{D},\theta) = p(\mathcal{D}|\theta)p(\theta)$ is the joint distribution.

Computing this gradient directly is often challenging due to the expectation over the variational distribution. Two primary techniques are employed to address this challenge:

3.3.3.2 *Reparameterization trick*

For certain distributions (e.g., Gaussian), we can employ the reparameterization trick. For a Gaussian variational distribution with parameters μ and σ, we write:

$$\theta = \mu + \sigma\epsilon, \quad \epsilon \sim \mathcal{N}(0,1). \tag{3.9}$$

This transformation converts the stochastic expectation into a deterministic function:

$$\mathbb{E}_{q(\theta;\phi)}[f(\theta)] = \mathbb{E}_\epsilon[f(\mu+\sigma\epsilon)], \tag{3.10}$$

enabling straightforward backpropagation through the expectation.

3.3.3.3 *Monte Carlo approximation*

When the reparameterization trick is not applicable, Monte Carlo sampling is used to approximate the gradient:

$$\nabla_\phi \mathcal{L}(\phi) \approx \frac{1}{S}\sum_{s=1}^{S} \nabla_\phi[\log p(\mathcal{D}, \theta^{(s)}) - \log q(\theta^{(s)}; \phi)], \tag{3.11}$$

where $\theta^{(s)} \sim q(\theta; \phi)$ are samples from the variational distribution.

3.3.3.4 *Parameter estimation from variational distributions*

Once the optimal variational distribution $q^*(\theta; \phi^*)$ is obtained through ELBO maximization, parameter estimation can proceed using several approaches that leverage the tractable nature of the variational approximation:

Maximum a Posteriori (MAP) estimation:

$$\theta^* = \arg\max_\theta q^*(\theta; \phi^*). \tag{3.12}$$

This approach finds the mode of the variational distribution, which serves as a point estimate of the parameters. For Gaussian variational distributions, this simply corresponds to the mean parameter.

Posterior mean estimation:

$$\hat{\theta} = \mathbb{E}_{q^*(\theta;\phi^*)}[\theta]. \tag{3.13}$$

The posterior mean provides a Bayes-optimal point estimate under squared loss and can be computed analytically for many variational families.

3.3.4 *Mean-field approximation*

To further simplify the optimization, the variational distribution is often assumed to factorize across parameter groups:

$$q(\theta; \phi) = \prod_{i=1}^{d} q_i(\theta_i; \phi_i), \tag{3.14}$$

where $\theta = (\theta_1, \ldots, \theta_d)$ and each q_i is a simpler distribution. This mean-field approximation reduces the complexity of the optimization problem but may underestimate the posterior uncertainty by ignoring parameter correlations.

3.3.5 *Extensions and advanced topics*

3.3.5.1 *Hierarchical variational Bayes*

Variational Bayes extends naturally to hierarchical models where hyperparameters ζ govern the parameter distribution:

$$p(\theta, \mathcal{D}, \zeta) = p(\mathcal{D}|\theta)p(\theta|\zeta)p(\zeta). \tag{3.15}$$

The variational approximation then targets the joint posterior $p(\theta, \zeta|\mathcal{D})$ using a factorized approximation $q(\theta; \phi_\theta)q(\zeta; \phi_\zeta)$.

3.3.5.2 *Variational autoencoders*

In variational autoencoders (VAEs), variational approaches estimate parameters of a latent generative model. The ELBO for VAEs is:

$$\mathcal{L}(\phi, \theta) = \mathbb{E}_{q_\phi(z|x)}[\log p_\theta(x|z)] - \mathrm{KL}(q_\phi(z|x)\|p(z)), \tag{3.16}$$

where z represents latent variables, x denotes observed data, $q_\phi(z|x)$ is the encoder network, and $p_\theta(x|z)$ is the decoder network.

3.3.5.3 *Black-box variational inference*

Black-box variational inference provides a general framework for variational inference that can be applied to any differentiable model. It approximates gradients of the ELBO using stochastic gradient methods with Monte Carlo estimates, making it applicable to a wide range of models without requiring model-specific derivations.

3.3.5.4 *Neural variational inference*

In neural variational inference, the variational distribution $q(\theta; \phi)$ is parameterized by neural network weights ϕ. This approach enables flexible approximations and efficient parameter estimation across large datasets, particularly in deep learning applications.

3.4 Differences between parametric and non-parametric methods

The fundamental distinction between parametric and non-parametric methods lies in their assumptions about the underlying structure of the data and the flexibility of the models they employ. This section provides a comprehensive mathematical analysis of their key differences, theoretical properties, and practical implications.

3.4.1 *Model assumptions and structure*

3.4.1.1 *Parametric methods*

Parametric methods assume a specific finite-dimensional form for the underlying model. The model is characterized by a function $f(x;\theta)$ where the parameter vector θ belongs to a finite-dimensional parameter space:

$$f(x;\theta), \quad \theta \in \Theta \subseteq \mathbb{R}^p, \tag{3.17}$$

where p is the fixed dimension of the parameter space, independent of the sample size n.

The parametric approach embodies strong structural assumptions about the data-generating process. For instance, in linear regression, we assume:

$$f(x;\theta) = \theta^\top x = \theta_0 + \theta_1 x_1 + \cdots + \theta_p x_p. \tag{3.18}$$

The primary goal in parametric estimation is to determine the optimal parameter vector θ^* that best fits the observed data according to some criterion (e.g., maximum likelihood, least squares).

3.4.1.2 *Non-parametric methods*

Non-parametric methods do not impose a specific parametric form on the underlying function $f(x)$. Instead, $f(x)$ is treated as an element of an infinite-dimensional function space:

$$f \in \mathcal{F}, \tag{3.19}$$

where $\mathcal{F}$ represents a function space such as Sobolev spaces, reproducing kernel Hilbert spaces (RKHS), or spaces of smooth functions.

The key advantage of this approach is its flexibility—the model can adapt to the data's inherent structure without being constrained by rigid parametric assumptions. However, this flexibility comes at the cost of increased complexity and the need for regularization to prevent overfitting.

3.4.2 *Flexibility and model complexity*

3.4.2.1 *Parametric complexity*

In parametric methods, model complexity is determined by the number of parameters p, which remains fixed regardless of the sample size. This constraint provides several advantages:

1. **Interpretability:** The fixed number of parameters often corresponds to meaningful quantities in the problem domain.

2. **Computational efficiency:** Estimation involves optimization over a fixed-dimensional space.
3. **Statistical efficiency:** When the parametric assumption is correct, these methods achieve optimal convergence rates.

However, parametric methods suffer from potential model misspecification. This means that if the assumed functional form is incorrect, the method may perform poorly regardless of the sample size.

3.4.2.2 *Non-parametric complexity*

Non-parametric methods exhibit complexity that typically grows with the sample size n. This adaptive complexity allows the model to capture increasingly subtle patterns as more data becomes available.

Consider kernel density estimation (KDE) as an example:

$$\hat{f}(x) = \frac{1}{nh} \sum_{i=1}^{n} K\left(\frac{x - X_i}{h}\right), \tag{3.20}$$

where $K(\cdot)$ is a kernel function and h is the bandwidth parameter that controls the smoothness of the estimate.

The flexibility of non-parametric methods enables them to approximate any smooth function as $n \to \infty$, but this comes with the risk of overfitting, particularly in finite samples, necessitating careful regularization.

3.4.3 *Estimation techniques in non-parametric methods*

3.4.3.1 *Kernel methods*

Kernel methods form a cornerstone of non-parametric estimation. The general form of a kernel estimator is:

$$\hat{f}(x) = \frac{1}{n} \sum_{i=1}^{n} K_h(x - x_i), \tag{3.21}$$

where $K_h(x) = \frac{1}{h}K(x/h)$ is the scaled kernel and h is the bandwidth parameter.

The choice of kernel K and bandwidth h significantly affects the estimator's performance. Common kernels include the Gaussian kernel:

$$K(u) = \frac{1}{\sqrt{2\pi}} \exp\left(-\frac{u^2}{2}\right). \tag{3.22}$$

3.4.3.2 *Spline methods*

Spline methods balance data fidelity with smoothness by minimizing a penalized least squares objective:

$$\sum_{i=1}^{n}(y_i - f(x_i))^2 + \lambda \int \|f''(x)\|^2 dx, \tag{3.23}$$

where λ is the regularization parameter controlling the trade-off between fit and smoothness.

The solution can be expressed as a linear combination of basis functions:

$$\hat{f}(x) = \sum_{j=1}^{m} \beta_j B_j(x), \tag{3.24}$$

where $B_j(x)$ are the basis functions (e.g., B-splines).

3.4.3.3 *Gaussian processes*

Gaussian processes provide a probabilistic framework for non-parametric regression. Given training data (X, Y), the posterior mean estimate for a new point x is:

$$\hat{f}(x) = k(x, X)[K(X, X) + \sigma^2 I]^{-1} Y, \tag{3.25}$$

where $k(x, X)$ is the kernel vector between x and the training inputs, $K(X, X)$ is the Gram matrix, and σ^2 is the noise variance.

3.4.4 *Practical examples*

3.4.4.1 *Parametric examples*

Linear regression:

$$y = \beta_0 + \beta_1 x + \varepsilon, \quad \varepsilon \sim \mathcal{N}(0, \sigma^2). \tag{3.26}$$

Logistic regression:

$$P(y = 1 \mid x) = \frac{1}{1 + \exp(-\theta^\top x)}. \tag{3.27}$$

Polynomial regression:

$$y = \beta_0 + \beta_1 x + \beta_2 x^2 + \cdots + \beta_p x^p + \varepsilon. \tag{3.28}$$

3.4.4.2 *Non-parametric examples*

Kernel density estimation:

$$\hat{f}(x) = \frac{1}{nh}\sum_{i=1}^{n} K\left(\frac{x - X_i}{h}\right). \tag{3.29}$$

Nadaraya-Watson kernel regression:

$$\hat{f}(x) = \frac{\sum_{i=1}^{n} K_h(x - x_i)y_i}{\sum_{i=1}^{n} K_h(x - x_i)}. \tag{3.30}$$

k-nearest neighbors:

$$\hat{f}(x) = \frac{1}{k}\sum_{i\in\mathcal{N}_k(x)} y_i, \tag{3.31}$$

where $\mathcal{N}_k(x)$ denotes the set of k nearest neighbors of x.

The modern trend in machine learning often involves hybrid approaches that combine the benefits of both paradigms, such as semi-parametric models or parametric models with non-parametric components.

3.5 Gibbs measures in non-parametric estimation

The concept of Gibbs measures provides a principled probabilistic framework for non-parametric estimation, particularly in complex, high-dimensional systems. This connection has proven especially valuable in statistical mechanics-inspired methods, including Bayesian non-parametrics, spatial statistics, and image processing applications.

In the classical setting of statistical mechanics, a Gibbs measure μ on a configuration space $\mathcal{X}$ is characterized by its local conditional probabilities. However, in the context of non-parametric estimation, we consider a more general formulation where the configuration space $\mathcal{X}$ corresponds to a function space, such as spaces of density functions or regression functions, and the energy function $H(x)$ encodes prior beliefs and data constraints.

3.5.1 *Bayesian framework for non-parametric estimation*

Non-parametric estimation aims to estimate an unknown function $f \in \mathcal{F}$ based on observed data, where $\mathcal{F}$ represents an appropriate function space. Within a Bayesian framework, we can introduce a Gibbs measure to model the posterior distribution of f given the data:

$$\mu(df \mid \text{data}) \propto e^{-\beta H(f)}\mu_0(df), \tag{3.32}$$

where $\mu_0(df)$ is a prior measure on $\mathcal{F}$ (such as a Gaussian process prior), $\beta = 1/T > 0$ is an inverse temperature parameter, and $H(f)$ is the energy function that typically combines data fidelity and regularization terms.

The choice of function space $\mathcal{F}$ depends on the specific problem at hand. Common choices include Sobolev spaces $H^s(\Omega)$ when smoothness regularization is desired, or reproducing kernel Hilbert spaces (RKHS) $\mathcal{H}_K$ for kernel-based methods. The energy function $H(f)$ naturally decomposes into two components that balance model fit and complexity.

3.5.2 *Energy function decomposition*

The energy function $H(f)$ typically combines a data fidelity term with a regularization term. The data fidelity term measures how well the function f explains the observed data:

$$H_{\text{data}}(f) = -\sum_{i=1}^{n} \log p(y_i \mid f(x_i)), \tag{3.33}$$

where $p(y_i \mid f(x_i))$ represents the likelihood of observation y_i given the function value $f(x_i)$.

The regularization term imposes structural constraints or smoothness properties on the estimated function:

$$H_{\text{reg}}(f) = \lambda \|f\|_{\mathcal{H}}^2, \tag{3.34}$$

where $\|f\|_{\mathcal{H}}$ is an appropriate norm in the chosen function space $\mathcal{H}$, and $\lambda > 0$ is a regularization parameter that controls the trade-off between data fidelity and smoothness.

Combining these components, the posterior Gibbs measure becomes:

$$\mu(df \mid \text{data}) \propto e^{-\beta(H_{\text{data}}(f)+H_{\text{reg}}(f))}\mu_0(df). \tag{3.35}$$

3.5.3 *Estimation methods*

The Gibbs measure framework naturally leads to several estimation approaches. Maximum a posteriori (MAP) estimation seeks the mode of the posterior distribution:

$$f_{\text{MAP}} = \arg\max_f \mu(df \mid \text{data}), \tag{3.36}$$

which is equivalent to solving the optimization problem:

$$f_{\text{MAP}} = \arg\min_f \left[H_{\text{data}}(f) + H_{\text{reg}}(f)\right]. \tag{3.37}$$

Alternatively, full Bayesian estimation utilizes the entire posterior distribution through the posterior mean or sampling-based approaches, providing natural uncertainty quantification alongside point estimates.

3.5.4 *Computational approaches*

Since direct sampling from the posterior measure $\mu(df \mid \text{data})$ is typically intractable in high-dimensional function spaces, Markov chain Monte Carlo (MCMC) methods become essential. Gibbs sampling and related algorithms provide practical approaches for posterior exploration.

The Metropolis-Hastings algorithm can be adapted to propose changes to the function f and accept or reject these proposals based on energy differences. More sophisticated approaches include Hamiltonian Monte Carlo (HMC), which leverages gradient information of the energy function $H(f)$ to enable more efficient exploration of the posterior distribution.

3.6 Conclusion

In this chapter, we explored the role of Gibbs measures as a fundamental probabilistic tool for parameter estimation in various machine learning models. Beginning with a general introduction to parameter estimation, we illustrated how Gibbs distributions naturally arise in the estimation of hidden or latent variables in models inspired by statistical physics, such as the Ising model, the Solid-on-Solid model, and biological sequence models like DNA. These examples demonstrate the flexibility and physical interpretability of Gibbs measures in capturing dependencies and regularities in high-dimensional systems.

We also examined variational approaches to parameter estimation, which use Gibbs-like approximations for tractable inference. The distinction between parametric and non-parametric methods was discussed, highlighting the scenarios in which Gibbs measures provide a principled way to encode prior structure and enforce sparsity or smoothness in estimation.

Finally, we investigated the emerging role of Gibbs measures in non-parametric estimation, particularly through Markov random fields and Bayesian frameworks. The insights gained from statistical mechanics continue to influence new directions in machine learning, particularly in designing energy-based models and understanding the geometry of parameter spaces.

For further reading see [155], [342], [348], [352], [359], [397].

Chapter 4

Applications of Gibbs Measures in Bayesian Inference and Probabilistic Modeling

This chapter explores the integration of Gibbs measures into Bayesian inference and probabilistic modeling frameworks. Beginning with the foundations of Bayesian parameter estimation and hierarchical modeling, it illustrates how Gibbs measures naturally emerge in multi-level Bayesian structures. Special emphasis is placed on applications involving graphical models and energy-based representations. The chapter also introduces Ising models with hidden Markov structures, examining translation-invariant Gibbs measures and their interpretability in machine learning contexts. It concludes with discussions on Bayesian non-parametric methods, including the Dirichlet process, and on generalized Bayesian inference, highlighting the advantages of Gibbs-based formulations in complex hierarchical models.

4.1 Bayesian inference

Bayesian inference is a probabilistic learning technique (see [146] for detailed presentation) that provides a framework for defining and updating our beliefs in light of new evidence. This approach systematically incorporates new observations to refine our understanding or predictions.

In machine learning, the primary objective is to develop a model that accurately represents or predicts data, particularly unseen data. Often, we have some preliminary knowledge or assumptions about the model before observing the data. Bayesian inference leverages this prior knowledge by representing it as a probability distribution over possible models, known as the *prior distribution* or *prior belief*.

4.1.1 *Foundational principles*

The Bayesian paradigm begins by encoding prior knowledge about the system under investigation through a prior distribution. Consider the classical coin-flipping problem, where the system is characterized by p_h, the probability of obtaining heads. In the absence of specific prior information, we might express our initial uncertainty through a uniform distribution over the unit interval:

$$P_0(p_h) = \text{Uniform}(0, 1). \tag{4.1}$$

This prior reflects the principle of indifference, assigning equal probability to all possible values of p_h.

Upon observing data, Bayesian inference systematically updates this prior belief to yield the posterior distribution $P_{\text{post}}(p_h)$. This update mechanism is governed by Bayes' theorem:

$$P_{\text{post}}(p_h \mid \text{data}) = \frac{P(\text{data} \mid p_h)P_0(p_h)}{P(\text{data})}, \tag{4.2}$$

where $P(\text{data} \mid p_h)$ represents the likelihood function quantifying the probability of observing the data given parameter p_h, and $P(\text{data})$ serves as a normalization constant ensuring the posterior forms a valid probability distribution.

4.1.2 *Sequential updating through conjugate priors*

The power of Bayesian inference becomes evident through sequential updating. Consider observing a single heads outcome: the likelihood becomes $P(\text{heads} \mid p_h) = p_h$. Combined with the uniform prior, the unnormalized posterior is proportional to p_h, yielding a linear distribution over the unit interval after normalization.

This sequential process can be continued indefinitely. If the subsequent observation is tails, the likelihood $P(\text{tails} \mid p_h) = 1 - p_h$ is multiplied by the current posterior to produce an updated belief. This iterative procedure exemplifies the self-consistent nature of Bayesian learning.

The beta distribution emerges naturally as the conjugate prior for binomial likelihood functions. After observing n_h heads and n_t tails, the posterior distribution takes the form:

$$P_{\text{post}}(p_h) = \text{Beta}(n_h + 1, n_t + 1), \tag{4.3}$$

where the beta distribution with parameters α and β has density:

$$f(x; \alpha, \beta) = \frac{\Gamma(\alpha + \beta)}{\Gamma(\alpha)\Gamma(\beta)} x^{\alpha-1}(1 - x)^{\beta-1}. \tag{4.4}$$

For instance, after observing 26 heads and 21 tails, the posterior becomes Beta(27, 22).

The posterior distribution enables various forms of inference. We can assess whether the coin exhibits bias toward heads by computing $P(p_h > 0.5)$, integrating the posterior density over the appropriate region. Similarly, predictive probabilities for future events can be calculated by integrating over the posterior. The probability of observing five consecutive heads is given by:

$$P(5 \text{ consecutive heads}) = \int_0^1 p_h^5 P_{\text{post}}(p_h)\, dp_h. \tag{4.5}$$

4.1.3 *General Bayesian framework*

The coin-flipping example generalizes to a comprehensive framework for model selection and parameter estimation. Consider a stochastic process generating independent and identically distributed events E_n for $n = 1, 2, 3, \ldots$, where the underlying probability mechanism is unknown. Let the model space $\{M_m\}$ represent a collection of mutually exclusive and exhaustive hypotheses about this process.

Each model M_m is characterized by its conditional probabilities $P(E_n \mid M_m)$, while our initial beliefs are encoded through prior probabilities $P(M_m)$ satisfying the normalization constraint:

$$\sum_m P(M_m) = 1. \tag{4.6}$$

When an event E is observed, the prior distribution over models is updated to the posterior distribution using Bayes' theorem:

$$P(M_m \mid E) = \frac{P(E \mid M_m)P(M_m)}{\sum_{m'} P(E \mid M_{m'})P(M_{m'})}. \tag{4.7}$$

This framework naturally extends to sequences of observations. For a sequence of independent observations $\mathbf{E} = (e_1, e_2, \ldots, e_n)$, the posterior probability becomes:

$$P(M_m \mid \mathbf{E}) = \frac{P(\mathbf{E} \mid M_m)P(M_m)}{\sum_{m'} P(\mathbf{E} \mid M_{m'})P(M_{m'})}, \tag{4.8}$$

where the likelihood function factorizes due to independence:

$$P(\mathbf{E} \mid M_m) = \prod_{k=1}^{n} P(e_k \mid M_m). \tag{4.9}$$

This formulation provides a systematic approach to learning from accumulating evidence, with each new observation refining our beliefs about the underlying generative process. The framework naturally balances model complexity with data fit, automatically implementing a form of Occam's razor through the normalization constraint in Bayes' theorem.

4.2 Bayesian parameter estimation

The power of Bayesian inference becomes particularly evident in parameter estimation problems, where we seek to learn about unknown parameters of a probabilistic model from observed data. Rather than treating parameters as fixed but unknown constants, the Bayesian approach treats them as random variables with their own probability distributions. This perspective enables us to quantify uncertainty about parameter values and update our beliefs systematically as new data become available.

4.2.1 *The parameterized model framework*

Consider a statistical model parameterized by a vector $\boldsymbol{\theta}$ that spans the parameter space. By parameterizing the space of possible models through $\boldsymbol{\theta}$, we can update our beliefs about all models simultaneously in a single Bayesian update. The distribution of belief over the model space can then be interpreted as a distribution of belief over the parameter space itself.

Let the initial prior distribution over $\boldsymbol{\theta}$ be denoted $p(\boldsymbol{\theta} \mid \boldsymbol{\alpha})$, where $\boldsymbol{\alpha}$ represents a set of hyperparameters that characterize our prior beliefs. This prior distribution encodes our initial knowledge about the parameters before observing any data, which may arise from previous experiments, theoretical considerations, domain expertise, or subjective judgment. This is illustrated in the top-right panel of Figure 4.1, where the prior distribution over the 2D parameter space emphasizes plausible values before incorporating any data.

When we observe a sequence of data points $\mathbf{X} = (x_1, \ldots, x_n)$, where each observation x_i is distributed according to $p(x_i \mid \boldsymbol{\theta})$ for some true but unknown parameter value $\boldsymbol{\theta}$, we can apply Bayes' theorem to obtain the posterior distribution over the parameters.

4.2.2 *Posterior inference*

The posterior distribution is computed through Bayes' theorem:

$$p(\boldsymbol{\theta} \mid \mathbf{X}, \boldsymbol{\alpha}) = \frac{p(\mathbf{X} \mid \boldsymbol{\theta}) p(\boldsymbol{\theta} \mid \boldsymbol{\alpha})}{p(\mathbf{X} \mid \boldsymbol{\alpha})}, \tag{4.10}$$

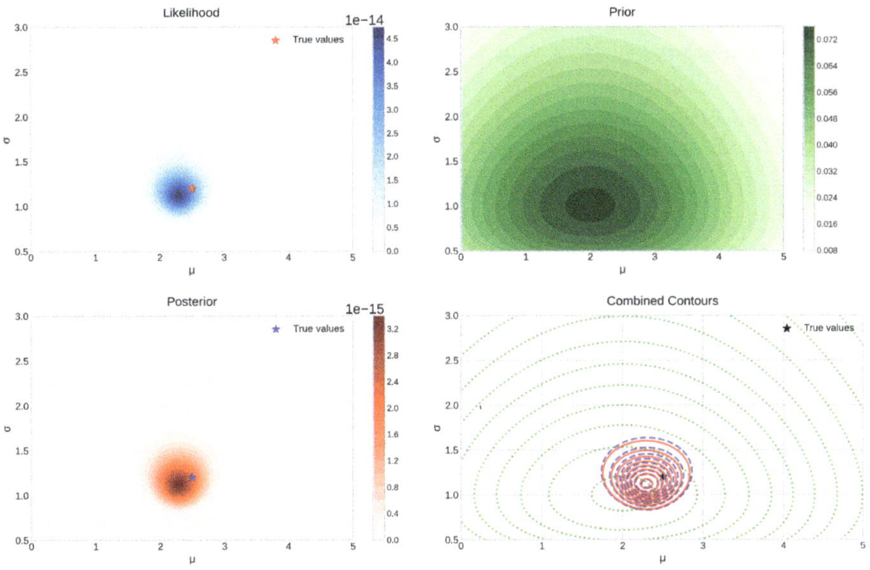

Fig. 4.1: 2D parameter space: Likelihood, prior, and posterior.

where the marginal likelihood or evidence is given by:

$$p(\mathbf{X} \mid \boldsymbol{\alpha}) = \int p(\mathbf{X} \mid \boldsymbol{\theta}) p(\boldsymbol{\theta} \mid \boldsymbol{\alpha}) \, d\boldsymbol{\theta}. \tag{4.11}$$

The likelihood function $p(\mathbf{X} \mid \boldsymbol{\theta})$ describes how the observed data are generated given the parameters. The top-left panel of Figure 4.1 visualizes this likelihood function over the two-dimensional parameter space. Brighter regions correspond to parameter values that make the observed data more probable. Under the standard assumption of independent and identically distributed observations, this likelihood factorizes as:

$$p(\mathbf{X} \mid \boldsymbol{\theta}) = \prod_{i=1}^{n} p(x_i \mid \boldsymbol{\theta}). \tag{4.12}$$

Since the marginal likelihood serves as a normalization constant when making inferences about $\boldsymbol{\theta}$, we often work with the proportional relationship:

$$p(\boldsymbol{\theta} \mid \mathbf{X}, \boldsymbol{\alpha}) \propto p(\mathbf{X} \mid \boldsymbol{\theta}) p(\boldsymbol{\theta} \mid \boldsymbol{\alpha}). \tag{4.13}$$

The marginal likelihood plays a crucial role beyond normalization. It provides a measure of how well the model explains the observed data by averaging the likelihood over all possible parameter values weighted by their prior probabilities. This quantity becomes essential for model comparison and selection. Finally, the bottom-right panel in Figure 4.1 overlays the prior, likelihood, and posterior contours, allowing direct comparison of how the posterior shifts from the prior toward regions supported by the data.

4.2.3 *Predictive inference*

A fundamental advantage of the Bayesian framework lies in its natural approach to prediction, which automatically accounts for parameter uncertainty. Rather than making predictions based on point estimates of parameters, Bayesian prediction integrates over the entire posterior distribution, providing a principled way to propagate uncertainty from parameters to predictions.

The posterior predictive distribution represents our beliefs about a new observation $\tilde{x}$ after having observed the data $\mathbf{X}$:

$$p(\tilde{x} \mid \mathbf{X}, \boldsymbol{\alpha}) = \int p(\tilde{x} \mid \boldsymbol{\theta}) p(\boldsymbol{\theta} \mid \mathbf{X}, \boldsymbol{\alpha}) \, d\boldsymbol{\theta}. \tag{4.14}$$

This distribution captures both the inherent variability in the data-generating process and the epistemic uncertainty in our parameter estimates. The integration over the posterior ensures that predictions account for all plausible parameter values weighted by their posterior probabilities, naturally providing uncertainty quantification for predictions.

The prior predictive distribution offers complementary insights by describing our beliefs about potential observations before seeing any data:

$$p(\tilde{x} \mid \boldsymbol{\alpha}) = \int p(\tilde{x} \mid \boldsymbol{\theta}) p(\boldsymbol{\theta} \mid \boldsymbol{\alpha}) \, d\boldsymbol{\theta}. \tag{4.15}$$

This distribution serves multiple important purposes in Bayesian analysis. It can be used to check the appropriateness of our prior assumptions through prior predictive checks, guide experimental design by identifying informative observations, and assess the informativeness of the data by comparing prior and posterior predictive distributions.

This framework for Bayesian parameter estimation provides a coherent and principled approach to learning from data while maintaining explicit quantification of uncertainty. The systematic treatment of parameters as random variables, combined with the coherent updating mechanism provided by Bayes' theorem, makes this approach particularly powerful for scientific inference and decision-making under uncertainty.

4.2.4 *Estimates of parameters and predictions*

Bayesian inference often involves estimating parameters or making predictions based on a posterior distribution. We now review several common estimators (extending those in Section 3.3.3.4) and the posterior predictive distribution, with illustrative examples.

Posterior Mean: When the posterior has a finite mean, the *posterior mean* provides the Bayes estimator under quadratic loss:

$$\tilde{\theta} = \mathbb{E}[\theta \mid \mathbf{X}, \alpha] = \int \theta \, p(\theta \mid \mathbf{X}, \alpha) \, d\theta.$$

This estimator minimizes the posterior expected squared error.

Posterior Median: In one-dimensional problems, the *posterior median* is defined by

$$P(\theta \leq \text{Median}) = \tfrac{1}{2},$$

and minimizes the posterior expected absolute loss. For symmetric, unimodal posteriors, the median coincides with the mean.

Maximum a Posteriori (MAP): The *MAP estimator* maximizes the posterior density:

$$\theta_{\text{MAP}} \in \arg\max_{\theta} \; p(\theta \mid \mathbf{X}, \alpha).$$

When the posterior is multimodal or unbounded, a unique MAP may not exist.

Bayes Estimators via Posterior Risk: More generally, given a loss function $L(\theta, \hat{\theta})$, the Bayes estimator minimizes

$$\hat{\theta}_{\text{Bayes}} = \arg\min_{\hat{\theta}} \int L(\theta, \hat{\theta}) \, p(\theta \mid \mathbf{X}, \alpha) \, d\theta.$$

- Quadratic loss $L(\theta, \hat{\theta}) = (\theta - \hat{\theta})^2 \implies$ posterior mean.
- Absolute loss $L(\theta, \hat{\theta}) = |\theta - \hat{\theta}| \implies$ posterior median.

Posterior Predictive Distribution: To predict a new observation $\tilde{x}$, integrate over the posterior:

$$p(\tilde{x} \mid \mathbf{X}, \alpha) = \int p(\tilde{x} \mid \theta)\, p(\theta \mid \mathbf{X}, \alpha)\, d\theta.$$

Example 22 (Cookie Problem)**.** *Two bowls contain plain and chocolate cookies:*

- *Bowl 1: 30 plain, 10 chocolate;*
- *Bowl 2: 20 plain, 20 chocolate.*

A cookie is drawn and is plain. What is $P(Bowl = 1 \mid plain)$*?*

Applying Bayes' theorem,

$$P(1 \mid E) = \frac{P(E \mid 1)\, P(1)}{P(E \mid 1)\, P(1) + P(E \mid 2)\, P(2)} = \frac{0.75 \cdot 0.5}{0.75 \cdot 0.5 + 0.5 \cdot 0.5} = 0.6.$$

Example 23 (Archaeological Dating)**.** *An archaeologist dates pottery fragments to century* $c \in [11, 16]$*. Let* $E =$ *glazed and decorated. Assume:*

$$P(glaze \mid c) = 0.01 + \frac{0.81 - 0.01}{5}(c - 11), \quad P(decor \mid c) = 0.5 - \frac{0.5 - 0.05}{5}(c - 11).$$

Then

$$P(E \mid c) = P(glaze \mid c)\, P(decor \mid c),$$

and with a uniform prior $f_C(c) = 1/5$*, Bayes' theorem gives*

$$f_C(c \mid E) = \frac{P(E \mid c)\, f_C(c)}{\displaystyle\int_{11}^{16} P(E \mid t)\, f_C(t)\, dt}. \tag{4.16}$$

These examples illustrate how Bayesian methods yield both point estimates of parameters and full predictive distributions. In contrast to point–estimation approaches (e.g. maximum likelihood), Bayesian inference emphasizes integrating over uncertainty, which is especially valuable in data-scarce or risk-sensitive settings.

4.3 Hierarchical modeling

Hierarchical models involve organizing parameters or variables into multiple levels, reflecting nested or layered dependencies.

4.3.1 *Bayesian hierarchical models*

In many applications one assumes that the observations $y_1, \ldots, y_n$ are exchangeable.[1]

Definition 23 (Finite Exchangeability). *A finite sequence* $\{y_i\}_{i=1}^n$ *is* exchangeable *if its joint density is invariant under any permutation* π*:*

$$p(y_1, \ldots, y_n) = p\big(y_{\pi(1)}, \ldots, y_{\pi(n)}\big).$$

De Finetti's theorem then implies that exchangeable data can be represented as i.i.d. conditional on some latent variable θ:

$$y_i \mid \theta \overset{\text{i.i.d.}}{\sim} f(y_i \mid \theta), \qquad \theta \sim p(\theta).$$

Example 24 (Urn Sampling). *An urn contains one red and one blue ball. Draw two balls without replacement and let*

$$Y_i = \begin{cases} 1, & \text{if the } i\text{th draw is red,} \\ 0, & \text{otherwise.} \end{cases}$$

Then

$$p(Y_1 = 1, Y_2 = 0) = p(Y_1 = 0, Y_2 = 1) = \tfrac{1}{2},$$

so (Y_1, Y_2) *is exchangeable but not independent.*

Definition 24 (Infinite Exchangeability). *An infinite sequence* $\{y_i\}_{i \geq 1}$ *is* exchangeable *if every finite subsequence is exchangeable.*

Bayesian hierarchical modeling incorporates *hyperparameters* ϕ (parameters of the prior) and *hyperpriors* $p(\phi)$. For instance, one might have

$$Y \mid \theta \sim \mathcal{N}(\theta, 1), \quad \theta \mid \mu \sim \mathcal{N}(\mu, 1), \quad \mu \sim \mathcal{N}(0, 1),$$

where μ is a hyperparameter with hyperprior $\mathcal{N}(0, 1)$.

Let y_j be observation j and θ_j its corresponding parameter. Assume the θ_j are generated exchangeably from a common population distribution governed by ϕ. Equivalently, one can view the model as a three-level hierarchy:

- *Stage 1 (Data level):* Given the individual parameter θ_j and hyperparameter ϕ, the data y_j are generated according to the likelihood

  $$y_j \mid \theta_j, \phi \;\sim\; P(y_j \mid \theta_j, \phi).$$

  This stage captures how the observed data depend on the latent parameters.

[1] `https://en.wikipedia.org/wiki/Bayesian_hierarchical_modeling`.

- *Stage 2 (Parameter level):* The individual parameters θ_j themselves are drawn from a common prior distribution indexed by ϕ:

$$\theta_j \mid \phi \sim P(\theta_j \mid \phi).$$

This "pooling" step shares information across groups through the hyperparameter.
- *Stage 3 (Hyperparameter level):* Finally, the hyperparameter ϕ has its own prior (the hyperprior):

$$\phi \sim P(\phi).$$

This stage encodes any prior beliefs about the overall population from which the θ_j are drawn.

Putting these together, the joint posterior over all unknowns is

$$p(\theta_{1:n}, \phi \mid y_{1:n}) \propto \left[\prod_{j=1}^{n} P(y_j \mid \theta_j, \phi)\right]\left[\prod_{j=1}^{n} P(\theta_j \mid \phi)\right] P(\phi).$$

Example 25 (SAT Score Estimation). *A student's GPA Y is modeled as $P(Y \mid \theta)$, where θ is their SAT score. The SAT score θ is drawn from a distribution $P(\theta \mid \phi)$ governed by high-school performance ϕ, which itself has prior $P(\phi)$. The posterior is*

$$p(\theta, \phi \mid Y) \propto P(Y \mid \theta)\, P(\theta \mid \phi)\, P(\phi).$$

Hierarchical models thus enable "partial pooling", improving estimation when group-level data are limited.

4.3.2 *Hierarchical clustering*

Hierarchical clustering is a clustering technique aimed at constructing a hierarchy of clusters.

We use notations:

Dendrogram: A dendrogram is a diagram representing a tree (see Fig. 4.2). This diagrammatic representation is frequently used in hierarchical clustering, it illustrates the arrangement of the clusters produced by the corresponding analyses. *Greed algorithm*: A greedy algorithm is any algorithm that follows the problem-solving heuristic of making the locally optimal choice at each stage.

Hierarchical clustering approach is divided into two main strategies:

1. *Agglomerative (Bottom-Up)*: Each observation begins as its own cluster, and clusters are successively merged as one ascends the hierarchy.

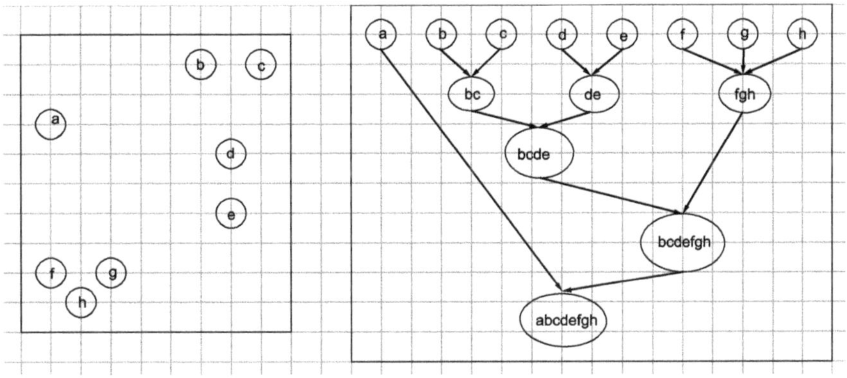

Fig. 4.2: An example of data is to be clustered, and the Euclidean distance is the distance metric. The left is data and the right is corresponding dendrogram.

2. *Divisive (Top-Down)*: All observations start in a single cluster, and recursive splits are performed as one descends the hierarchy.

The process of merging and splitting is typically done in a greedy manner. The output of hierarchical clustering is often visualized as a dendrogram.

The distance between two clusters, denoted $\mathcal{A}$ and $\mathcal{B}$, is commonly defined in one of the ways in Table 4.1.

In cases where there are tied minimum distances between pairs of clusters, one of the following approaches can be used:

1. A random choice between tied pairs, leading to multiple possible dendrograms.

2. Joining all tied pairs simultaneously, resulting in a unique dendrogram.

One of the key advantages of hierarchical clustering is its flexibility in using any valid distance metric. Additionally, it does not require the raw observations; a distance matrix is sufficient. However, for most algorithms, except for single-linkage clustering, finding the optimal solution is computationally infeasible due to the combinatorial complexity (e.g., exhaustive search requiring $\mathcal{O}(2^n)$ time).

Divisive clustering (DIANA Algorithm): The DIANA algorithm is a hierarchical divisive clustering approach. It starts with all data points grouped in a single cluster and progressively splits the largest cluster until each object forms its own singleton cluster. Since splitting a cluster involves

Table 4.1: Clustering Metrics and Their Formulas.

Names	**Formula**
Maximum or complete-linkage clustering	$\max_{a\in\mathcal{A},b\in\mathcal{B}} d(a,b)$
Minimum or single-linkage clustering	$\min_{a\in\mathcal{A},b\in\mathcal{B}} d(a,b)$
Unweighted average linkage clustering (or UPGMA)	$\frac{1}{\lvert\mathcal{A}\rvert\cdot\lvert\mathcal{B}\rvert}\sum_{a\in\mathcal{A}}\sum_{b\in\mathcal{B}} d(a,b)$
Weighted average linkage clustering (or WPGMA)	$\frac{d(i,k)+d(j,k)}{2}$
Centroid linkage clustering, or UPGMC	$\lVert\mu_A-\mu_B\rVert^2$ where μ_A and μ_B are the centroids of A resp. B.
Median linkage clustering, or WPGMC	$d(i\cup j,k)=d(m_i,m_j,m_k)$ where $m_i=\frac{1}{2}(m_i+m_j)$
Versatile linkage clustering	$\sqrt{\frac{1}{\lvert\mathcal{A}\rvert\cdot\lvert\mathcal{B}\rvert}\sum_{a\in\mathcal{A}}\sum_{b\in\mathcal{B}} d(a,b)^p}, p\neq 0$
Ward linkage: Minimum Increase of Sum of Squares (MISSQ)	$\frac{\lvert\mathcal{A}\rvert\cdot\lvert\mathcal{B}\rvert}{\lvert\mathcal{A}\cup\mathcal{B}\rvert}\lVert\mu_A-\mu_B\rVert^2 = \sum_{x\in\mathcal{A}\cup\mathcal{B}}\lVert x-\mu_{\mathcal{A}\cup\mathcal{B}}\rVert^2 - \sum_{x\in\mathcal{A}}\lVert x-\mu_{\mathcal{A}}\rVert^2 - \sum_{x\in\mathcal{B}}\lVert x-\mu_{\mathcal{B}}\rVert^2$
Minimum Error Sum of Squares (MSSQ)	$\sum_{x\in\mathcal{A}\cup\mathcal{B}}\lVert x-\mu_{\mathcal{A}\cup\mathcal{B}}\rVert^2$
Minimum Increase in Variance (MIVAR)	$\frac{1}{\lvert\mathcal{A}\cup\mathcal{B}\rvert}\sum_{x\in\mathcal{A}\cup\mathcal{B}}\lVert x-\mu_{\mathcal{A}\cup\mathcal{B}}\rVert^2 - \frac{1}{\lvert\mathcal{A}\rvert}\sum_{x\in\mathcal{A}}\lVert x-\mu_{\mathcal{A}}\rVert^2 - \frac{1}{\lvert\mathcal{B}\rvert}\sum_{x\in\mathcal{B}}\lVert x-\mu_{\mathcal{B}}\rVert^2 = \mathrm{Var}(\mathcal{A}\cup\mathcal{B})-\mathrm{Var}(\mathcal{A})-\mathrm{Var}(\mathcal{B})$
Minimum Variance (MNVAR)	$\frac{1}{\lvert\mathcal{A}\cup\mathcal{B}\rvert}\sum_{x\in\mathcal{A}\cup\mathcal{B}}\lVert x-\mu_{\mathcal{A}\cup\mathcal{B}}\rVert^2=\mathrm{Var}(\mathcal{A}\cup\mathcal{B})$
Hausdorff linkage	$\max_{x\in\mathcal{A}\cup\mathcal{B}}\min_{y\in\mathcal{A}\cup\mathcal{B}} d(x,y)$
Minimum Sum Medoid linkage	$\min_{m\in\mathcal{A}\cup\mathcal{B}}\sum_{y\in\mathcal{A}\cup\mathcal{B}} d(m,y)$ such that m is the medoid of the resulting cluster

evaluating $\mathcal{O}(2^n)$ possibilities, the algorithm employs heuristics to make this process computationally feasible.

Heuristic splitting process: DIANA selects the object with the maximum average dissimilarity to other objects in the cluster. This object becomes the seed for a new cluster (splinter group). Subsequently, objects that are more similar to the new cluster than to the original cluster are reassigned to it.

Rather than simply "dividing" clusters, DIANA operates more like "hollowing out". At each iteration:

1. A cluster (initially the entire dataset) is selected.

2. A nested cluster is formed within it by moving the most dissimilar objects.

3. This process continues until the parent cluster contains no loose objects.

Over time, the original cluster is "hollowed out" into smaller, nested subclusters.

Algorithm 6 DIANA (Divisive Analysis) Algorithm

Initialization:
$C_0 \leftarrow \{1, \ldots, n\}$ {Initial cluster containing all objects}
$\mathcal{C} \leftarrow \{C_0\}$ {Collection of formed clusters}

while $|\mathcal{C}| < n$ **do**
  Find the largest cluster based on diameter:
  $C_* \leftarrow \arg\max_{C \in \mathcal{C}} \max_{i_1, i_2 \in C} \delta(i_1, i_2)$

  Identify the most dissimilar object in C_*:
  $i_* \leftarrow \arg\max_{i \in C_*} \frac{1}{|C_*|-1} \sum_{j \in C_* \setminus \{i\}} \delta(i, j)$
  $C_{\text{new}} \leftarrow \{i_*\}$ {Create new cluster with this object}

  Populate the new cluster C_{new}:
  repeat
    for each object $i \in C_*$ **do**
      Compute $D(i) = \frac{1}{|C_*|-1} \sum_{j \in C_* \setminus \{i\}} \delta(i, j) - \frac{1}{|C_{\text{new}}|} \sum_{j \in C_{\text{new}}} \delta(i, j)$
    end for
    $i_* \leftarrow \arg\max_{i \in C_*} D(i)$
    if $D(i_*) > 0$ **then**
      Migrate i_* from C_* to C_{new}
    end if
  until $D(i_*) \leq 0$ or no more objects can be migrated

  Update cluster collection:
  $\mathcal{C} \leftarrow \mathcal{C} \cup \{C_{\text{new}}\}$
end while

Note that the measure $D(i)$ captures the "desire" of an object to leave its current cluster. However, it penalizes the migration if the object does not fit well into the new cluster. Objects with weak fit to both clusters will likely start their own splinter groups in future iterations.

Dendrogram construction. A dendrogram is constructed by treating each splinter group C_{new} as a child of the hollowed-out cluster C_*. The root of the tree is the initial cluster C_0, and the leaves are the final singleton clusters, representing individual objects.

4.4 Application of Gibbs measures in hierarchical models

Gibbs measures are particularly well-suited for hierarchical modeling, where parameters are organized into multiple levels of hierarchy. In this setting, Gibbs measures allow us to sample from the joint distribution of all parameters, taking into account the dependencies between them at different levels of the hierarchy. This enables us to capture complex relationships in the data and make more accurate predictions.

Gibbs measures play a central role in hierarchical modeling in the following ways:

4.4.1 *Graphical representations*

Hierarchical models are often represented as factor graphs or Bayesian networks. Gibbs measures provide a probabilistic foundation for these representations:

- Nodes represent variables or parameters.

- Edges represent dependencies, which are encoded in the energy function H.

Representing hierarchical models with factor graphs:

A factor graph is a bipartite graph consisting of:

- Variable nodes $X_1, X_2, \ldots, X_n$, representing random variables in the model.

-Factor nodes $f_1, f_2, \ldots, f_m$, representing factors (local dependencies or interactions).

To explain the mathematical representation consider the joint probability distribution of the system expressed as:

$$P(X_1, X_2, \ldots, X_n) = \frac{1}{Z} \prod_{i=1}^{m} f_i(\mathbf{X}_i),$$

where - $\mathbf{X}_i \subseteq \{X_1, X_2, \ldots, X_n\}$ are the variables associated with the factor f_i.

- $f_i(\mathbf{X}_i) \geq 0$ is the factor function encoding the interactions between the variables.

- Z is the partition function ensuring normalization:

$$Z = \sum_{\mathbf{X}} \prod_{i=1}^{m} f_i(\mathbf{X}_i).$$

Bayesian networks. A Bayesian network is a directed acyclic graph (DAG) where

- Nodes represent random variables.
- Edges represent conditional dependencies between variables.

The joint distribution is factorized according to the graph structure:

$$P(X_1, X_2, \ldots, X_n) = \prod_{i=1}^{n} P(X_i \mid \text{Pa}(X_i)),$$

where $\text{Pa}(X_i)$ are the parent nodes of X_i in the DAG.

Hierarchical models are often connected to Gibbs measures, which provide the statistical mechanics framework for encoding dependencies.

Recall that a Gibbs measure is a probability distribution defined in terms of an energy function H:

$$P(\mathbf{X}) = \frac{1}{Z} e^{-H(\mathbf{X})}.$$

Here $\mathbf{X} = (X_1, X_2, \ldots, X_n)$ is the configuration of variables and Z is the partition function, ensuring normalization.

The energy function $H(\mathbf{X})$ is typically expressed as a sum of local interaction terms:

$$H(\mathbf{X}) = \sum_{i=1}^{m} H_i(\mathbf{X}_i),$$

where $H_i(\mathbf{X}_i)$ encodes the interaction for a subset $\mathbf{X}_i$ of variables.

This structure aligns with factor graphs, where $H_i(\mathbf{X}_i) = -\log f_i(\mathbf{X}_i)$ relates factors to energy terms.

Example. Consider a simple system with three variables X_1, X_2, X_3 and interactions encoded as:

$$H(\mathbf{X}) = -\log f_1(X_1, X_2) - \log f_2(X_2, X_3).$$

The factor graph has variable nodes X_1, X_2, X_3 and factor nodes f_1, f_2 connected to the relevant variables.

The corresponding Gibbs measure is:

$$P(X_1, X_2, X_3) = \frac{1}{Z} f_1(X_1, X_2) f_2(X_2, X_3).$$

This representation encodes the dependencies in the hierarchical model.

4.4.2 Multi-level energy functions

For hierarchical systems, the Hamiltonian H can be decomposed into contributions from different levels:

$$H = \sum_{l=1}^{L} H_l,$$

where H_l represents the interactions at the l-th level of the hierarchy. This decomposition allows for:

- Modeling interactions within each level.
- Capturing dependencies between levels.

A concrete example can be found in the Dyson hierarchical Ising model [45, 54, 141] or related systems, which are used in statistical mechanics to study phase transitions.

Let the system consist of a binary spin configuration $\sigma = (\sigma_1, \sigma_2, \ldots, \sigma_N)$ with $\sigma_i \in \{-1, 1\}$ and a Hamiltonian $H(\sigma)$ expressed as a sum of contributions from multiple hierarchical levels.

The Hamiltonian H is decomposed as:

$$H(\sigma) = \sum_{l=1}^{L} H_l(\sigma),$$

where $H_l(\sigma)$ is the contribution at level l. Each level corresponds to interactions involving groups of spins defined hierarchically.

At each level l, the Hamiltonian has the form:

$$H_l(\sigma) = -J_l \sum_{B \in \mathcal{B}_l} \left(\frac{1}{|B|} \sum_{i \in B} \sigma_i \right)^2,$$

where

- $\mathcal{B}_l$ is the set of spin blocks at level l.
- $|B|$ is the number of spins in block B.
- J_l is the coupling constant at level l, typically decaying with l, e.g., $J_l = J_0 r^l$ with $0 < r < 1$.

Hierarchical Organization:

- At level 1, interactions involve small spin groups (e.g., pairs or triplets).
- At higher levels, interactions involve larger blocks of spins, reflecting coarse-graining.

Mathematical properties:

1. *Recursive structure*: The energy contributions at level l can often be computed recursively:

$$H_l(\sigma) = H_{l-1}(\sigma') + \text{higher-order corrections},$$

where σ' represents the effective spins of aggregated blocks from the previous level.

2. *Scaling of couplings*: The couplings J_l typically decrease exponentially with l, ensuring convergence of the sum:

$$\sum_{l=1}^{\infty} J_l < \infty.$$

3. *Additive and hierarchical structure*: Each level contributes independently, making the system suitable for renormalization group analysis. Specifically, coarse-graining the system results in a self-similar effective Hamiltonian.

4. *Energy landscape*: The hierarchical decomposition creates an energy landscape with multi-scale interactions. This is useful in studying systems with long-range correlations or fractal-like structures.

Example 1.: Dyson hierarchical Ising model. In this model spins are organized in a binary tree of depth L. At each level l, interactions involve pairs of subtrees of depth $l-1$, with coupling J_l.

The Gibbs measures for the hierarchical Hamiltonian

$$H(\sigma) = -\sum_{l=1}^{L} J_l \sum_{B \in \mathcal{B}_l} \sigma_B \sigma_{B'},$$

where σ_B and $\sigma_{B'}$ represent effective spins of subtrees B and B', respectively, exhibit interesting mathematical properties due to the hierarchical structure.

Due to the hierarchical nature of the Hamiltonian, Gibbs measures can be constructed recursively:

- At the finest level ($l = 1$), local interactions determine the Gibbs measure for small spin groups.

- At higher levels ($l > 1$), effective spins σ_B are defined as averages or majority spins of subgroups at the previous level.

This recursive structure makes the system amenable to renormalization group methods, where the effective Hamiltonian at level $l+1$ is obtained from level l.

The following results are known (see [45], [69], [70], [113] and references therein) for this Hamiltonian:

1. *Existence and uniqueness of Gibbs measures*: For sufficiently high temperatures (small interaction strengths relative to the temperature), the Gibbs measure is unique. This follows from the Dobrushin uniqueness

condition (see e.g. [343]), which can be adapted to the hierarchical setup. The decay of coupling constants J_l with increasing l also ensures weak correlations at higher temperatures.

2. *Phase transition at low temperatures*: For low temperatures (or large J_l relative to temperature), the system exhibits a phase transition. Multiple Gibbs measures exist, corresponding to different macrostates (e.g., spontaneous magnetization).

3. *Critical temperature*: A critical temperature T_c can be defined, below which symmetry breaking occurs. The value of T_c depends on the sequence J_l, often scaling with the hierarchical depth L.

4. *Exponential decay of correlations*: At high temperatures, the correlation between two spins decays exponentially with their hierarchical distance. Specifically:

$$\mathrm{Cov}(\sigma_i, \sigma_j) \sim \exp(-\kappa d(i, j)),$$

where $d(i, j)$ is the hierarchical distance, and $\kappa > 0$ is the decay constant.

5. *Spontaneous magnetization*: Below T_c, the magnetization M is nonzero

$$M = \lim_{N \to \infty} \frac{1}{N} \sum_{i=1}^{N} \langle \sigma_i \rangle.$$

6. *Scaling of magnetization*: Near the critical temperature, M follows a power law

$$M \sim (T_c - T)^{\beta},$$

where β is a critical exponent depending on J_l and the hierarchical structure.

Example 2. Following [71] we give another example of hierarchical model: one dimensional Ising spin systems with long range interactions, where a geometric description of the spin configurations in terms of triangles which play the role of contours. Construction of such triangular contour may be applicable to several problems in machine learning. Moreover, this yields a direct proof of the well known result by Dyson about phase transitions at low temperatures (see [71] and [70] for details).

The Hamiltonian of the model (an Ising ferromagnet on a one dimensional lattice) is given by

$$h(\sigma) = \frac{1}{2} \sum_{x, y \in \mathbb{Z}} \frac{1}{|x - y|^{2-\alpha}} \mathbf{1}_{\sigma(x) \neq \sigma(y)},$$

where $\sigma(x) \in \{-1, 1\}$ and $\alpha \in [0, 1/2]$.

A graphical description of a spin configuration $\sigma \in \{-1,1\}^{\mathbb{Z}}$ in terms of a configuration of triangles allows to introduce the notion of internal and external interfaces. The description is as follows.

Take a finite subset of $\mathbb{Z}$ and consider homogeneous boundary conditions (outside of the finite set), i.e. the spins in the boundary are either all $+1$ or all -1. By the spin flip symmetry, we restrict to the former, so that we will only study configurations $\sigma = \{\sigma_x, x \in \mathbb{Z}\} \in \mathcal{X}_+$, namely such that $\sigma_x = 1$ for all $|x|$ large enough. The goal here is to reconstruct an image in which the configurations are represented by a set of interfaces. In one dimensions, an interface at $(x, x+1)$ means $\sigma_x \sigma_{x+1} = -1$. The precise location of the interface in the interval is immaterial and we will use it to our advantage by choosing a point in each interval $(x + 1/2) \pm 1/100$, $x \in \mathbb{Z}$, with the property that for any four distinct points r_i, $i = 1, .., 4$, $|r_1 - r_2| \neq |r_3 - r_4|$.

By definition, each interface point signifies a phase change. As a result, after the first interface point (approaching from the left), the second point indicates a return to the original phase, and this pattern continues. However, this viewpoint is not the most effective for analyzing spin configurations. Instead, we employ a criterion based on minimal distance, which is made geometrically intuitive through graphical representation. In this approach, each spin configuration is mapped to a set of triangles, with the vertices of these triangles representing pairs of coupled interface points.

Given the specified boundary conditions, any $\sigma \in \mathcal{X}_+$ contains a finite, even number of interface points. We then allow each interface point to evolve into two trajectories, represented in the (r, t) plane by the lines $r \pm t$ for $t \geq 0$. This results in a collection of growing v-shaped lines, each emanating from an interface point. When two v-lines meet, they are frozen and cease to grow, while the remaining v-lines continue to propagate undisturbed. This choice of interface point locations ensures that these collisions occur one at a time, making the definition unambiguous.

Graphically, the collision of two points in the (r, t) plane is represented by a triangle. The base of this triangle corresponds to the line segment joining the two interface points, while its sides are formed by the arms of the v-lines that intersect at the moment of collision.

Triangles will be denoted by T and we will write

$$|T| = \text{ cardinality of } \; T \sqcap \mathbb{Z}, \qquad \text{dist}(T, T') = \text{cardinality of } \; I \sqcap \mathbb{Z},$$

where I is the interval between T and T' if T and T' are disjoint; if T and T' are one contained in the other (no other possibility may arise in the above construction) then I denotes the minimal interval between the two.

We have thus represented a configuration $\sigma \in \mathcal{X}_+$ as a collection $\underline{T} = (T_1, .., T_n)$ of triangles in the (r, t) plane. The set of configurations of triangles obtained in this way are denoted by $\{\underline{T}\}$, and the above construction defines a one to one map from $\mathcal{X}_+$ onto $\{\underline{T}\}$. It is easy to see that a triangle configuration $\underline{T}$ belongs to $\{\underline{T}\}$ iff for any pair T and T' in $\underline{T}$

$$\mathrm{dist}(T, T') \geq \min\{|T|, |T'|\} \tag{4.17}$$

The two endpoints of a triangle play the role which has the interface in higher dimensions and we thus have, also in $d = 1$, a notion of external and internal interfaces. (see Fig. 4.3).

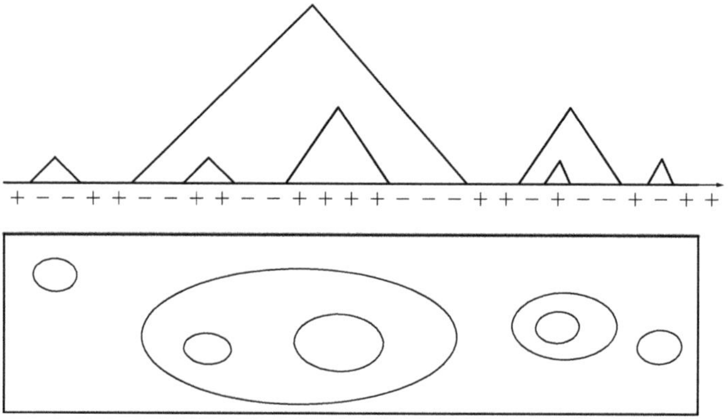

Fig. 4.3: External and internal interfaces in $d = 1$ and $d = 2$.

Thus there are finitely many phase changes and it is proved in [71] that each one of them has a small Gibbs weight.

Writing

$$H(\underline{T}) = h(\sigma), \qquad \sigma \in \mathcal{X}_+ \Leftrightarrow \underline{T} \in \{\underline{T}\} \tag{4.18}$$

and calling $\underline{T} = (T_1, .., T_n)$ with $|T_i| \leq |T_{i+1}|$, we have

$$H(\underline{T}) = H(T_1 \,|\, \underline{T} \setminus T_1) + H(\underline{T} \setminus T_1), \qquad H(\underline{S} \,|\, \underline{T}) := H(\underline{S} \sqcup \underline{T}) - H(\underline{T}) \tag{4.19}$$

In fact if $\underline{T} \in \{\underline{T}\}$, and $T \in \underline{T}$, then $\underline{T} \setminus T$ obviously satisfies (4.17) and therefore it is in $\{\underline{T}\}$. $\underline{T} \setminus T$ is obtained from the configuration σ corresponding to $\underline{T}$ by flipping all the spins inside the basis of T. By iteration,

$$H(\underline{T}) = \sum_{i=1}^{n} H(T_i \,|\, \underline{T} \setminus [T_1 \sqcup \cdots \sqcup T_i]) \tag{4.20}$$

Lemma 4. *For any* i,

$$H\big(T_i \,\big|\, \underline{T} \setminus [T_1 \sqcup \cdots \sqcup T_i]\big) \geq W(|T_i|) \tag{4.21}$$

where

$$W(L) = \sum_{x=1}^{L} \left(\sum_{\substack{y\in[L+1,2L] \\ y\in[-L+1,0]}} J(|x-y|) - \sum_{\substack{y\in[2L+1,\infty] \\ y\in[-\infty,-L]}} J(|x-y|) \right) \tag{4.22}$$

Proof. Call $I_i^{\pm}$ the two intervals in $\mathbb{Z}$ which are to the right and to the left of T_i, each one consisting of $|T_i|$ sites. There is no interface point inside T_i, and inside $I_i^{\pm}$ as well, because $|T_i|$ is the minimal length in $\underline{T}\setminus[T_1\sqcup\cdots\sqcup T_{i-1}]$ and all T_j, $j > i$, have distance from T_i which is $\geq |T_i|$. Then, if σ corresponds to $\underline{T} \setminus [T_1 \sqcup \cdots \sqcup T_{i-1}]$, the spins in $T_i \sqcap \mathbb{Z}$ are all equal to each other and opposite to those in $I_i^{\pm}$. Instead, in the configuration σ' which corresponds to $\underline{T} \setminus [T_1 \sqcup \cdots \sqcup T_i]$, the spins are all the same in $(T_i \sqcap \mathbb{Z}) \sqcup I_i^+ \sqcup I_i^-$. By (4.19), $H\big(T_i \,\big|\, \underline{T} \setminus [T_1 \sqcup \cdots \sqcup T_i]\big) = h(\sigma) - h(\sigma')$, so that (4.21) and the lemma are proved. □

In [71] it is proved that if $J(1)$ large enough, there is $\zeta > 0$ such that

$$W(L) \geq \zeta h_\alpha(L), \quad \text{where} \quad h_\alpha(L) := \begin{cases} L^\alpha & \alpha \in (0, 1/2] \\ \ln L + 4 & \alpha = 0. \end{cases} \tag{4.23}$$

Thus

$$H(\underline{T}) \geq \zeta \sum_{i=1}^{n} |T_i|^\alpha, \qquad \underline{T} = (T_1, .., T_n) \tag{4.24}$$

The inequality can be interpreted as an analogue of the Peierls estimate in dimensions $d > 1$, where the excess energy of a configuration of interfaces is bounded from below in proportion to the surface area of these interfaces. Here, $|T_i|$ represents the volume enclosed by the interface, and the parameter α corresponds to the ratio $(d-1)/d$, with d serving as an "effective dimension" of the system.

However, this remains only an analogy. To truly implement a Peierls bound within our framework, it is necessary to "localize the estimates," meaning we must calculate the weight of an individual triangle within a generic configuration. The earlier bound was straightforward because it allowed us to estimate the weights of triangles sequentially, in the order of their lengths. For a generic triangle T in a configuration $\underline{T}$, however, $|T|$ may not be the smallest length, introducing situations where other triangles exist within $T \sqcup I^+ \sqcup I^-$ (see Lemma 4 for notation). Indeed, smaller triangles T' could be added to $T \sqcup I^+ \sqcup I^-$ without violating inequality (4.17). The approach involves the following steps:

1. Connecting triangles that are "dangerously close" to one another,
2. Defining contours as "connected clusters" of triangles, and
3. Calculating probabilities of contours rather than individual triangles.

To compute the probability of a contour, we begin by ordering the triangles within the contour in increasing order of their lengths. This enables a generalization of the earlier argument, utilizing the property that triangles not part of the contour are "sufficiently far away" due to the manner in which contours are defined.

In [71] the triangles are clustered into contours and then extend Lemma 4 to contours, thus concluding the analysis of the energy of contours. After that entropy bounds (on the number of contours) are obtained, which show that the temperature low enough, energy wins against entropy. These results allow to use the classical argument for existence of a phase transition. We think that this argument can be used in machine learning problems, to classify the complex systems.

4.5 Ising Models with Hidden Markov Structure

As deep learning continues to evolve (see Chapter 7), exploring the full potential of Gibbs measures in enhancing model flexibility, scalability, and interpretability remains an exciting research frontier. Addressing challenges related to efficient inference, scalability to large datasets, and incorporating structured priors will be key areas of focus for advancing the use of Gibbs measures in deep learning applications.

This section is based on [179].

4.5.1 *Definitions*

Here we investigate tree-indexed Markov chains (Gibbs measures) defined by a Hamiltonian that couples two Ising layers: hidden spins $s(x) \in \{\pm 1\}$ and observed spins $\sigma(x) \in \{\pm 1\}$ on a Cayley tree. The Hamiltonian incorporates Ising interactions within each layer and site-wise emission couplings between layers, extending hidden Markov models to a bilayer Markov random field. Specifically, we explore translation-invariant Gibbs measures (TIGM) of this Hamiltonian on Cayley trees.

The motivations behind this study are twofold. First, hierarchical models have gained prominence in machine learning due to their ability to represent complex, multi-level structures in data. Hidden Markov models (HMMs) have been widely used to model such hierarchical systems, where the data is assumed to be generated by underlying hidden states (see [151] and the references therein). However, inference in these models, especially when dealing with spatial or temporal correlations (as seen in image or sequence data), often becomes intractable. By introducing the Ising interactions between the hidden states, we aim to explore more effective ways to model these dependencies while maintaining computational feasibility.

Second, probabilistic graphical models have shown tremendous success in machine learning, particularly in areas such as generative modeling, denoising, and anomaly detection (see [41], [301], [378]). However, a key challenge remains: how to efficiently perform inference when the underlying data is noisy or missing. The incorporation of a hidden Markov structure within the Ising framework provides a principled way to handle noisy or incomplete observations, as the model's energy function captures both the relationship between hidden and observed variables as well as their dependencies.

Let us recall some definitions. A Cayley tree $\Gamma^k = (V, L)$ (where V is the set of vertices and L is the set of edges) with branching factor $k \geq 1$ is a connected infinite graph, every vertex of which has exactly $k+1$ neighbors. The graph Γ^k is acyclic, meaning it has no loops or cycles.

Fix a vertex $x^0 \in V$, interpreted as the *root* of the tree. We say that $y \in V$ is a *direct successor* of $x \in V$ if x is the penultimate vertex on the unique path leading from the root x^0 to the vertex y; that is, $d(x^0, y) = d(x^0, x) + 1$ and $d(x, y) = 1$. The set of all direct successors of $x \in V$ is denoted by $S(x)$.

For a fixed $x^0 \in V$ we set $W_n = \{x \in V \mid d(x, x^0) = n\}$,

$$V_n = \{x \in V \mid d(x, x^0) \leq n\}, \quad L_n = \{l = \langle x, y\rangle \in L \mid x, y \in V_n\}. \quad (4.25)$$

For $x \in W_n$ the set $S(x)$ then has the form

$$S(x) = \{y \in W_{n+1} : \langle x, y\rangle\}. \quad (4.26)$$

In this section we consider Ising model's spins $s(x) \in I = \{-1, 1\}$ which are assigned to the vertices of Cayley tree.

A configuration s on V is then defined as a function $x \in V \mapsto s(x) \in I$; the set of all configurations is $\Omega := I^V$.

We consider a hidden configuration $s \in \Omega$ and observed configuration $\sigma \in \Omega$ and formulate a Hamiltonian that depends on both configurations in the context of the hidden Markov model (HMM) applied to the Ising model.

The HMM assumes that the observed configuration $\{\sigma(x), x \in V\}$ depends on the hidden spin configuration $\{s(x), x \in V\}$ in some probabilistic manner. So, the observed configuration is a noisy or indirect reflection of the actual spins.

Hamiltonian for an HMM: Hidden Markov model (HMM) can be described in statistical mechanics terms using a Hamiltonian that captures the joint distribution of hidden states and observations (see [151], [383]). In this context, the hidden states $s(x)$ play the role of spins, and the observations $\sigma(x)$ are the emitted (visible) symbols or measurements.

Let us define a general Hamiltonian for the HMM on a graph (e.g., a Cayley tree or lattice), where $x \in V$ an vertex, $s(x) \in \mathcal{S}$ are hidden states (e.g., spins), $\sigma(x) \in \Sigma$ are observed symbols.

Then the Hamiltonian is:

$$H(s, \sigma) = -\sum_{\langle x,y \rangle} E(s(x), s(y)) - \sum_{x} p(s(x) \mid \sigma(x)), \tag{4.27}$$

here interaction term (hidden states): $-\sum_{\langle x,y \rangle} E(s(x), s(y))$ represents the Markov property: hidden states interact locally (e.g., adjacent in a graph). $E(s(x), s(y))$ is the interaction energy between neighboring hidden states.

Emission term (observation likelihood): $-\sum_x p(s(x) \mid \sigma(x))$. This captures the probability of emitting observation $\sigma(x)$ given the hidden state $s(x)$. Here $p(s \mid \sigma) = \log P(\sigma \mid s)$ is a log-likelihood function of the emission probability.

The model: Consider Ising spins $s(x) \in I = \{-1, 1\}$ (i.e. $\mathcal{S} = \Sigma = I$) which are assigned to the vertices of a Cayley tree.

Considered Hamiltonian, $H(s, \sigma)$, which defines the energy of a given configuration pair (s, σ), incorporating both the Ising interactions and the relationship between the hidden states s and the observations σ given as follows:

$$H(s, \sigma) = -J \sum_{\langle x,y \rangle} (s(x)s(y) - \sigma(x)\sigma(y)) - \sum_{x \in V} p(\sigma(x)|s(x)) \tag{4.28}$$

Here, $J \in \mathbb{R}$.

Remark 4. *In a more general setting, the Hamiltonian (4.28) can be written as*

$$H(s, \sigma) = -J_1 \sum_{\langle x,y \rangle} s(x)s(y) + J_2 \sum_{\langle x,y \rangle} \sigma(x)\sigma(y) - \sum_{x \in V} p(\sigma(x) \mid s(x)), \tag{4.29}$$

where $J_1, J_2 \in \mathbb{R}$. *In this case, one can also prove an analogue of Theorem 17 (see below); however, the corresponding system of equations becomes more complicated due to the presence of an additional parameter. For this reason, we restrict ourselves to the simpler case* $J_1 = J_2 = J$.

Remark 5. *Comparing Hamiltonians (4.27) and (4.28) one can see that (4.28) has (additional) the antiferromagnetic coupling term* $J\sum_{\langle x,y\rangle}\sigma(x)\sigma(y)$, *coupling the observable* σ *spins.*

1. *The first part,* $s(x)s(y)$, *as in (4.27), represents the natural interaction between spins on neighbors* x *and* y *in the Ising model (favoring alignment of spins with coupling strength* J*). The second part,* $\sigma(x)\sigma(y)$, *introduces a discrepancy between the observed configurations, which could indicate how much the observations diverge from the hidden states. The difference* $(s(x)s(y)-\sigma(x)\sigma(y))$ *essentially* **penalizes mismatches** *between the hidden spin configuration and the observed configuration in the Ising framework.*
2. *The second term in (4.28) is similar to second term of (4.27) reflects the probabilistic nature of the observations. It models how likely the observed spin* $\sigma(x)$ *is, given the true (hidden) spin* $s(x)$. *The probability* $p(\sigma(x)|s(x))$ *can be thought of as a noise model for the observation, where different types of distributions (e.g., Gaussian, Bernoulli) can be chosen depending on the nature of the observations.*
3. *The Hamiltonian (4.28) does* **not** *define an HMM because it includes a direct interaction term for observations* $(\sigma(x)\sigma(y))$, *which violates the fundamental HMM property that observations should be conditionally independent given the hidden states. Instead, (4.28) describes a more general coupled system, deviating from the standard HMM framework defined by (4.27). While standard HMMs enforce conditional independence of observations given hidden states, (4.28) introduces observation correlations, generalizing to a bilayer Markov chain indexed by the tree. Therefore the model constructed by (4.28) can be called as "HMM with Ising-structured observations".*

Remark 6. *In [122], a variant of (4.28)—lacking the direct observation coupling term* $J\sum_{\langle x,y\rangle}\sigma(x)\sigma(y)$ *—was analyzed. Unlike our approach, [122] assumes no interactions between observed spins, instead modeling* *noisy observations of coupled hidden spins via site-wise (time-dependent) couplings between $\sigma(x)$ and $s(x)$. The authors studied homogeneous Gibbs measures (HGMs) on a Cayley tree under infinite-temperature Glauber dynamics, demonstrating distinct non-Gibbsian behaviors in the intermediate state (e.g., free-boundary Gibbs state) versus extremal states ($\pm$). While [122]'s model holds relevance for machine learning (via HMMs), our work focuses exclusively on characterizing HGMs for Hamiltonian (4.28) without Glauber dynamical evolution.*

Motivation for adding the second term to Hamiltonian (4.28): *The inclusion of the antiferromagnetic term in Hamiltonian (4.28) serves not only to facilitate the study of translation-invariant Gibbs measures but also offers significant advantages for machine learning applications. Specifically, we propose using the expression $(s(x)s(y) - \sigma(x)\sigma(y))$—which represents the difference between the Ising model energies of the hidden and observed configurations—as a* **loss function** *for machine learning tasks (see [233] for the representation of a Hamiltonian as an energy-loss function). This formulation is motivated by its ability to directly quantify the discrepancy in pairwise spin interactions, a key feature of Ising-type models. Minimizing this loss function encourages the hidden configuration to replicate the statistical behavior of the observed one in terms of energy, thereby leading to more interpretable and physically meaningful learning outcomes—especially in settings where spin correlations are of fundamental importance.*

Interpretation in Machine Learning:

- Energy minimization: The Hamiltonian $H(s, \sigma)$ essentially represents the total energy of a configuration. In machine learning, the goal would typically be to minimize this energy function (or equivalently maximize the likelihood of the observed data given the hidden states).

- Inference: Given observed data σ, the task might be to infer the most likely hidden configuration s that generated the observations, which is a typical problem in HMMs and probabilistic graphical models. The Hamiltonian provides a framework for this by considering both the structure of the hidden system and the noise in the observations.

This setup aligns with standard approaches in machine learning, where the goal is often to infer hidden structures (like the spins s) from noisy or indirect observations (like σ), with the Hamiltonian serving as the energy function to minimize during the inference process. Below we construct some Gibbs measures of the Hamiltonian (4.28) and then use these measures to solve above mentioned problems of Machine learning.

4.5.2 *The compatibility of measures*

Define a finite-dimensional distribution of a probability measure μ in the volume V_n as

$$\mu_n(s_n,\sigma_n) = Z_n^{-1}\exp\left\{-\beta H_n(s_n,\sigma_n) + \sum_{x\in W_n} h_{s(x),\sigma(x),x}\right\}, \tag{4.30}$$

where $\beta = 1/T$, and $T > 0$ is the temperature, Z_n^{-1} is the normalizing factor,

$$\{h_x = (h_{-1,-1,x}, h_{-1,1,x}, h_{1,-1,x}, h_{1,1,x}) \in \mathbb{R}^4, x \in V\} \tag{4.31}$$

is a collection of vectors and

$$H_n(s_n,\sigma_n) = -J\sum_{\langle x,y\rangle\in L_n}(s(x)s(y)-\sigma(x)\sigma(y)) - \sum_{x\in V_n} p(\sigma(x)|s(x)).$$

Definition 25. *We say that the probability distributions (4.30) are compatible if for all $n \geq 1$ and $s_{n-1} \in I^{V_{n-1}}$, $\sigma_{n-1} \in I^{V_{n-1}}$:*

$$\sum_{(w_n,\omega_n)\in I^{W_n}\times I^{W_n}} \mu_n(s_{n-1}\vee w_n, \sigma_{n-1}\vee\omega_n) = \mu_{n-1}(s_{n-1},\sigma_{n-1}). \tag{4.32}$$

Here symbol $\vee$ denotes the concatenation (union) of the configurations.

In this case, by Kolmogorov's theorem ([136, p.251]) there exists a unique measure μ on $I^V \times I^V$ such that, for all n and $s_n, \sigma_n \in I^{V_n}$,

$$\mu(\{(s,\sigma)|_{V_n} = (s_n,\sigma_n)\}) = \mu_n(s_n,\sigma_n).$$

Such a measure is called a *splitting Gibbs measure* (SGM) corresponding to the Hamiltonian (4.28) and vector-valued functions (4.31).

The following theorem describes conditions on h_x guaranteeing compatibility of $\mu_n(s_n,\sigma_n)$.

Theorem 17. *Probability distributions $\mu_n(s_n,\sigma_n)$, $n \in \mathbb{N}$, in (4.30) are compatible iff for any $x \in V \setminus \{x^0\}$ the following equation holds:*

$$z_{-1,1,x} = \prod_{y\in S(x)} \frac{\theta + az_{-1,1,y} + bz_{1,-1,y} + \theta^{-1}cz_{1,1,y}}{1+\theta az_{-1,1,y} + \theta^{-1}bz_{1,-1,y} + cz_{1,1,y}}, \tag{4.33}$$

$$z_{1,-1,x} = \prod_{y\in S(x)} \frac{\theta^{-1} + az_{-1,1,y} + bz_{1,-1,y} + \theta cz_{1,1,y}}{1+\theta az_{-1,1,y} + \theta^{-1}bz_{1,-1,y} + cz_{1,1,y}}, \tag{4.34}$$

$$z_{1,1,x} = \prod_{y\in S(x)} \frac{1 + \theta^{-1}az_{-1,1,y} + \theta bz_{1,-1,y} + cz_{1,1,y}}{1+\theta az_{-1,1,y} + \theta^{-1}bz_{1,-1,y} + cz_{1,1,y}}, \tag{4.35}$$

where

$$z_{\epsilon,\delta,x} = \exp(h_{\epsilon,\delta,x} - h_{-1,-1,x}),\ \epsilon,\delta = -1,1, \quad \theta = \exp(2J\beta), \tag{4.36}$$

$$a = \exp[\beta(p(-1|1) - p(-1|-1))],$$

$$b = \exp[\beta(p(1|-1) - p(-1|-1))],$$

$$c = \exp[\beta(p(1|1) - p(-1|-1))],$$

and $S(x)$ is the set of direct successors of x.

Proof. Below we use the following equalities:

$$V_n = V_{n-1} \cup W_n, \quad W_n = \bigcup_{x \in W_{n-1}} S(x).$$

Necessity. Assume that (4.32) holds, we shall prove (4.33). Substituting (4.30) into (4.32), obtain that for any configurations (s_n, σ_{n-1}): $x \in V_{n-1} \mapsto (s_n(x), \sigma_{n-1}(x)) \in I \times I$:

$$\frac{Z_{n-1}}{Z_n} \sum_{(w_n,\omega_n)\in I^{W_n}\times I^{W_n}} \exp\left(\sum_{x\in W_{n-1}} \sum_{y \in S(x)} (J\beta(s_{n-1}(x)w_n(y) - \sigma_{n-1}(x)\omega_n(y))\right.$$

$$\left.+\beta p(\omega_n(y)|w_n(y)) + h_{w_n(y),\omega_n(y),y})\right) = \exp\left(\sum_{x\in W_{n-1}} h_{s_{n-1}(x),\sigma_{n-1}(x),x}\right). \tag{4.37}$$

From (4.37) we get:

$$\frac{Z_{n-1}}{Z_n} \sum_{(w_n,\omega_n)\in I^{W_n}\times I^{W_n}} \prod_{x\in W_{n-1}} \prod_{y\in S(x)} \exp\left(J\beta(s_{n-1}(x)w_n(y) - \sigma_{n-1}(x)\omega_n(y)\right)$$

$$+\beta p(\omega_n(y)|w_n(y)) + h_{w_n(y),\omega_n(y),y}) = \prod_{x\in W_{n-1}} \exp\left(h_{s_{n-1}(x),\sigma_{n-1}(x),x}\right).$$

Fix $x \in W_{n-1}$ and rewrite the last equality for $(s_{n-1}(x), \sigma_{n-1}(x)) = (\epsilon, \delta)$, $\epsilon, \delta = -1, 1$, keeping the configurations unchanged on $W_{n-1} \setminus \{x\}$. Then dividing each of equalities to equation of the case $(s_{n-1}(x), \sigma_{n-1}(x)) = (-1, -1)$, we get

$$\prod_{y\in S(x)} \frac{\sum_{(j,u)\in I\times I} \exp\left(J\beta(\epsilon j - \delta u) + \beta p(u|j) + h_{j,u,y}\right)}{\sum_{(j,u)\in I\times I} \exp\left(-J\beta(j-u) + \beta p(u|j) + h_{j,u,y}\right)} = \exp\left(h_{\epsilon,\delta,x} - h_{-1,-1,x}\right), \tag{4.38}$$

where $\epsilon, \delta = -1, 1$.

Now by using notations (4.36), from (4.38) we get (4.33)–(4.35). Note that $z_{-1,-1,x} \equiv 1$.

Sufficiency. Suppose that (4.33)–(4.35) hold. It is equivalent to the representations

$$\prod_{y\in S(x)} \sum_{(j,u)\in I\times I} \exp\left(J\beta(\epsilon j - \delta u) + \beta p(u|j) + h_{j,u,y}\right) = a(x)\exp\left(h_{\epsilon,\delta,x}\right),$$

$$\epsilon, \delta = -1, 1 \tag{4.39}$$

for some function $a(x) > 0, x \in V$. We have

$$\text{LHS of (4.32)} = \frac{1}{Z_n}\exp(-\beta H(s_{n-1}, \sigma_{n-1}))$$

$$\times \prod_{x\in W_{n-1}} \prod_{y\in S(x)} \sum_{(\epsilon,u)\in I\times I} \exp\left(J\beta(s_{n-1}(x)\epsilon - \sigma_{n-1}(x)u)\right.$$

$$\left. + \beta p(u|\epsilon) + h_{\epsilon,u,y}\right). \tag{4.40}$$

Substituting (4.5.2) into (4.40) and denoting $A_n = \prod_{x\in W_{n-1}} a(x)$, we get

$$\text{RHS of (4.40)} = \frac{A_{n-1}}{Z_n}\exp(-\beta H(s_{n-1}, \sigma_{n-1}))$$

$$\times \prod_{x\in W_{n-1}} \exp(h_{s_{n-1}(x),\sigma_{n-1}(x),x}). \tag{4.41}$$

Since μ_n, $n \geq 1$ is a probability, we should have

$$\sum_{(s_{n-1},\sigma_{n-1})\in I^{V_{n-1}}\times I^{V_{n-1}}} \sum_{(w_n,\omega_n)\in I^{W_n}\times I^{W_n}} \mu_n(s_{n-1}\vee w_n, \sigma_{n-1}\vee \omega_n) = 1.$$

Hence from (4.41) we get $Z_{n-1}A_{n-1} = Z_n$, and (4.32) holds.

□

From Theorem 17, it follows that for any $z = \{z_{\epsilon,\delta,x},\ x \in V\}$ that satisfies the system of functional equations (4.33)–(4.35), there exists a unique SGM μ, and conversely. Therefore, the core problem of describing Gibbs measures reduces to solving the system of functional equations (4.33)–(4.35).

However, solving this system is particularly challenging due to its nonlinear nature, multidimensional structure, and the fact that the unknown functions are defined on a tree. Even the task of determining all constant functions (which are independent of the tree's vertices) is complex. Here we present a class of such constant solutions and explore how the corresponding Gibbs measures can be applied in machine learning.

4.5.3 *Translation-invariant Gibbs measures*

Now we focus on translation-invariant Gibbs measures, which correspond to solutions of the form:

$$z_{\epsilon,i,x} \equiv z_{\epsilon,i}, \quad \text{for all } x \in V. \tag{4.42}$$

Denote

$$u = az_{-1,1}, \quad v = bz_{1,-1}, \quad w = cz_{1,1}.$$

Substituting this into equation (4.33)-(4.35) gives:

$$\begin{aligned} u &= a\left(\tfrac{\theta+u+v+\theta^{-1}w}{1+\theta u+\theta^{-1}v+w}\right)^k \\ v &= b\left(\tfrac{\theta^{-1}+u+v+\theta w}{1+\theta u+\theta^{-1}v+w}\right)^k \\ w &= c\left(\tfrac{1+\theta^{-1}u+\theta v+w}{1+\theta u+\theta^{-1}v+w}\right)^k. \end{aligned} \tag{4.43}$$

4.5.3.1 *Case:* $a = b = c = 1$

In this case write the system (4.43) as an equation of fixed point $F(t) = t$, where $t = (u, v, w)$ and the operator $F : \mathbb{R}^3_+ \to \mathbb{R}^3_+$ is given by

$$\begin{aligned} u' &= \left(\tfrac{\theta+u+v+\theta^{-1}w}{1+\theta u+\theta^{-1}v+w}\right)^k \\ v' &= \left(\tfrac{\theta^{-1}+u+v+\theta w}{1+\theta u+\theta^{-1}v+w}\right)^k \\ w' &= \left(\tfrac{1+\theta^{-1}u+\theta v+w}{1+\theta u+\theta^{-1}v+w}\right)^k. \end{aligned} \tag{4.44}$$

Consider sets

$$I_1 = \{t = (u, v, w) \in \mathbb{R}^3_+ : u = 1, v = w\}.$$

$$I_2 = \{t = (u, v, w) \in \mathbb{R}^3_+ : u = w, v = 1\}.$$

$$I_3 = \{t = (u, v, w) \in \mathbb{R}^3_+ : u = v, w = 1\}.$$

The following lemma is straightforward:

Lemma 5. *Each set* I_i, $i = 1, 2, 3$ *is invariant with respect to* F, *i.e.* $F(I_i) \subset I_i$.

Now we shall find fixed points of F on each invariant set I_i. Fixed points of restricted operator F on each of the invariant set, will be an equation the form

$$x = f(x, \gamma) := \left(\frac{1 + \gamma x}{\gamma + x}\right)^k. \tag{4.45}$$

Lemma 6. *Let $k \geq 2$. The equation (4.45) has unique solution $x = 1$ if $0 < \gamma \leq \frac{k+1}{k-1}$ and it has three solutions if $\gamma > \frac{k+1}{k-1}$.*

Proof. This is known see [172]. Here we give the proof because we will use this lemma several times. Note that $x = 1$ is a solution of (4.45) independently on parameters.

Case: $\gamma < 1$. In this case (4.45) has unique solution, because $f(x,\gamma)$ is decreasing in $(0, +\infty)$.

Case: $\gamma > 1$. Denote $u = \sqrt[k]{x}$. Rewrite (4.45) as

$$u^{k+1} - \gamma u^k + \gamma u - 1 = 0. \tag{4.46}$$

By Descartes rule (see[2] [303], p. 28), the equation (4.46) has up to 3 positive roots. We show that, under suitable condition on parameters, it has exactly three roots.

The function $f(x,\gamma)$ for $x > 0$ is increasing and bounded. We have

$$\frac{d}{dx} f(1,\gamma) = f'(1,\gamma) = k\frac{\gamma - 1}{\gamma + 1} > 0 \tag{4.47}$$

and

$$f'(1,\gamma) = 1, \quad \text{gives} \quad \gamma_c = \frac{k+1}{k-1}.$$

If $0 < \gamma \leq \gamma_c$ then $f'(1,\gamma) < 1$, the solution $x = 1$ is a stable fixed point of the map $f(x,\gamma)$, and $\lim_{n\to\infty} f^{(n)}(x,\gamma) = 1$, for any $x > 0$. Here, $f^{(n)}$ is the nth iterate of the above map $f(x,\gamma)$. Therefore, 1 is the unique positive solution. On the other hand, under $\gamma > \gamma_c$, the fixed point 1 is unstable. Iterates $f^{(n)}(x,\gamma)$ remain for $x > 1$, monotonically increasing and hence converge to a limit, $z^* \geq 1$ which solves (4.45). However, $z^* > 1$ as 1 is unstable. Similarly, for $x < 1$ one gets a solution $0 < z_* < 1$. This completes the proof. □

4.5.3.2 *Case:* I_1

Restricting (4.44) on the set I_1, the equation $t = F(t)$ becomes as

$$v = \left(\frac{\theta^{-1} + 1 + (1+\theta)v}{1 + \theta + (1+\theta^{-1})v}\right)^k = \left(\frac{1+\theta v}{\theta + v}\right)^k. \tag{4.48}$$

[2]The Descartes rule states that if the nonzero terms of a single-variable polynomial with real coefficients are ordered by descending variable exponent, then the number of positive roots of the polynomial is either equal to the number of sign changes between consecutive (nonzero) coefficients, or is less than it by an even number. A root of multiplicity n is counted as n roots. In particular, if the number of sign changes is zero or one, the number of positive roots equals the number of sign changes.

Applying Lemma 6 to (4.48) with $\gamma = \theta$ implies that

Result 1. The operator F on invariant set I_1 has unique fixed point $\mathbf{1} := (1,1,1)$, i.e., $u = v = w = 1$ if $0 < \theta \leq \frac{k+1}{k-1}$ and it has three fixed points $\mathbf{1}$, $(1, v_i, v_i)$, $i = 1, 2$ if $\theta > \frac{k+1}{k-1}$.

For example, if $k = 2$ then we have v_i, $i = 1, 2$ are

$$\begin{aligned} v_1 &:= v_1(\theta) = \tfrac{1}{2}(\theta^2 - 2\theta - 1 - (\theta - 1)\sqrt{(\theta+1)(\theta-3)}), \\ v_2 &:= v_2(\theta) = \tfrac{1}{2}(\theta^2 - 2\theta - 1 + (\theta - 1)\sqrt{(\theta+1)(\theta-3)}), \quad \theta > 3 \end{aligned} \tag{4.49}$$

4.5.3.3 *Case: I_2*

Restricting (4.44) on the set I_2, gives

$$u = \left(\frac{1 + \theta^{-1}u}{\theta^{-1} + v}\right)^k. \tag{4.50}$$

Now by Lemma 6, from (4.50) with $\gamma = 1/\theta$ we get

Result 2. The operator F on invariant set I_2 has unique fixed point $\mathbf{1} := (1,1,1)$, if $\theta \geq \frac{k-1}{k+1}$ and it has three fixed points $\mathbf{1}$, $(u_i, 1, u_i)$, $i = 1, 2$ if $0 < \theta < \frac{k-1}{k+1}$.

4.5.3.4 *Case: I_3*

Restricting (4.44) on the set I_3, gives

$$u = \left(\frac{1 + \gamma u}{\gamma + u}\right)^k, \quad \text{with} \quad \gamma = \frac{2}{\theta + \theta^{-1}}. \tag{4.51}$$

Since $\frac{2}{\theta+\theta^{-1}} < 1$, by Lemma 6, from (4.51) we get

Result 3. The operator F on invariant set I_3 has unique fixed point $\mathbf{1} := (1,1,1)$, for any $\theta > 0$.

We summarize the above mentioned Results 1-3, applying Theorem 17 in the following

Theorem 18. *If $k \geq 2$, $a = b = c = 1$ then for Hamiltonian*[3] *(4.28) there exists at least one SGM if $\theta \in (\frac{k-1}{k+1}, \frac{k+1}{k-1})$ and at least three SGMs if $\theta \in (0, \frac{k-1}{k+1}] \cup [\frac{k+1}{k-1}, +\infty)$.*

Remark 7. *Note that even when $a = b = c = 1$, the analysis of the system (4.43) is quite complicated for unknowns outside the invariant sets I_i. However, based on numerical (computer) analysis, we* **conjecture** *that there are no positive solutions outside these invariant sets.*

[3]The condition $a = b = c = 1$ implies $p(-1|1) = p(1|-1) = p(1|1) = p(-1|-1)$.

Remark 8. *It is well-known ([149]) that each Gibbs measure defines a state of the system determined by a Hamiltonian. The existence of certain parameter values (such as $\theta_c = \frac{k-1}{k+1}$) at which the uniqueness of the Gibbs measure transitions to non-uniqueness is interpreted as a phase transition.*

Phase transitions in Hidden Markov Models (HMMs) also refer to sudden or qualitative changes in the system's behavior, often driven by the model's parameters and the number of hidden states. These transitions can manifest as a shift from predictable, stable behavior to erratic or oscillatory behavior, as well as a transition between underfitting and overfitting as the number of hidden states changes.

Understanding these phase transitions is crucial for tuning HMMs and ensuring the model performs optimally for a given dataset. In the following, we will clarify this point specifically in the context of the non-uniqueness of Gibbs measures.

4.5.3.5 *Case:* $k = 1,\ a = b,\ c = 1.$

Here we assume that

$$p(1|1) = p(-1|-1), \quad p(1|-1) = p(-1|1). \tag{4.52}$$

In this case from (4.43) we get

$$\begin{aligned} u &= a\left(\frac{\theta+u+v+\theta^{-1}w}{1+\theta u+\theta^{-1}v+w}\right) \\ v &= a\left(\frac{\theta^{-1}+u+v+\theta w}{1+\theta u+\theta^{-1}v+w}\right) \\ w &= \left(\frac{1+\theta^{-1}u+\theta v+w}{1+\theta u+\theta^{-1}v+w}\right). \end{aligned} \tag{4.53}$$

Proposition 5. *For* $k = 1$, $a = b$, $c = 1$ *the system (4.43) (i.e. (4.53)) has unique positive solution.*

Proof. By subtracting the second equation from the first equation of (4.53), and then subtracting 1 from the third equation, we obtain:

$$\begin{aligned} u - v &= \frac{a(\theta^{-1}-\theta)}{1+\theta u+\theta^{-1}v+w}(w-1), \\ w - 1 &= \frac{\theta^{-1}-\theta}{1+\theta u+\theta^{-1}v+w}(u-v). \end{aligned} \tag{4.54}$$

From (4.54) we get

$$u = v, \quad w = 1 \tag{4.55}$$

or

$$u \neq v, \quad w \neq 1, \quad (1+\theta u+\theta^{-1}v+w)^2 = a(\theta^{-1}-\theta)^2. \tag{4.56}$$

Subcase: Consider the case (4.55), then system of equations (4.43) is reduced to

$$u = a\left(\frac{\theta+\theta^{-1}+2u}{2+(\theta+\theta^{-1})u}\right),$$

that can be rewritten as

$$u^2+(1-a)\Theta u - a = 0, \quad \text{with } \Theta = \frac{2}{\theta+\theta^{-1}}.$$

We are interested to positive solutions u. It is easy to see that for any $a > 0$ and $\theta > 0$, the last equation has unique positive solution:

$$u = u_1 := \frac{1}{2}\left((a-1)\Theta + \sqrt{((a-1)\Theta)^2+4a}\right). \tag{4.57}$$

Subcase: Let us consider the case (4.56), then from (4.53) we get

$$\begin{cases} |\theta-\theta^{-1}|u = \alpha\left(\theta+u+v+\theta^{-1}w\right) \\ |\theta-\theta^{-1}|v = \alpha\left(\theta^{-1}+u+v+\theta w\right) \\ \alpha|\theta-\theta^{-1}|w = 1+\theta^{-1}u+\theta v+w \end{cases} \tag{4.58}$$

where $\alpha = \sqrt{a}$.

Adding the first and second equations of (4.58) and adding its third equation to the equation given in (4.56) one gets

$$\begin{cases} (2\alpha - M)s + \alpha(\theta+\theta^{-1})\tau = 0 \\ (\theta+\theta^{-1})s + (2-\alpha M)\tau = 0 \end{cases} \tag{4.59}$$

where

$$M = |\theta-\theta^{-1}|,\ s = u+v,\ \tau = w+1.$$

Since $\theta \neq 1$ we have

$$\begin{vmatrix} 2\alpha - M & \alpha(\theta+\theta^{-1}) \\ \theta+\theta^{-1} & 2-\alpha M \end{vmatrix} = -2M(\alpha^2+1) \neq 0.$$

Therefore, the system (4.59) has unique solution $s = 0$, $\tau = 0$, which gives $u = -v$, $w = -1$, but we only interested to positive solutions. Thus system (4.53) has unique positive solution $(u_1, u_1, 1)$, where u_1 is given by (4.57). □

4.5.3.6 *Case:* $k \geq 2$, $a = b$, $c = 1$.

In this case we give analysis of (4.43).

Subcase: $x = y$, $z = 1$. The system is reduced to

$$\frac{1}{a}x = \varphi(x, \Theta) := \left(\frac{1+\Theta x}{\Theta + x}\right)^k, \ \Theta = \frac{2}{\theta + \theta^{-1}}. \tag{4.60}$$

Proposition 6. *For any $k \geq 2$, $a > 0$, $\theta > 0$ the equation (4.60) has unique solution.*

Proof. Note that $\Theta < 1$, because $\theta + \theta^{-1} > 2$, for $\theta > 0$, $\theta \neq 1$. Consequently, the function $\varphi(x, \Theta)$ is monotone decreasing function of x for all $\Theta \in (0, 1)$. But the function $\frac{x}{a}$ in the RHS of (4.60) is increasing. Therefore, equation (4.60) has unique solution. □

From (4.43), denoting $x = \sqrt[k]{u}$, $y = \sqrt[k]{v}$ and $z = \sqrt[k]{w}$ we get

$$\begin{aligned} x &= a \cdot \tfrac{\theta + x^k + y^k + \theta^{-1}z^k}{1+\theta x^k + \theta^{-1}y^k + z^k} \\ y &= a \cdot \tfrac{\theta^{-1} + x^k + y^k + \theta z^k}{1+\theta x^k + \theta^{-1}y^k + z^k} \\ z &= \tfrac{1+\theta^{-1}x^k + \theta y^k + z^k}{1+\theta x^k + \theta^{-1}y^k + z^k} \end{aligned} \tag{4.61}$$

Lemma 7. *In system (4.61) $x = y$ iff $z = 1$.*

Proof. Rewrite (4.61) as

$$\begin{aligned} x - y &= \tfrac{a(\theta - \theta^{-1})}{1+\theta x^k + \theta^{-1}y^k + z^k}(1 - z^k) \\ 1 - z &= \tfrac{(\theta - \theta^{-1})}{1+\theta x^k + \theta^{-1}y^k + z^k}(x^k - y^k) \end{aligned} \tag{4.62}$$

If $x = y$ then from the second equation we get $z = 1$. If $z = 1$ then from the first equation we get $x = y$. □

Subcase: $x \neq y$, $z \neq 1$. For simplicity we consider the case $k = 2$ and from (4.62) we obtain

$$(1 + \theta x^2 + \theta^{-1}y^2 + z^2)^2 = a(\theta - \theta^{-1})^2(x + y)(1 + z).$$

But this relation does not simplify the system (4.61). It still is very complicated therefor in the Table we give some numerical results.

k	θ	a	x	y	z
2	0.1	2	1.268048128 0.2005870619 192.3741268	1.268048128 1.263964993 3.052913735	1 0.6570348177 152.1989357
2	0.1	0.5	0.7886136007 0.005198204231 49.85366406	0.7886136007 0.7911611517 0.3275559308	1 0.1586966909 63.01328616
2	0.1	1	1 0.1010204092 98.98989796	1 1 1	1 0.1010204092 98.98989796
2	1.3	1	1	1	1
2	1.3	0.5	0.5376526550	0.5376526550	1

4.5.4 *Interpretations of results and their applications in machine learning*

4.5.4.1 *Prediction of a hidden configuration based on the observed one*

For a given sequence of observed data (configuration) $\sigma = (\sigma(x)$ over vertices x in the tree, one of main mathematical problem in HMM on the Ising model is to infer the hidden states (the spin configurations) from these observations. This can be formalized by calculation the conditional probability $\mu(s_n|\sigma_n)$ with condition (observed) configuration σ_n. We start by considering the joint distribution $\mu_n(s_n, \sigma_n)$ and the Hamiltonian $H_n(s_n, \sigma_n)$.

Expanding the Hamiltonian, we separate terms involving $s(x)$ and $\sigma(x)$. The terms involving $\sigma(x)$ are constants when conditioning on σ_n, so they can be factored out into the normalization constant. The remaining terms form an effective Hamiltonian for the s_n spins, which includes the interaction between s spins, the local terms $\beta p(\sigma(x)|s(x))$, and the boundary fields $h_{s(x),\sigma(x),x}$.

The final conditional probability is given by:

$$\mu(s_n|\sigma_n) = \frac{1}{Z(\sigma_n)} \exp\left(\beta J \sum_{\langle x,y\rangle} s(x)s(y) + \sum_{x\in V_n} \beta p(\sigma(x)|s(x)) \right.$$
$$\left. + \sum_{x\in W_n} h_{s(x),\sigma(x),x}\right)$$

$$= \frac{1}{Z(\sigma_n)} \prod_{\langle x,y \rangle} \theta^{\frac{s(x)s(y)}{2}} \prod_{x \in V_n} \exp(\beta p(\sigma(x)|s(x))) \prod_{x \in W_n} z_{s(x),\sigma(x),x}, \tag{4.63}$$

where $Z(\sigma_n)$ is the normalization factor dependent on σ_n.

Remark 9. *It follows from formula (4.63) that to define the conditional probability $\mu(s_n|\sigma_n)$, it suffices to know the conditional probabilities on each edge. Therefore, we examine these probabilities on each edge. Since the Gibbs measures we have derived above are translation-invariant, in formula (4.63), we have $z_{s(x),\sigma(x),x} = z_{s(x),\sigma(x)}$, meaning that it does not depend on the vertex x, but depends on the values of configurations at the vertex.*

Recall (see Theorem 17):

$$z_{\epsilon,\delta,x} = \exp(h_{\epsilon,\delta,x} - h_{-1,-1,x}), \ \epsilon, \delta = -1, 1.$$

Without loss of generality we assume $h_{-1,-1,x} \equiv 0$, then for each translation-invariant solution (4.42) we have

$$h_{\epsilon,\delta,x} \equiv \log z_{\epsilon,\delta}, \quad \epsilon, \delta = -1, 1. \tag{4.64}$$

Thus, on Cayley trees, belief propagation (BP) computes marginals $\mu(s_n|\sigma_n)$ efficiently by (4.63). The observed σ act as fixed boundary conditions, reducing degeneracy in hidden states.

Case: $a = b = c = 1$: Let us illustrate this for configurations on a fixed edge $\ell_0 = \langle x, y \rangle$ and for three distinct Gibbs measures of Result 1: μ_0 corresponding to $u = v = w = 1$ and μ_i, $i = 1, 2$ corresponding to solutions $(1, v_i, v_i)$ with v_i given in (4.49):

Measure μ_0: In case of Result 1 we have condition $a = b = c = 1$ that is

$$p(-1|1) = p(1|1) = p(1|-1) = p(-1|-1) \equiv \frac{1}{2}. \tag{4.65}$$

Since $u = v = w = 1$ from (4.64) we get

$$h_{\epsilon,\delta,x} \equiv 0, \quad \epsilon, \delta = -1, 1.$$

Consequently, for fixed edge $\ell_0 = \langle x, y \rangle$ we have

$$\mu_0(s_{\ell_0}|\sigma_{\ell_0}) = \frac{\theta^{\frac{1+s(x)s(y)}{2}}}{2(1+\theta)}. \tag{4.66}$$

By this formula, one can see that with respect to measure μ_0 the conditional probability does not depend on condition (observed) configuration. Moreover, hidden configuration on the end-points of edge ℓ_0 has equal values with probability $\theta/(1+\theta)$ and distinct values with probability $1/(1+\theta)$.

Measure μ_1: In this case we have

$$\mu_1(s_{\ell_0}|\sigma_{\ell_0}) = \frac{\theta^{\frac{1+s(x)s(y)}{2}} z_{s(x),\sigma(x)} z_{s(y),\sigma(y)}}{\sum_{\epsilon,\delta\in\{-1,1\}} \theta^{\frac{1+\epsilon\delta}{2}} z_{\epsilon,\sigma(x)} z_{\delta,\sigma(y)}}. \tag{4.67}$$

For solution $(1, v_i, v_i)$ this conditional probability also do not depend on condition σ_{ℓ_0}. For example,

$$\mu_1((1,1)|\sigma_{\ell_0}) = \frac{\theta v_1(\theta)^2}{\theta v_1(\theta)^2 + 2v_1(\theta) + \theta}$$

Measure μ_2: For the measure μ_2 similarly to the case μ_1 we obtain

$$\mu_2((1,1)|\sigma_{\ell_0}) = \frac{\theta v_2(\theta)^2}{\theta v_2(\theta)^2 + 2v_2(\theta) + \theta} \tag{4.68}$$

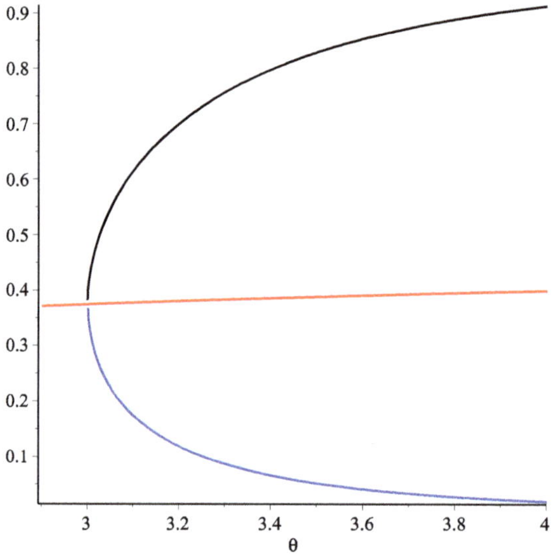

Fig. 4.4: The graph of $\mu_0((1,1)|(\sigma(x),\sigma(y)))$ (shown in the red), $\mu_1((1,1)|(\sigma(x),\sigma(y)))$ (blue) and $\mu_2((1,1)|(\sigma(x),\sigma(y)))$ (black) as functions of parameter θ.

Remark 10. *By formulas (4.66)–(4.68) (see also Figure 4.4), it is clear that the system described by the Hamiltonian (4.28) with the condition (4.65) has three distinct equilibrium states:*

- *For the state corresponding to μ_0, the hidden configurations at the endpoints of each edge of the tree appear with equal values with probability $\frac{\theta}{1+\theta}$ and with distinct values with probability $\frac{1}{1+\theta}$. Moreover, as $\theta \to \infty$,* *the hidden configuration is observed to be either all* $+1$ *or all* -1*, with probability* 1 *for the* μ_0 *state.*

- *For the state corresponding to* μ_1*, the hidden configurations at the endpoints of each edge of the tree are most likely to be* $+1$*, with the highest probability corresponding to* μ_1*.*
- *For the state corresponding to* μ_2*, the hidden configurations at the endpoints of each edge of the tree are most likely to be* -1*, with the highest probability corresponding to* μ_2*.*

Case $k = 1$**,** $a = b$**,** $c = 1$**:** In this case by Proposition 5 we have unique Gibbs measure, denote it by μ^*. From (4.63) we obtain (recall u_1 given in (4.57)) for condition $\sigma_{\ell_0} = (1, 1)$:

$$\mu^*((1,1)|(1,1)) = \frac{\theta}{2\theta + u_1 + u_1^2}, \quad \mu^*((-1,1)|(1,1)) = \frac{u_1}{2\theta + u_1 + u_1^2}, \tag{4.69}$$

$$\mu^*((1,-1)|(1,1)) = \frac{\theta}{2\theta + u_1 + u_1^2}, \quad \mu^*((-1,-1)|(1,1)) = \frac{u_1^2}{2\theta + u_1 + u_1^2}.$$

and for condition $\sigma_{\ell_0} = (-1, 1)$:

$$\mu^*((1,1)|(-1,1)) = \frac{u_1}{\theta + (1+\theta)u_1 + u_1^2},$$

$$\mu^*((-1,1)|(-1,1)) = \frac{\theta}{\theta + (1+\theta)u_1 + u_1^2}, \tag{4.70}$$

$$\mu^*((1,-1)|(-1,1)) = \frac{u_1^2}{\theta + (1+\theta)u_1 + u_1^2},$$

$$\mu^*((-1,-1)|(-1,1)) = \frac{\theta u_1}{\theta + (1+\theta)u_1 + u_1^2}.$$

Remark 11. *From formulas (4.69)–(4.70) (see also Figure 4.5), it is evident that the system described by the Hamiltonian (4.28) on a one-dimensional lattice, with the condition (4.52), has a unique equilibrium state corresponding to* u_1*. In this case, the conditional probability of predicting a hidden configuration (at the endpoints of each edge of the 1D tree) depends on the conditioned observed configuration. For instance, the hidden configuration with the highest* μ^**-probability coincides with the (conditioned) observed configuration.*

Case $k = 2$**,** $a = b$**,** $c = 1$**:** In this case we consider $a = 2$, $\theta = 0.1$ and choose solution given in the second row of the Table: $u = x^2 \approx 0.04$,

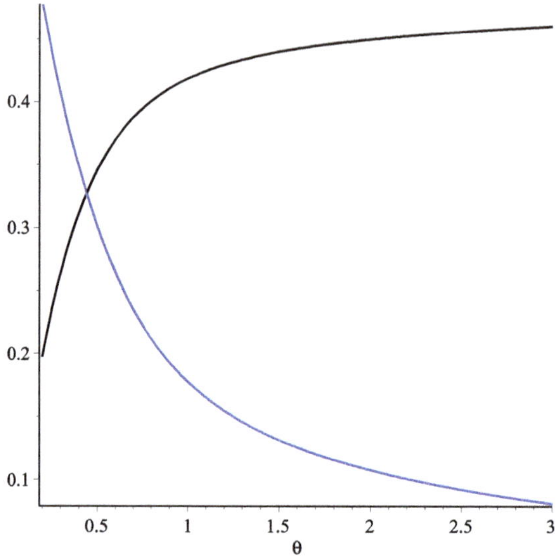

Fig. 4.5: The graph of $\mu^*((1,1)|(1,1))$ (black) and $\mu^*((1,1)|(-1,1))$ (blue) at $a = 0.3$ as functions of parameter θ.

$v = y^2 \approx 1.6$, $w = z^2 \approx 0.432$. Denote by μ_3 the corresponding Gibbs measure, then for these values from (4.63) we get

$$\mu_3((1,1)|(1,1)) \approx 0.347, \quad \mu_3((-1,1)|(1,1)) \approx 0.324,$$

$$\mu_3((1,-1)|(1,1)) \approx 0.324, \quad \mu_3((-1,-1)|(1,1)) \approx 0.005.$$

For condition $\sigma_{\ell_0} = (-1,1)$ we have

$$\mu_3((1,1)|(-1,1)) \approx 0.003, \quad \mu_3((-1,1)|(-1,1)) \approx 0.86,$$

$$\mu_3((1,-1)|(-1,1)) \approx 0.127, \quad \mu_3((-1,-1)|(-1,1)) \approx 0.01.$$

Thus with respect to measure μ_3 the conditional probability of predicting a hidden configuration depends on the conditioned observed configuration. For instance, the hidden configuration with the highest μ_3-probability coincides with the (conditioned) observed configuration.

4.5.4.2 *Applications to machine learning*

Let us now discuss some relations of our results in machine learning:

- In tasks where observed data (e.g., pixel intensities in images, word sequences) have intrinsic correlations, the term $\sigma(x)\sigma(y)$ in the

Hamiltonian (4.28) allows the model to capture dependencies in the observations (as explained in the previous subsection) while inferring the hidden structure s. For instance, in image denoising, the observed pixels σ are noisy, and the hidden spins s represent the clean image labels. The model learns to recover the true signal from noisy observations by leveraging the correlation between the noisy pixels and the hidden clean labels (e.g., [61]).

- In weakly supervised learning, the Hamiltonian's mismatch penalty $(s(x)s(y) - \sigma(x)\sigma(y))$ enforces consistency between local predictions and global correlations in the observed data σ. This penalty term ensures that the model respects both the local relationships (captured by s) and the global structure in the data. It plays a critical role in training models where supervision is limited (e.g., [400]).
- In graphical model learning, the couplings J and emission parameters (embedded in $p(\sigma|s)$) via contrastive divergence are essential for controlling the dynamics of the system related to the model. The penalty $(s(x)s(y) - \sigma(x)\sigma(y))$ depends on the sign of J. Specifically, for $J > 0$, the model penalizes configurations where $s(x)s(y) < \sigma(x)\sigma(y)$, while for $J < 0$, the opposite holds, meaning the model favors configurations where $s(x)s(y) > \sigma(x)\sigma(y)$. The tree structure allows for exact calculations of critical parameters (e.g., [378]).
- In anomaly detection, outliers are detected by identifying configurations where $s(x)s(y)$ deviates significantly from $\sigma(x)\sigma(y)$. Such mismatches signal discrepancies between expected and observed correlations, which often correspond to unusual or anomalous data points (e.g., [77]).
- Belief Propagation: On tree-like structures, belief propagation (or the sum-product algorithm) is exact. This means that the messages passed between neighboring vertices x and y in the tree structure can be used to compute marginals of the spin configuration at each vertex. Belief propagation updates the beliefs about the state of each spin based on the observed data and the messages received from neighboring vertices. The algorithm for the Ising model on a Cayley tree iterates over the tree, updating the probability of the spin at each vertex based on its neighbors:

$$\mu_{x\to y}(s(x)) = \sum_{s(y)} p(s(x)|s(y))p(\sigma(x)|s(x)) \prod_{z\in\mathcal{N}(x)\setminus y} \mu_{z\to x}(s(z)),$$

where $\mathcal{N}(x)$ is the set of all neighbors of x

This iterative process continues until convergence, and the marginal distribution at each vertex provides an estimate for the hidden spin configuration (e.g., [301], [394]).

4.6 Bayesian non-parametric modeling

Gibbs measures are also widely used in Bayesian nonparametric modeling, where the number of parameters is not fixed but can grow with the data. By defining a prior distribution over an infinite-dimensional parameter space, Gibbs measures allow us to sample from the posterior distribution over this space, capturing the complexity of the data without specifying a fixed number of parameters.

Definitions. Bayesian non-parametric (BNP) modeling provides a flexible framework for statistical inference where the complexity of the model grows with the data. Unlike traditional parametric models with a fixed, finite number of parameters, BNP models assume an infinite-dimensional parameter space. This flexibility allows BNP models to adapt to data without requiring a rigid structure.

BNP modeling operates within the Bayesian paradigm, where the goal is to infer a posterior distribution over an unknown quantity given observed data. Formally:

$$p(\theta \mid \mathcal{D}) = \frac{p(\mathcal{D} \mid \theta)p(\theta)}{p(\mathcal{D})},$$

where θ is the unknown parameter, possibly infinite-dimensional, $p(\theta)$ is the prior distribution, $p(\mathcal{D} \mid \theta)$ is the likelihood of the observed data $\mathcal{D}$ given θ, and $p(\mathcal{D}) = \int p(\mathcal{D} \mid \theta)p(\theta)d\theta$ is the marginal likelihood. In BNP models, θ often represents functions, distributions, or other infinite-dimensional objects.

BNP models use priors over infinite-dimensional spaces, such as:

- Gaussian Processes (GPs) where priors over functions, used for regression and classification. A GP is defined as:

$$f(x) \sim \mathcal{GP}(m(x), k(x, x')),$$

where $m(x)$ is the mean function and $k(x, x')$ is the covariance kernel.

- Dirichlet Processes (DPs) where priors over distributions, used in clustering and mixture models. A DP is defined as:

$$G \sim \mathrm{DP}(\alpha, G_0),$$

where α is the concentration parameter and G_0 is the base measure.

- Indian Buffet Process (IBP) where a prior over infinite binary matrices, used for latent feature models.

These priors allow the complexity of the model to grow as more data is observed.

Dirichlet process and stick-breaking construction. The Dirichlet Process is a cornerstone of BNP modeling. It is a distribution over distributions and is defined as:

$$G \sim \mathrm{DP}(\alpha, G_0),$$

where G_0 is the base measure (mean of the DP) and α controls the variability around G_0.

The stick-breaking construction provides a constructive definition of a DP:

$$G = \sum_{k=1}^{\infty} \pi_k \delta_{\theta_k},$$

where $\pi_k = V_k \prod_{j=1}^{k-1}(1 - V_j)$ and $V_k \sim \mathrm{Beta}(1, \alpha)$, and $\theta_k \sim G_0$.

This construction illustrates how a DP generates a discrete distribution.

Chinese restaurant process (CRP) representation. The CRP provides an intuitive representation of clustering induced by a DP. Given n data points:

- The probability of assigning a new point to an existing cluster k is proportional to the number of points in that cluster:

$$p(z_{n+1} = k \mid \mathbf{z}) \propto n_k,$$

where n_k is the size of cluster k.

- The probability of starting a new cluster is proportional to α:

$$p(z_{n+1} = \text{new} \mid \mathbf{z}) \propto \alpha.$$

Bayesian non-parametric modeling provides a principled approach to modeling uncertainty and complexity, making it an essential tool in modern statistics and machine learning.

4.7 Generalized Bayesian inference

4.7.1 *Definitions and examples*

Let us first introduce notation of Gibbs posterior.

A Gibbs posterior is a probability distribution used in statistical learning when the likelihood is replaced by a loss function $L(\theta; D)$, where θ

represents model parameters and D is the data. It assigns higher probabilities to parameter values with lower loss, that is,

$$\pi(\theta|D) \propto \exp\left(-\lambda L(\theta; D)\right) \pi_0(\theta),$$

where $\lambda > 0$ is a parameter that controls the influence of the loss function relative to the prior $\pi_0(\theta)$.

As one can see from the formulas, both concepts Gibbs measure and Gibbs posterior involve exponential weighting (via the Boltzmann/Gibbs factor), but they differ fundamentally:

- Gibbs measures model physical systems and equilibrium states, and are tied to spatial structures (e.g., lattices or trees).
- Gibbs posteriors focus on parameter estimation and learning under uncertainty, using a "loss-driven" approach instead of likelihood-based Bayesian inference.

Generalized Bayesian inference (GBI) (see [81, 192, 203, 357]) refers to extensions or modifications of classical Bayesian inference to address limitations or adapt to specific challenges in data analysis, modeling, or decision-making.

As discussed in the previous sections, classical Bayesian inference updates prior beliefs about parameters using observed data and a likelihood function, in accordance with Bayes' rule. GBI extends this framework in several ways while preserving the fundamental principle of belief updating.

Instead of strictly following Bayes' rule, GBI defines a posterior that incorporates alternative likelihood models, loss functions, or updating schemes. This allows for more flexibility in handling data that do not conform to traditional probabilistic assumptions.

A generalized posterior might take the form:

$$\pi(\theta|x) \propto \pi(\theta) \exp(-\alpha \cdot \mathcal{D}(x, \theta)),$$

where

- $\mathcal{D}(x, \theta)$ is a discrepancy measure (not necessarily a log-likelihood),
- $\alpha > 0$ controls the strength of the updating.

Note that the loss function measures compatibility between the parameters and observed data, and this compatibility is translated into a posterior using the Gibbs-Boltzmann distribution. This approach allows practitioners to model problems with greater flexibility, especially in cases where the likelihood function is difficult to specify or overly restrictive.

For example, by selecting a loss function tailored to specific preferences or penalties, one can address decision-theoretic problems more effectively. This framework has been applied in machine learning, energy-based modeling, and robust statistics. The Gibbs posterior can recover traditional Bayesian inference by choosing the negative log-likelihood as the loss function, making it a strict generalization of Bayesian methods.

Furthermore, its flexibility has drawn attention in contexts where classical likelihood-based approaches struggle, such as complex natural phenomena or energy-based models in machine learning.

Jones's explanation of Gibbs posteriors and their flexibility in inference[4]: This source provides insights into how Gibbs posteriors use loss functions to generalize Bayesian approaches. It also links this concept to energy-based models in machine learning, which share similar formulations using energy functions as a basis for probabilities.

Let us give some examples of generalized Bayesian methods (see [43], [168] for more detailed material):

1. Bayesian updating with divergences: Instead of using a likelihood, a divergence $\mathcal{D}$ between the observed data and the model predictions is used:

$$\pi(\theta|x) \propto \pi(\theta) \exp(-\alpha \mathcal{D}(P_x, P_\theta)),$$

where P_x and P_θ are empirical and model distributions.

2. PAC-Bayesian inference: The PAC-Bayesian framework combines Bayesian reasoning with a bound on the generalization error. This is widely used in machine learning for robust model selection and parameter tuning.

3. Wasserstein-Bayes: The Wasserstein distance is used in place of the likelihood to create a generalized posterior:

$$\pi(\theta|x) \propto \pi(\theta) \exp(-\alpha W(P_x, P_\theta)),$$

where W is the Wasserstein distance between empirical and model distributions.

4.7.2 *Key advantages in hierarchical contexts*

Advantages in hierarchical contexts are as follows (see e.g. [25], [146], [149], [323], [324]).

- Flexibility: Gibbs measures adapt to various interaction structures, including trees, grids, and hypergraphs.

[4] https://andrewcharlesjones.github.io/journal/ebms.html.

- Markov property: The local conditional independence property simplifies computations.

- Statistical mechanics analogy: Hierarchical models often resemble systems in statistical mechanics (e.g., Ising and Potts models), where Gibbs measures describe equilibrium states.

- Applications:

Statistical mechanics on trees: Gibbs measures on Cayley trees or hierarchical structures have been widely studied for Ising and Potts models, where parameters reflect hierarchical interactions. For example, recursive construction of Gibbs measures for spin systems on trees (see [323], [324] and references therein).

Hierarchical Bayesian models: In Bayesian statistics, Gibbs sampling is a key method for computing posterior distributions, particularly in hierarchical models where parameters are modeled conditionally (see [6], [146], [236], [238]).

Machine learning: Gibbs measures are fundamental in probabilistic graphical models, such as hidden Markov models and Gaussian mixture models, which have hierarchical interpretations (see [41], [201], [278], [378]).

Following [76] (and references therein), we note that Bayesian inference often faces significant challenges due to the necessity of specifying a likelihood function. These challenges are particularly pronounced in practical applications, where concerns about potential model misspecification can undermine the validity of the statistical analysis.

Non-parametric Bayesian methods [76] address the challenge of methodological inflexibility by enhancing flexibility and improving the robustness of statistical inferences. However, fully adopting non-parametric approaches introduces additional complexity. Evaluating the trade-offs between this added complexity and the resulting robustness necessitates careful, context-specific considerations.

As an alternative, generalized Bayesian inference [81,192,203,357] offers a model-free framework designed to mitigate the risks of model misspecification and overly complex formulations.

Gibbs posteriors have found applications in diverse fields [39, 203, 312, 357, 358]. Recently, their theoretical basis as rational belief updates—qualifying them as valid posterior distributions—has been rigorously established [43]. In [191], introduced important advancements in methods for selecting the temperature parameter λ.

4.8 Conclusion

In this chapter, we explored the deep interconnections between Gibbs measures and modern Bayesian inference techniques. Starting with foundational concepts in Bayesian interfaces and parameter estimation, we developed a layered understanding of hierarchical modeling both in classical and modern forms. We saw how Gibbs measures naturally arise in hierarchical structures and multi-level energy models, serving as powerful tools for representing complex dependencies and latent variables.

The integration of Gibbsian frameworks with graphical models and hidden Markov structures, especially, within the Ising model—demonstrated the strength of these methods in capturing spatial and temporal dependencies. The investigation of translation-invariant Gibbs measures in hidden Markov contexts illuminated their utility in machine learning applications such as pattern recognition, denoising, and unsupervised classification.

Furthermore, the chapter addressed the frontiers of Bayesian nonparametric modeling via Dirichlet processes and stick-breaking constructions, emphasizing their flexibility in high-dimensional and data-rich environments. The section on generalized Bayesian inference highlighted recent trends in robust and scalable inference procedures, particularly in hierarchical contexts.

For further reading see [22], [86], [87] (and references therein), [268] (for an introduction to probabilistic programming that frequently uses Gibbs sampling and hierarchical models), [151], [152] (These works are devoted to probabilistic programming, Bayesian optimization, data compression and automatic model discovery).

Chapter 5

Gibbs Measures for Unsupervised Learning

This chapter investigates the role of Gibbs measures in clustering and unsupervised learning. It begins by defining clustering within the machine learning context and introduces fairness-aware methods, including generalized Bayesian fair clustering. The chapter then addresses the challenges of clustering high-dimensional data using Bayesian techniques, emphasizing the advantages of Gibbs-based approaches. Further sections explore how Gibbs measures support unsupervised learning tasks and serve as a foundation for probabilistic generative models. Throughout, the chapter highlights how principles from statistical physics enrich the design and interpretation of clustering algorithms in modern data analysis.

5.1 Clustering with Gibbs measures

5.1.1 *Definition of clustering in machine learning*

Clustering is a unsupervised learning technique in machine learning that involves grouping a set of data points into subsets (clusters) based on their similarity [60, 137, 290]. The goal is to partition the data such that points within the same cluster are more similar to each other (according to some distance or similarity measure) than to points in different clusters.

By defining joint probability distributions over the data and cluster assignments, Gibbs measures facilitate the inference of cluster parameters and the assignment of data points to clusters in a probabilistic manner. This approach allows for robust and interpretable clustering results, even in the presence of complex data structures.

Clustering is commonly used in applications such as image segmentation, document categorization, customer segmentation, and anomaly detection.

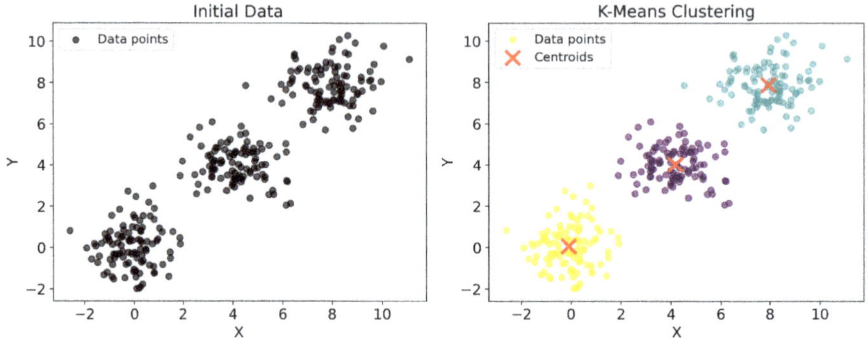

Fig. 5.1: K-means clustering of input data

Mathematically, clustering can be formulated as an optimization problem or a probabilistic modeling task, depending on the specific algorithm used. The following are key mathematical concepts:

Dataset representation: Let the dataset consist of n data points:

$$\mathcal{X} = \{x_1, x_2, \ldots, x_n\}, \quad x_i \in \mathbb{R}^d$$

where d is the number of features (dimensionality of the data).

Clustering objective: The objective is to partition $\mathcal{X}$ into k clusters $\mathcal{C} = \{C_1, C_2, \ldots, C_k\}$ such that:

- $\bigcup_{i=1}^{k} C_i = \mathcal{X}$ (all data points are assigned to a cluster),
- $C_i \cap C_j = \emptyset$ for $i \neq j$ (clusters do not overlap).

Similarity measure: A similarity or distance function $d(x_i, x_j)$ quantifies how similar or different two data points are. Commonly used measures include Euclidean distance $d(x_i, x_j) = \|x_i - x_j\|_2$ and cosine similarity $\cos(x_i, x_j) = \frac{x_i^\top x_j}{\|x_i\|_2 \|x_j\|_2}$.

Clustering optimization problems: Different clustering algorithms solve clustering as different optimization problems:

(1) K-Means clustering minimizes within-cluster variance:

$$\arg \min_{\mu_1, \mu_2, \ldots, \mu_k} \sum_{i=1}^{k} \sum_{x \in C_i} \|x - \mu_i\|_2^2$$

where $\mu_i = \frac{1}{|C_i|} \sum_{x \in C_i} x$ is the centroid of cluster C_i (see Figure 5.1).

(2) Hierarchical clustering optimizes cluster merging/splitting based on linkage criteria:

$$\arg \min_{C_1, \ldots, C_k} \sum_{i<j} d(C_i, C_j) \cdot \mathbf{1}[\text{merge}(C_i, C_j)]$$

where $d(C_i, C_j)$ is defined by single linkage $\min_{x\in C_i, y\in C_j} d(x, y)$, complete linkage $\max_{x\in C_i, y\in C_j} d(x, y)$, or average linkage $\frac{1}{|C_i|\cdot|C_j|} \sum_{x\in C_i, y\in C_j} d(x, y)$.

(3) DBSCAN density-based clustering maximizes cluster density:

$$\arg\max_{\{C_i\}} \sum_i |C_i|^2 \text{ subject to } \min_{x\in C_i} |N_\varepsilon(x)| \geq \text{minPts}$$

where $N_\varepsilon(x) = \{y : d(x, y) \leq \varepsilon\}$.

(4) EM for GMMs maximizes log-likelihood:

$$\arg \max_{\{\pi_i, \mu_i, \Sigma_i\}} \sum_{j=1}^{n} \log \left(\sum_{i=1}^{k} \pi_i \mathcal{N}(x_j \mid \mu_i, \Sigma_i) \right)$$

where $\sum_{i=1}^{k} \pi_i = 1$ and $\pi_i \geq 0$.

Thus a clustering is a versatile unsupervised learning approach rooted in the optimization of similarity measures, probabilistic modeling, or density estimation. Its choice depends on the structure and properties of the data.

5.1.2 *Fairness in clustering*

Fairness in clustering in machine learning is built upon ensuring that the clusters produced by an algorithm are equitable with respect to sensitive attributes (e.g., gender, race, or age). It involves creating constraints or objectives to prevent unfair bias while preserving the core properties of clustering.

A clustering algorithm partitions a dataset $\mathcal{X} = \{x_1, x_2, \ldots, x_n\} \subseteq \mathbb{R}^d$ into k clusters:

$$\mathcal{C} = \{\mathcal{C}_1, \mathcal{C}_2, \ldots, \mathcal{C}_k\}, \quad \bigcup_{i=1}^{k} \mathcal{C}_i = \mathcal{X}, \quad \mathcal{C}_i \cap \mathcal{C}_j = \emptyset \text{ for } i \neq j.$$

Each data point x_i is associated with a sensitive attribute $a_i \in \mathcal{A}$, where $\mathcal{A}$ represents the set of possible values of the sensitive attribute (e.g., $\mathcal{A} = \{\text{male, female}\}$).

The clustering objective is to minimize within-cluster dissimilarity (e.g., k-means, k-median) or maximize between-cluster separation, typically expressed as:

$$\text{Minimize: } \sum_{i=1}^{k} \sum_{x\in \mathcal{C}_i} \mathrm{d}(x, c_i),$$

where c_i is the center or representative of cluster $\mathcal{C}_i$, and $\mathrm{d}(\cdot,\cdot)$ is a distance metric.

Fairness in clustering introduces additional constraints to ensure equitable representation of sensitive groups within clusters. The most common mathematical formulations include proportional representation and fairness-aware balance constraints. In proportional representation, each cluster $\mathcal{C}_i$ should contain a proportion of points from each group (e.g., by sensitive attribute a) that approximates the global distribution. Let π_a denote the global fraction of points with attribute a:

$$\pi_a = \frac{|\{x \in \mathcal{X} : a_x = a\}|}{|\mathcal{X}|}.$$

For fairness, each cluster $\mathcal{C}_i$ should satisfy:

$$\frac{|\{x \in \mathcal{C}_i : a_x = a\}|}{|\mathcal{C}_i|} \approx \pi_a, \quad \forall a \in \mathcal{A}, \forall i.$$

Fairness-aware balance constraints provide a strict version of proportional representation that enforces balance within each cluster:

$$\alpha \cdot \pi_a \cdot |\mathcal{C}_i| \leq |\{x \in \mathcal{C}_i : a_x = a\}| \leq \beta \cdot \pi_a \cdot |\mathcal{C}_i|,$$

where α, β are user-defined constants (e.g., $\alpha = 0.9, \beta = 1.1$).

The optimization problem for fair clustering can be formulated as a constrained optimization problem:

$$\text{Minimize: } \sum_{i=1}^{k} \sum_{x \in \mathcal{C}_i} \mathrm{d}(x, c_i),$$

subject to:

$$\frac{|\{x \in \mathcal{C}_i : a_x = a\}|}{|\mathcal{C}_i|} \approx \pi_a \quad \forall a \in \mathcal{A}, \forall i.$$

This problem can also be framed in terms of regularization:

$$\text{Minimize: } \sum_{i=1}^{k} \sum_{x \in \mathcal{C}_i} \mathrm{d}(x, c_i) + \lambda \cdot \text{Fairness Penalty}(\mathcal{C}),$$

where λ is a regularization parameter balancing fairness and clustering objectives.

Common fairness penalty examples include representation error and entropy-based fairness. The representation error penalty is defined as:

$$\text{Fairness Penalty}(\mathcal{C}) = \sum_{i=1}^{k} \sum_{a \in \mathcal{A}} \left| \frac{|\{x \in \mathcal{C}_i : a_x = a\}|}{|\mathcal{C}_i|} - \pi_a \right|.$$

Entropy-based fairness encourages clusters with uniform distribution over $\mathcal{A}$ using entropy:

$$\text{Fairness penalty}(\mathcal{C}) = -\sum_{i=1}^{k}\sum_{a\in\mathcal{A}} p_i(a)\log(p_i(a)),$$

where $p_i(a) = \frac{|\{x\in C_i : a_x = a\}|}{|C_i|}$.

Fair clustering algorithms adjust traditional clustering methods to include fairness constraints. By combining optimization, regularization, and algorithm design, fairness in clustering provides a rigorous and equitable approach to partitioning data.

We shall now formally introduce (see [76] and references therein) the notion of balance in the context of fair clustering.

For a positive integer t, denote $[t] := \{1,\ldots,t\}$. Suppose that we observe data $\{(x_i,a_i)\}_{i=1}^N$, where x_i denotes the d-variate observation for the i-th data unit, and a_i the label of the protected attribute. For each a, let $\{x_i^{(a)}\}_{i=1}^{N_a}$ denote the observations corresponding to the a-th level of the protected attribute, where

$$N_a = \sum_{i=1}^{N} 1(a_i = a), \quad \text{and} \quad \sum_{a=1}^{r} N_a = N.$$

The goal of fair clustering is to assign the data points $\{(x_i,a_i)\}_{i=1}^N$ into clusters $C = (C_1,\ldots,C_K)$, such that:

$$\bigcup_{k=1}^{K} C_k = [N],$$

while respecting the notion of balance [82].

Definition 26. (*Balance*). *Given* $\{(x_i,a_i)\in\mathcal{X}\times[r], i\in[N]\}$, *such that* $a_i = a$ *for* $i = \sum_{j=0}^{a-1} N_j + 1,\ldots,\sum_{j=1}^{a} N_j$, *where* $a\in[r]$ *and* $N_0 = 0$, *the balance in cluster* C_k *is defined as:*

$$Balance(C_k) = \min_{1\le j_1 < j_2 \le r} \min\left\{\frac{|C_k^{(j_1)}|}{N_{j_1}}, \frac{|C_k^{(j_2)}|}{N_{j_2}}\right\},$$

where $|C_k^{(j)}|$ *denotes the number of observations in* C_k *with* $a = j$.

The overall balance of the clustering is:

$$Balance(C) = \min_{k=1,\ldots,K} Balance(C_k).$$

The higher this measure is for a clustering configuration, the fairer the clustering is.

Given the aforementioned definition of balance, in [82] the authors introduced the notion of *fairlets* as minimal fair sets that approximately maintain the selected clustering objective. The authors illustrated that addressing any fair clustering problem involves initially obtaining a fairlet decomposition of the data through the solution of a minimum-cost flow problem.

5.1.3 *Generalized Bayesian fair clustering*

Following [76] we consider Bayesian fair clustering. For the observed data $\{(x_i, a_i) \in \mathcal{X} \times [2], i \in [N]\}$, let $\mathcal{U} \subseteq \mathcal{X}^m$ be the class of all "m fairlet centers."

Now we place a uniform prior on the space of all possible fairlet decompositions $\mathcal{U}$, i.e., assume

$$\pi(u) = \frac{1}{|\mathcal{U}|}, \quad u \in \mathcal{U}. \tag{5.1}$$

Then, the generalized Bayes posterior for fairlet decomposition takes the form:

$$\pi(u \mid \lambda_f, \{(x_i, a_i)\}_{i=1}^N) \propto \frac{\exp\left(-\lambda_f L_f(u)\right)}{\sum_{u \in \mathcal{U}} \exp\left(-\lambda_f L_f(u)\right)},$$

where λ_f is a temperature parameter, and $L_f(u)$ is the loss function.

Given a fairlet decomposition $u \in \mathcal{U}$, which may differ from the optimal fairlet decomposition $u^\star$, a typical Bayesian model for clustering is based on the assumption that observations follow:

$$u_i \mid \theta_k, i \in C_k \overset{\text{ind}}{\sim} \pi(u_i \mid \theta_k), \quad \theta_k \overset{\text{i.i.d.}}{\sim} \pi(\theta), \, k \in [K]. \tag{5.2}$$

Under the above model and prior specification, the posterior distributions of clustering configurations take the form:

$$\pi(C \mid u) \propto \pi(C) \prod_{k=1}^{K} \int_{\Theta} \prod_{i \in C_k} \pi(u_i \mid \theta)\pi(\theta)\, d\theta, \tag{5.3}$$

where $\pi(C)$ is the prior probability of C, $\pi(u \mid \theta)$ is the within-cluster likelihood, and $\pi(\theta)$ is the prior distribution on the cluster-specific parameters.

While equation (5.3) serves as the foundation for an extensive body of literature on Bayesian clustering, it gives rise to significant practical challenges. The integral often does not admit a closed-form expression, introducing computational complexities. Furthermore, the posterior of clustering configurations is highly sensitive to the precise specifications of the data-generating mechanism $\pi(u \mid \theta)$.

Such model-based clustering frameworks (5.2) are routinely criticized for various aspects. Firstly, clustering may simply serve as a convenient preprocessing step, and there might not be distinct groups present in the data. Moreover, even if such groups exist, the distribution of the data within each cluster is unlikely to precisely adhere to the chosen distribution.

In terms of computational aspects, even when the data accurately conform to the selected mixture distribution, we frequently encounter substantial computational bottlenecks, especially in high dimensions (see [76], [266] and references therein).

To address these brittleness issues, the paper [312] implemented a Gibbs posterior framework tailored for clustering, aiming to navigate around these challenges.

Given the optimal fairlet decomposition of the observed data

$$u^\star = (u_1^\star, \dots, u_m^\star)^T \in \mathcal{U},$$

the paper [82] proposed to invoke existing machinery for traditional clustering algorithms to cluster the m-fairlet centers into K groups. In [76] the authors study the flexible class of clustering mechanisms characterized by a factorized loss $L_c(C \mid u^\star)$. Assume that the number of clusters K is fixed, e.g., it is known or has been selected in an exploration phase. Let $u^\star_{(k)}$ denote the center of the fairlet centers $\{u_i^\star : i \in C_k\}$ belonging to cluster C_k, for $k \in [K]$. Then, the clustering is characterized by a factorized loss of the form:

$$L_c(C \mid u^\star) = \sum_{k=1}^{K} \sum_{i \in C_k} D(u_i^\star, u^\star_{(k)}), \quad C : |C| = K, \tag{5.4}$$

where $D(u_i^\star, u^\star_{(k)}) \geq 0$ is a function of $u_i^\star$ and $u^\star_{(k)}$ that quantifies the discrepancy of the i-th unit from the k-th cluster.

Let the loss function $L(C \mid u)$ be as in (5.4). We are interested in employing uniform clustering priors of the form:

$$\pi(C) = \frac{1}{S(n, K)}, \quad C : |C| = K, \tag{5.5}$$

where

$$S(n, K) = \frac{1}{K!} \sum_{k=0}^{K} (-1)^{K-k} K! \left[(K-k)!k!\right]^{-1} k^n$$

is the Stirling number of the second kind. Prior (5.5) is uniform over partitions having K components. Although the framework readily accommodates more intricate clustering priors, our emphasis in this paper remains on the uniform case.

The generalized Bayes posterior under a generalized Bayes product partition model has the form:

$$\pi(C \mid \lambda_c, u) \propto \prod_{k=1}^{K} \exp\left(-\lambda_c \sum_{i \in C_k} D(u_i, u^{(k)})\right), \quad C : |C| = K,$$

where λ_c is a temperature parameter and $\{u^{(k)}, k \in [K]\}$ are the K cluster centers.

Now we are ready to introduce a generalized Bayesian posterior framework for the fair clustering problem. Consider the uniform clustering prior in (5.5) and utilize the uniform prior on the space of all possible fairlet decompositions $\mathcal{U}$, introduced in (5.1). Then, the generalized Bayes posterior for fair clustering takes the form:

$$\pi(C, u \mid (\lambda_f, \lambda_c), \{(x_i, a_i)\}_{i=1}^{N}) \propto \frac{\exp(-\lambda_f L_f(u))}{\sum_{u \in \mathcal{U}} \exp(-\lambda_f L_f(u))} \times \exp\left(-\lambda_c L_c(C \mid u)\right).$$

Under the assumption of factorized clustering loss, the posterior simplifies to:

$$\pi(C, u \mid (\lambda_f, \lambda_c), \{(x_i, a_i)\}_{i=1}^{N})$$

$$\propto \frac{\exp(-\lambda_f L_f(u))}{\sum_{u \in \mathcal{U}} \exp(-\lambda_f L_f(u))} \times \prod_{k=1}^{K} \exp\left(-\lambda_c \sum_{i \in C_k} D(u_i, u^{(k)})\right), \tag{5.6}$$

such that $C : |C| = K$.

The paper [76] employed the methodology proposed in [191] for the selection of the temperature parameters (λ_f, λ_c).

A natural competitor for the proposed methodology based on the joint posterior in (5.6) is the fair clustering with fairlets [82].

Let $\pi(C, u \mid (\lambda_f, \lambda_c), \{(x_i, a_i)\}_{i=1}^{N})$ denote the joint posterior of the fairlet decompositions of the observed data and clustering configurations in (5.6). Then, we denote the posterior mode by:

$$(C_{\text{MAP}}, u_{\text{MAP}}) = \arg\max_{C,u} \pi(C, u \mid (\lambda_f, \lambda_c), \{(x_i, a_i)\}_{i=1}^{N}).$$

Proposition 7 (76). *The clustering configuration C_{MAP} does not coincide with the optimal clustering obtained via fair clustering with fairlets [82].*

Proof. From Equation (5.6), we note that the quantities $(C_{\text{MAP}}, u_{\text{MAP}})$ are computed via solving

$$\arg\max_{C,u} \pi(C, u \mid (\lambda_f, \lambda_c), \{(x_i, a_i)\}_{i=1}^N)$$

$$= \arg\min_{C,u} \left[\lambda_f L_f(u) + \lambda_c \sum_{k=1}^{K} \sum_{i \in C_k} D(u_i, u_{(k)}) \right],$$

where $\lambda_f \geq 0$ and $\lambda_c \geq 0$.

On the other hand, the optimal clustering obtained via fair clustering with fairlets is calculated in a two-step process:

1. First, the optimal fairlet decomposition $u^\star \in \mathcal{U}$ is obtained by:

$$u^\star = \arg\min_{u \in \mathcal{U}} L_f(u),$$

2. Then, the optimal clustering is obtained by minimizing the factorized loss given $u^\star$:

$$C_{\text{FCF}} = \arg\min_{C} L_c(C \mid u^\star) = \arg\min_{C:|C|=K} \sum_{k=1}^{K} \sum_{i \in C_k} D(u_i^\star, u_{(k)}^\star).$$

This completes the proof. □

For the purpose of point estimation, we argue that C_{MAP} should be preferred over C_{FCF}, since it is obtained via maximizing the posterior with respect to the fairlet decomposition and clustering configuration simultaneously. The C_{FCF} estimator, on the other hand, should be considered as an approximation of our C_{MAP} estimator, as C_{FCF} is practically obtained via the following two-step process:

1. Solve:

$$u^\star = \arg\max_{u \in \mathcal{U}} \exp(-\lambda_f L_f(u)),$$

2. Then, given the optimal fairlet decomposition $u^\star$ of the observed data, solve:

$$C_{\text{FCF}} = \arg\max_{C} \prod_{k=1}^{K} \exp\left(-\lambda_c \sum_{i \in C_k} D(u_i^*, u_{(k)}^*) \right).$$

By numerical studies, one can see that the maximum a posteriori estimate of the proposed fair k-means clustering via the Gibbs posterior turns out to be very similar to the fair clustering with fairlets estimator. The proposed Gibbs posterior-based approach, unlike a traditional optimization-based approach [123], also provides uncertainty quantification associated with the clustering configurations, along with the maximum a posteriori estimate.

5.2 Bayesian clustering of high-dimensional data

Following [153] (see also references therein) we note that Model-based clustering of moderate (or large-dimensional data) is a challenging task, especially in the presence of model misspecification. In [153] the authors propose a model for simultaneous dimensionality reduction and clustering by assuming a mixture model for a set of latent scores, which are then linked to the observations through a Gaussian latent factor model.

To address this issue, a repulsive point process is introduced prior for the component-specific means of the mixture for the latent scores, leading to a more robust model that outperforms the standard mixture model in several simulated scenarios. This repulsive point process is specifically designed to be anisotropic to favor well-separated clusters of data.

To connect this with Gibbs measures, we interpret the repulsive point process as a Gibbs point process, where the interactions between points (i.e., latent scores) are governed by a potential function that favors well-separated clusters. The Gibbs measure framework allows for a probabilistic interpretation of the repulsion between points, and its properties ensure that the posterior distribution is well-behaved, allowing for efficient sampling and inference. A general construction for anisotropic determinantal point processes provides an efficient and interpretable mechanism for clustering in this setting.

Bayesian clustering via latent mixtures. Let $y = (y_1, \ldots, y_n)$, $y_i \in \mathbb{R}^p$, $\Lambda \in \mathbb{R}^{p\times d}$ be the matrix of factor loadings, $\eta_1, \ldots, \eta_n \in \mathbb{R}^d$ be a set of latent factors, and $\Psi = \text{diag}(\sigma_1^2, \ldots, \sigma_p^2)$, $\sigma_j > 0$, be a diagonal covariance matrix. Let $\mathcal{N}_p(a, B)$ denote the p-dimensional Gaussian distribution with mean a and covariance matrix B. A latent factor mixture model assumes that:

$$y_i \Big| \eta_i, \Lambda, \Psi \overset{\text{ind}}{\sim} \mathcal{N}_p\left(\Lambda\eta_i, \Psi\right), \quad \eta_i \Big| w, \mu, \Gamma \overset{\text{i.i.d.}}{\sim} \sum_{h=1}^{m} w_h \mathcal{N}_d\left(\mu_h, \Gamma_h\right) \tag{5.7}$$

Extensions to other kernel distributions for the mixture model for latent factors and observations are straightforward. The model is completed by assigning priors to $w = (w_1, \ldots, w_m)$, $\mu = (\mu_1, \ldots, \mu_m)$, $\Gamma = (\Gamma_1, \ldots, \Gamma_m)$, and σ_j^2. In repulsive mixtures m is also random, allowing the number of components to be learned from the data.

Introducing latent cluster indicators c_i with $\Pr(c_i = h \mid w) = w_h$, the prior for η_i in (5.7) can be restated as $\eta_i \mid c_i = h, \mu, \Gamma \overset{\text{ind}}{\sim} \mathcal{N}_d(\mu_h, \Gamma_h)$. Thus, clustering of y_i is inherited from the latent η_i: y_i and y_j belong to the same cluster if $c_i = c_j$. Clusters correspond to allocated components $h \in \{1, \ldots, m\}$ where $c_i = h$ for some i.

Repulsive priors and Gibbs measures. Well-separated clusters require a prior that encourages regular configurations of cluster centers μ_h. Repulsive mixture models achieve this by specifying a Gibbs measure as the prior for μ_h. A Gibbs measure, defined via an energy function encoding interactions between points, assigns higher probability to configurations where points satisfy spatial regularity. Specifically, the prior for $\{\mu_1, \ldots, \mu_m\}$ is modeled as a pairwise interaction Gibbs process with density relative to a Poisson point process:

$$p(\{\mu_1, \ldots, \mu_m\}) = \frac{1}{Z} \prod_{j=1}^{m} \phi_1(\mu_j) \prod_{1 \leq h < k \leq m} \phi_2(\|\mu_h - \mu_k\|), \tag{5.8}$$

where ϕ_1 (bounded) and ϕ_2 (nondecreasing) define attractive and repulsive interactions, respectively, and Z is the intractable normalizing constant (partition function) of the Gibbs measure. This formulation ensures that configurations with closely spaced μ_h are penalized, promoting separation. The connection to Gibbs measures highlights the role of ϕ_2 as an interaction potential in the energy function $-\log p(\{\mu_h\})$, which governs the spatial arrangement of clusters.

However, ensuring separation in μ_h does not guarantee well-separated data clusters. By the Gaussian structure,

$$\{y_i : c_i = h\} \mid c, \mu, \Lambda, \Psi \overset{\text{i.i.d.}}{\sim} \mathcal{N}_p(\Lambda\mu_h, \Lambda\Gamma_h\Lambda^\top + \Psi),$$

the effective separation between clusters depends on $\|\Lambda(\mu_h - \mu_k)\|$, not merely $\|\mu_h - \mu_k\|$. Thus, repulsion must be carefully calibrated to Λ. Modifying (5.8) to $\phi_2(\|\Lambda(\mu_h - \mu_k)\|)$ introduces Λ-dependent repulsion, but the normalization constant Z becomes a function of Λ, exacerbating computational challenges.

Posterior computation and Gibbs measures. Sampling from the posterior under this Gibbs measure prior requires addressing the intractable Z. The exchange algorithm (see [277]), which circumvents Z via auxiliary variable sampling, has been used for such models (see [34]). However, preliminary experiments show poor mixing when updating high-dimensional Λ, as the exchange step relies on perfect sampling from (5.8), which becomes inefficient in multi-dimensional settings.

In [153] the authors define an anisotropic determinantal point process (DPP) as the prior for the component-specific latent factor centers $\{\mu_1, \ldots, \mu_m\}$, where the matrix of factor loadings drives the anisotropy.

This DPP on $\{\mathbb{R}^d, \mathcal{B}(\mathbb{R}^d)\}$, (where $\mathcal{B}(\mathbb{R}^d)$ denotes the Borel σ-algebra), defined as a random subset $\{\mu_1, \ldots, \mu_m\} \subset \mathbb{R}^d$ characterized by a continuous complex-valued covariance kernel $K : \mathbb{R}^d \times \mathbb{R}^d \to \mathbb{C}$. The probability distribution of a DPP is uniquely determined by K through its m-factorial moment measure (see [153] and references therein).

While the general definition involves technicalities, we focus on DPPs restricted to a compact region $R \subset \mathbb{R}^d$, where the density with respect to a unit-rate Poisson process is expressible via Mercer's theorem (see e.g. [154, Chapter 3]). This theorem guarantees a spectral decomposition of the kernel:

$$K(x, y) = \sum_{j \geq 1} \gamma_j \xi_j(x) \overline{\xi_j(y)},$$

where $\{\xi_j\}$ form an orthonormal basis of $L^2(R; \mathbb{C})$, and $\{\gamma_j\}$ are nonnegative, summable eigenvalues. For $\gamma_j < 1$, the DPP density on R is:

$$p(\{\mu_1, \ldots, \mu_m\}) = e^{|R|-D} \det \{C(\mu_h, \mu_k)\}_{h,k=1,\ldots,m},$$

where

$$C(x, y) = \sum_{j \geq 1} \frac{\gamma_j}{1 - \gamma_j} \xi_j(x) \overline{\xi_j(y)}, \quad D = -\sum_{j \geq 1} \log(1 - \gamma_j), \quad |R| = \int_R dx. \tag{5.9}$$

The determinant $\det\{C(\mu_h, \mu_k)\}$ enforces repulsion by vanishing as points μ_h and μ_k approach each other, a hallmark of DPPs.

DPPs share conceptual ties with Gibbs measures, which model particle systems with interaction energies. While Gibbs measures typically specify interactions via explicit potentials (e.g., pairwise repulsion $U(\mu_h - \mu_k)$), DPPs encode repulsion globally through the determinant, which arises from the antisymmetric nature of fermionic systems in quantum mechanics. The density $p(\{\mu_1, \ldots, \mu_m\})$ can be interpreted as a Gibbs distribution with Hamiltonian $H(\{\mu_h\}) = -\log \det\{C(\mu_h, \mu_k)\}$, where the determinant encapsulates a many-body interaction structure. Unlike classical Gibbs measures, DPPs offer tractable normalization constants ($e^{|R|-D}$) due to their determinantal form, avoiding the computational hurdles of intractable partition functions common in Gibbsian models ([34], [154, Chapter 18]).

The following are results of [153]. An anisotropic DPP is constructed on $\mathbb{R}^d$ using a stationary kernel $K_0(x - y)$, parameterized by a matrix $\Sigma \in \mathbb{R}^{p \times d}$, a random scale $W > 0$, and intensity ρ. The kernel's Fourier transform $\phi(x)$ is derived explicitly, ensuring spectral positivity for valid DPP existence. Moreover, it is shown that the anisotropic DPP is equivalent to an isotropic DPP on the transformed space $\widetilde{\mu} = \Sigma\mu$, with repulsion governed by W. This equivalence underscores how anisotropy in repulsion (guided by Σ) mirrors spatially varying interaction ranges in anisotropic Gibbs measures.

Gaussian-like DPP: For $W = |\Sigma^T\Sigma|^{1/d}c^{-2/d}$, the kernel becomes:

$$K_0(x) = \rho \exp\left(-\frac{\|\Sigma x\|^2}{2|\Sigma^T\Sigma|^{1/d}c^{-2/d}}\right),$$

with Fourier transform

$$\phi(x) = \frac{\rho}{c}(2\pi)^{d/2} \exp\left(-2\pi^2|\Sigma^T\Sigma|^{1/d}c^{-2/d}x^T(\Sigma^T\Sigma)^{-1}x\right).$$

The intensity bound $\rho_{\max} = c(2\pi)^{-d/2}$ ensures the DPP's existence. The pair correlation function $g(x) = 1 - \exp\left(-\frac{\|\Sigma x\|^2}{|\Sigma^T\Sigma|^{1/d}c^{-2/d}}\right)$ quantifies anisotropic repulsion, analogous to orientation-dependent potentials in Gibbs frameworks.

In [153], the anisotropic repulsive point process latent mixture (APPLAM) model is introduced. The APPLAM model employs the anisotropic DPP as a prior for latent cluster centers $\{\mu_h\}$, inducing repulsion akin to a Gibbs measure with hard-core interactions.

The key point in this model is the prior specification for the latent factor centers $\{\mu_1, \ldots, \mu_m\}$. To introduce repulsiveness between the points $\{\Lambda\mu_h\}_{h=1}^m$, we first fix a compact $R \subset \mathbb{R}^d$, and assume that the locations $\{\mu_1, \ldots, \mu_m\}$ are distributed as an anisotropic DPP on R, conditioned on $m \geqslant 1$. Conditioning on the DPP being nonempty is necessary for this formulation of the model, as (5.7) is not well defined if $m = 0$. Therefore, consider the following density for the locations $\{\mu_1, \ldots, \mu_m\}$ and their cardinalities:

$$p(\{\mu_1, \ldots, \mu_m\} \mid \Lambda) = \frac{e^{|R|-D}}{1-e^{-D}} \det\{C(\mu_h, \mu_k)\}_{h,k=1,\ldots,m}, \tag{5.10}$$

where $m \geqslant 1, \mu_1, \ldots, \mu_m \in R$, and $p(\emptyset \mid \Lambda) = 0$, with D and C defined in (5.9).

Conditioning on $m \geq 1$, the density (5.10) ensures well-separated μ_h, critical for interpretable clusters. The prior structure parallels Gibbsian models where repulsive potentials discourage particle overlap, but with computational advantages from the DPP's analytic normalizer.

The compact support $R = [-10, 10]^d$ and parameterization $\rho = s\rho_{\max}$ (for $s \in (0,1)$) balance intensity and repulsion strength. The repulsion range, governed by c and Σ, aligns with interaction radii in Gibbs measures, while the global repulsiveness index $p_0 = \rho\pi^{d/2}c^{-1}$ reflects system-wide exclusion, similar to mean-field approximations.

Thus by framing DPPs within the Gibbs measure paradigm, we highlight their role as analytically tractable repulsive point processes. The anisotropic extension generalizes traditional isotropic Gibbs interactions, enabling direction-sensitive repulsion crucial for high-dimensional latent variable models. This connection enriches the understanding of DPPs as specialized Gibbs measures with inherently global, determinantal interactions.

5.3 Unsupervised learning with Gibbs measures

In unsupervised learning tasks, where the goal is to discover underlying patterns and relationships in data without labeled information, Gibbs measures play a crucial role in providing a probabilistic framework for modeling data. By incorporating latent variables and hierarchical structures, Gibbs measures enable the exploration of hidden patterns and dependencies within the data, leading to enhanced understanding and predictive capabilities.

Unsupervised learning represents a fundamental machine learning paradigm where algorithms are provided with unlabeled data and tasked with identifying patterns or structures within the data. To establish the foundation for understanding how Gibbs measures contribute to this field, we first present the core definitions, models, and principles of unsupervised learning.

5.3.1 *Problem setup and main objectives*

Consider a dataset $\mathcal{X} = \{x_1, x_2, \ldots, x_N\}$ consisting of N samples, where each $x_i \in \mathbb{R}^d$ is a d-dimensional vector. In unsupervised learning, the primary goal is to learn the underlying structure or distribution $p(x)$ of the data without any explicit labels.

The objectives in unsupervised learning encompass several interconnected tasks. Clustering involves partitioning the data into groups based on similarity, as discussed in previous sections. Dimensionality reduction seeks to reduce the dimensionality of the data while preserving its intrinsic structure. Density estimation focuses on estimating the probability density $p(x)$ underlying the data. Finally, generative modeling aims to learn a model that can generate new samples similar to the given data.

5.3.2 *Clustering approaches*

We recall the clustering tasks that were discussed in the previous sections as special cases of unsupervised learning. Two fundamental clustering approaches demonstrate the breadth of unsupervised learning techniques. K-means clustering aims to partition the data into k clusters by minimizing the within-cluster variance:

$$\min_{\{C_1,\ldots,C_k\}} \sum_{i=1}^{k} \sum_{x \in C_i} \|x - \mu_i\|^2,$$

where C_i represents the set of points in the i-th cluster and $\mu_i = \frac{1}{|C_i|} \sum_{x \in C_i} x$ is the centroid of C_i.

A more sophisticated probabilistic approach is provided by Gaussian mixture models (GMMs), which assume the data is generated from a mixture of Gaussian distributions:

$$p(x) = \sum_{k=1}^{K} \pi_k \mathcal{N}(x \mid \mu_k, \Sigma_k),$$

where π_k are the mixing coefficients satisfying $\sum_k \pi_k = 1$ and $\mathcal{N}(x \mid \mu_k, \Sigma_k)$ is the Gaussian distribution with mean μ_k and covariance Σ_k.

The parameters of GMMs are learned using the expectation-maximization (EM) algorithm, which alternates between two steps. The E-step estimates the posterior probabilities $r_{ik} = p(z_i = k \mid x_i)$, while the M-step updates the parameters (μ_k, Σ_k, π_k) by maximizing the expected log-likelihood.

5.3.3 *Dimensionality reduction techniques*

Dimensionality reduction techniques serve to discover lower-dimensional representations of high-dimensional data. Principal component analysis (PCA) finds such a representation by projecting the data onto the directions of maximum variance:

$$\max_{W \in \mathbb{R}^{d \times k}} \operatorname{tr}(W^\top \Sigma W), \quad \text{subject to } W^\top W = I_k,$$

where Σ is the covariance matrix of the data, and W contains the top k eigenvectors of Σ.

Autoencoders provide a neural network-based approach to dimensionality reduction. These networks are trained to reconstruct their input through a bottleneck layer:

$$\hat{x} = g(f(x; \theta); \phi),$$

where $f(x; \theta)$ is the encoder, $g(z; \phi)$ is the decoder, and the objective is to minimize the reconstruction loss:

$$\min_{\theta, \phi} \frac{1}{N} \sum_{i=1}^{N} \|x_i - \hat{x}_i\|^2.$$

5.3.4 *Density estimation methods*

Density estimation techniques aim to model the underlying probability distribution of the data. Kernel density estimation (KDE) provides a non-parametric approach by estimating the probability density function $p(x)$ as:

$$\hat{p}(x) = \frac{1}{Nh^d} \sum_{i=1}^{N} K\left(\frac{x - x_i}{h}\right),$$

where K is a kernel function (e.g., Gaussian), and $h > 0$ is the bandwidth parameter.

Variational autoencoders (VAEs) represent a more sophisticated probabilistic approach to density estimation. VAEs are probabilistic generative models that maximize a variational lower bound on the log-likelihood of the data:

$$\log p(x) \geq \mathbb{E}_{q(z|x)} [\log p(x \mid z)] - D_{\mathrm{KL}}(q(z \mid x) \| p(z)),$$

where $q(z \mid x)$ is the variational posterior, $p(z)$ is the prior, and D_{KL} denotes the Kullback-Leibler divergence.

5.3.5 *Generative models*

Generative models form a crucial category of unsupervised learning techniques that can create new data samples. Generative adversarial networks (GANs) represent a particularly influential approach, consisting of two competing neural networks. The generator $G(z; \theta_g)$ maps noise $z \sim p(z)$ to data, while the discriminator $D(x; \theta_d)$ distinguishes real data from generated data. The training objective is formulated as a min-max optimization problem:

$$\min_G \max_D \mathbb{E}_{x \sim p_{\mathrm{data}}} [\log D(x)] + \mathbb{E}_{z \sim p(z)} [\log(1 - D(G(z)))].$$

Restricted Boltzmann machines (RBMs) provide another important class of generative models based on energy functions. RBMs are energy-based models with a joint probability distribution:

$$p(x, h) = \frac{1}{Z} e^{-E(x,h)},$$

where $E(x, h)$ is the energy function, Z is the partition function, and h are hidden variables. This formulation naturally connects to the Gibbs measure framework, providing a bridge between classical statistical mechanics and modern machine learning approaches.

5.3.6 *The role of Gibbs measures*

Gibbs measures lie at the heart of many unsupervised learning techniques, providing a principled link between energy functions, probabilistic reasoning, and structured dependencies. In energy-based models (EBMs), each configuration x is assigned a scalar energy $E_\theta(x)$, and the corresponding Gibbs distribution

$$P_\theta(x) = \frac{e^{-E_\theta(x)}}{Z_\theta}, \qquad Z_\theta = \int e^{-E_\theta(x)}\, dx$$

defines a flexible family of probability densities over the data space. Lower-energy states receive higher probability, making Gibbs measures an ideal framework for capturing complex data structures [233].

The learning objective is to adjust θ so that $P_\theta(x)$ approximates the true data distribution $P_{\text{data}}(x)$. Equivalently, one maximizes the log-likelihood

$$\mathcal{L}(\theta) = \mathbb{E}_{x\sim P_{\text{data}}}\big[-E_\theta(x) - \log Z_\theta\big],$$

or, minimizing the Kullback–Leibler divergence

$$\text{KL}(P_{\text{data}} \| P_\theta) = \mathbb{E}_{x\sim P_{\text{data}}}\big[\log P_{\text{data}}(x) - \log P_\theta(x)\big].$$

A major computational hurdle is the intractability of Z_θ in high dimensions. Gibbs sampling, a Markov Chain Monte Carlo method, addresses this by iteratively resampling each coordinate x_i from its conditional distribution

$$P(x_i \mid x_{-i}) = \frac{\exp\big(-E_\theta(x_i, x_{-i})\big)}{\displaystyle\int \exp\big(-E_\theta(x_i', x_{-i})\big)\, dx_i'}.$$

Moreover, introducing latent variables h enriches the model's expressivity:

$$P_\theta(x,h) = \frac{e^{-E_\theta(x,h)}}{Z_\theta}, \quad P_\theta(x) = \sum_h \frac{e^{-E_\theta(x,h)}}{Z_\theta},$$

and gives rise to the free energy

$$F(x;\theta) = -\log \sum_h e^{-E_\theta(x,h)},$$

which serves as a surrogate energy for marginalizing over hidden structures. Minimizing free energy aligns the model with observed data while implicitly regularizing complexity.

Practical applications of this framework abound. LeCun et al. [233] laid the foundations of EBMs, elucidating how Gibbs measures underpin the modeling of dependencies in unsupervised learning. Ranzato et al. [308] further demonstrated how energy-based density models and the associated Gibbs sampling algorithms can derive efficient codes and restore corrupted data, illustrating both theoretical depth and algorithmic utility.

5.4 Normalizing flows

NFs are a powerful framework in machine learning for building complex probability distributions through a sequence of invertible transformations applied to simple base distributions [297,311]. Gibbs measures have notable connections to NFs when modeling distributions with specific constraints or regularities.

Let us explore the theory linking Gibbs measures and normalizing flows. Transform a simple distribution $p_z(z)$ (for example, Gibbs or Gaussian) into a complex distribution $p_\theta(x)$ using an invertible function f_θ.

The following is a standard formula for the change of variables

$$p_\theta(x) = p_z(f_\theta(x)) \cdot \left| \det \frac{\partial f_\theta}{\partial x} \right|.$$

The problem is to maximize:

$$\mathcal{L}(\theta) = \frac{1}{N} \sum_{i=1}^{N} \left[\log p_z(f_\theta(x_i)) + \log \left| \det \frac{\partial f_\theta}{\partial x} \right| \right].$$

In the RealNVP (Real-valued Non-Volume Preserving) model, a type of normalizing flow, the transformation between variables is designed to be invertible and efficient for both forward and reverse computations. The transformation is achieved using a series of affine transformations that split the input data into two parts. The key idea is to apply an invertible transformation to part of the data while keeping the rest unchanged, making it computationally efficient for Jacobian computation and likelihood estimation.

The transformation equations are as follows:

$$\mathbf{x}'_{1:d} = \mathbf{x}_{1:d},$$

$$\mathbf{x}'_{d+1:D} = \mathbf{x}_{d+1:D} \odot \exp(s(\mathbf{x}_{1:d})) + t(\mathbf{x}_{1:d}),$$

where

- $\mathbf{x} = (\mathbf{x}_{1:d}, \mathbf{x}_{d+1:D})$ represents the input data vector, split into two parts: $\mathbf{x}_{1:d}$ and $\mathbf{x}_{d+1:D}$.

- $\mathbf{x}' = (\mathbf{x}'_{1:d}, \mathbf{x}'_{d+1:D})$ is the transformed output vector.

- $s(\mathbf{x}_{1:d})$ and $t(\mathbf{x}_{1:d})$ are learnable scalar functions, often parameterized by neural networks. These functions are applied to the first part of the data $\mathbf{x}_{1:d}$.

- $\odot$ denotes the element-wise (Hadamard) product.

- $\exp(s(\mathbf{x}_{1:d}))$ ensures invertibility by guaranteeing positivity for scaling factors.

The transformation is invertible because the mapping for $\mathbf{x}'_{d+1:D}$ can be reversed:

$$\mathbf{x}_{d+1:D} = (\mathbf{x}'_{d+1:D} - t(\mathbf{x}_{1:d})) \odot \exp(-s(\mathbf{x}_{1:d})).$$

The Jacobian matrix for the transformation is triangular, making the log-determinant computation straightforward:

$$\log\det\left|\frac{\partial \mathbf{x}'}{\partial \mathbf{x}}\right| = \sum_{i=d+1}^{D} s(\mathbf{x}_{1:d})_i.$$

The flexibility of $s(\cdot)$ and $t(\cdot)$ enables the model to adapt to complex data distributions.

RealNVP typically applies multiple such transformations in a hierarchical manner to model complex, high-dimensional distributions effectively. This framework makes RealNVP suitable for tasks such as density estimation and generative modeling, as it enables both efficient sampling (via inversion) and likelihood computation.

Consider configurations $x \in \mathcal{X}$, parameterized by a Hamiltonian $H(x)$.

In machine learning, $H(x)$ often encodes constraints or specific properties of the data. Gibbs measures serve as a natural target for NFs in applications where constraints or equilibrium properties are essential.

Learning Gibbs measures with NFs:

- The flow learns a transformation f that maps the base distribution $p_u(z)$ to the Gibbs measure $\mu(x)$.

- The log-likelihood of the Gibbs measure is expressed as:

$$\log \mu(x) = \log p_u(f^{-1}(x)) - \log|\det J_f^{-1}(x)|,$$

where J_f is the Jacobian matrix of f.

Note that NFs provide efficient sampling from Gibbs measures since the invertible mappings enable direct sampling and density estimation.

As an applications of Gibbs measures in NFs one can see an image and signal processing, where image denoising and super-resolution tasks often require sampling from a Gibbs measure to ensure that the solution is consistent with physical or statistical constraints.

Incorporating constraints into machine learning models (e.g., smoothness, sparsity) can be achieved by designing Gibbs-like target distributions.

The partition function (given in definition of Gibbs measure) is typically intractable for high-dimensional systems, but normalizing flows sidestep this by directly learning the transformation between base and target distributions.

When the target Gibbs measure has sharp energy landscapes, flow-based models may struggle with optimization.

Non-normalized densities [194]:

In many cases, probabilistic models in machine learning, statistics, or signal processing are defined in terms of non-normalized probability densities. These models involve an unknown normalization constant, the computation of which is often infeasible in practice.

Assume we observe a random vector $\mathbf{x} \in \mathbb{R}^n$ with a probability density function denoted by $p_{\mathbf{x}}(\cdot)$. We consider a parametrized density model $p(\cdot;\boldsymbol{\theta})$, where $\boldsymbol{\theta} \in \mathbb{R}^m$ is a vector of parameters. The objective is to estimate the parameter $\boldsymbol{\theta}$ from the data $\mathbf{x}$, i.e., to approximate $p_{\mathbf{x}}(\cdot)$ by $p(\cdot;\hat{\boldsymbol{\theta}})$, where $\hat{\boldsymbol{\theta}}$ is the estimated parameter value. Consider the case of continuous-valued random variables.

The problem arises when the probability density function provided by the model can only be computed up to a multiplicative constant $Z(\boldsymbol{\theta})$:

$$p(\boldsymbol{\xi};\boldsymbol{\theta}) = \frac{1}{Z(\boldsymbol{\theta})} q(\boldsymbol{\xi};\boldsymbol{\theta}),$$

where $q(\boldsymbol{\xi};\boldsymbol{\theta})$ is a known function, but the normalization constant $Z(\boldsymbol{\theta})$ is difficult to compute. The constant $Z(\boldsymbol{\theta})$ is given by the integral:

$$Z(\boldsymbol{\theta}) = \int_{\mathbb{R}^n} q(\boldsymbol{\xi};\boldsymbol{\theta})\, d\boldsymbol{\xi},$$

which is often analytically intractable. For higher dimensions (specifically, for almost any $n > 2$), numerical computation of this integral is also practically infeasible.

Typically, one has to resort to Markov Chain Monte Carlo methods, or approximations of the normalization constant. In [194] the author proposes that such models can be estimated by minimizing the expected squared distance between the gradient of the log-density given by the model and the gradient of the log-density of the observed data. Let us give some details of this work.

Estimation by score matching. In the following, we extensively use the gradient of the log-density with respect to the data vector. For simplicity, we call this the score function. For the model density, the score function is denoted by $\boldsymbol{\psi}(\boldsymbol{\xi};\boldsymbol{\theta})$ and is defined as follows:

$$\boldsymbol{\psi}(\boldsymbol{\xi};\boldsymbol{\theta}) = \begin{pmatrix} \frac{\partial \log p(\boldsymbol{\xi};\boldsymbol{\theta})}{\partial \xi_1} \\ \vdots \\ \frac{\partial \log p(\boldsymbol{\xi};\boldsymbol{\theta})}{\partial \xi_n} \end{pmatrix} = \begin{pmatrix} \psi_1(\boldsymbol{\xi};\boldsymbol{\theta}) \\ \vdots \\ \psi_n(\boldsymbol{\xi};\boldsymbol{\theta}) \end{pmatrix} = \nabla_{\boldsymbol{\xi}} \log p(\boldsymbol{\xi};\boldsymbol{\theta}).$$

The key benefit of using the score function is that it does not depend on the normalization constant $Z(\boldsymbol{\theta})$. Specifically, we have:

$$\boldsymbol{\psi}(\boldsymbol{\xi};\boldsymbol{\theta}) = \nabla_{\boldsymbol{\xi}} \log q(\boldsymbol{\xi};\boldsymbol{\theta}).$$

Likewise, let $\boldsymbol{\psi}_x(\cdot) = \nabla_{\boldsymbol{\xi}} \log p_x(\cdot)$ denote the score function of the observed data distribution p_x. In principle, $\boldsymbol{\psi}_x(\cdot)$ could be estimated by computing the gradient of the logarithm of a non-parametric estimate of the pdf. However, as we will show below, no such computation is necessary.

Score functions are mappings from $\mathbb{R}^n$ to $\mathbb{R}^n$. We propose estimating the model by minimizing the expected squared distance between the model score function $\boldsymbol{\psi}(\cdot;\boldsymbol{\theta})$ and the data score function $\boldsymbol{\psi}_x(\cdot)$. The squared distance is defined as:

$$J(\boldsymbol{\theta}) = \frac{1}{2}\int_{\mathbb{R}^n} p_x(\boldsymbol{\xi}) \left\|\boldsymbol{\psi}(\boldsymbol{\xi};\boldsymbol{\theta}) - \boldsymbol{\psi}_x(\boldsymbol{\xi})\right\|^2 d\boldsymbol{\xi}.$$

Thus, the score matching estimator of $\boldsymbol{\theta}$ is given by:

$$\hat{\boldsymbol{\theta}} = \arg\min_{\boldsymbol{\theta}} J(\boldsymbol{\theta}).$$

The motivation for this estimator is that the score function $\boldsymbol{\psi}(\boldsymbol{\xi};\boldsymbol{\theta})$ can be directly computed from $q(\boldsymbol{\xi};\boldsymbol{\theta})$, as shown above, without requiring the computation of $Z(\boldsymbol{\theta})$. While this may seem to involve estimating the data score function $\boldsymbol{\psi}_x(\cdot)$, which would typically require non-parametric estimation, such computation is unnecessary. Using a simple partial integration trick, the objective function $J(\boldsymbol{\theta})$ can be computed easily, as shown in the following theorem:

Theorem 19. *[194] Assume the model score function $\boldsymbol{\psi}(\boldsymbol{\xi};\boldsymbol{\theta})$ is differentiable and satisfies weak regularity conditions:*

a. The data probability density function $p_x(\boldsymbol{\xi})$ is differentiable.

b. The expectations $\mathbb{E}_{\mathbf{x}}\left\{\|\boldsymbol{\psi}(\mathbf{x};\boldsymbol{\theta})\|^2\right\}$ and $\mathbb{E}_{\mathbf{x}}\left\{\|\boldsymbol{\psi}_x(\mathbf{x})\|^2\right\}$ are finite for any $\boldsymbol{\theta}$.

c. $p_x(\boldsymbol{\xi})\boldsymbol{\psi}(\boldsymbol{\xi};\boldsymbol{\theta}) \to 0$ for any $\boldsymbol{\theta}$ as $\|\boldsymbol{\xi}\| \to \infty$.

Then, the objective function $J(\boldsymbol{\theta})$ can be expressed as:

$$J(\boldsymbol{\theta}) = \int_{\mathbb{R}^n} p_x(\boldsymbol{\xi}) \left(\sum_{i=1}^{n} \partial_i \psi_i(\boldsymbol{\xi}; \boldsymbol{\theta}) + \frac{1}{2} \psi_i(\boldsymbol{\xi}; \boldsymbol{\theta})^2 \right) d\boldsymbol{\xi} + const,$$

where the constant does not depend on $\boldsymbol{\theta}$. Here:
- $\psi_i(\boldsymbol{\xi}; \boldsymbol{\theta}) = \frac{\partial \log q(\boldsymbol{\xi}; \boldsymbol{\theta})}{\partial \xi_i}$ is the i-th element of the model score function.
- $\partial_i \psi_i(\boldsymbol{\xi}; \boldsymbol{\theta}) = \frac{\partial \psi_i(\boldsymbol{\xi}; \boldsymbol{\theta})}{\partial \xi_i} = \frac{\partial^2 \log q(\boldsymbol{\xi}; \boldsymbol{\theta})}{\partial \xi_i^2}$ is the partial derivative of the i-th element of the model score function with respect to ξ_i.

In practice, we have T observations of the random vector $\mathbf{x}$, denoted by $\mathbf{x}(1), \ldots, \mathbf{x}(T)$. The sample version of J is naturally derived from Theorem 19 as:

$$\tilde{J}(\theta) = \frac{1}{T} \sum_{t=1}^{T} \sum_{i=1}^{n} \left(\partial_i \psi_i(\mathbf{x}(t); \theta) + \frac{1}{2} \psi_i(\mathbf{x}(t); \theta)^2 \right) + \text{const.} \tag{5.11}$$

By the law of large numbers, $\tilde{J}(\theta)$ is asymptotically equivalent to $J(\theta)$. To estimate the model, we propose minimizing $\tilde{J}(\theta)$ in the case of a finite sample.

One may wonder whether minimizing $J(\theta)$ is sufficient for estimating the model, or whether the score functions can be zero for different parameter values. If the model is degenerate, such that two different values of θ yield the same probability density function, then it is impossible to estimate θ. However, under the assumption that the model is non-degenerate and that $q > 0$ everywhere, the following theorem ensures local consistency:

Theorem 20 (194). *Assume that the probability density function of $\mathbf{x}$ follows the model $p_{\mathbf{x}}(\cdot) = p(\cdot; \theta^*)$ for some θ^*. Further assume that no other parameter value yields the same probability density function, i.e., $p(\cdot; \theta) \neq p(\cdot; \theta^*)$ for $\theta \neq \theta^*$, and that $q(\xi; \theta) > 0$ for all ξ and θ. Then:*

$$J(\theta) = 0 \iff \theta = \theta^*.$$

Under the assumptions of theorems, the score-matching estimator obtained by minimizing $\tilde{J}(\theta)$ is consistent. That is, the estimator converges in probability to the true value of θ as the sample size approaches infinity, provided the optimization algorithm identifies the global minimum. This follows from the law of large numbers. As the sample size grows, $\tilde{J}(\theta)$ converges to $J(\theta)$ in probability. Therefore, the estimator converges to a point where $J(\theta)$ is globally minimized. By Theorem 20, the global minimum is unique and occurs at θ^*. Since $J(\theta) \geq 0$, the global minimum satisfies $J(\theta^*) = 0$.

This result assumes that the global minimum of $\tilde{J}(\theta)$ is found by the optimization algorithm. In practice, this assumption may not hold due to the existence of multiple local minima. Thus, the consistency is local, meaning that the estimator is consistent if the optimization process is initialized sufficiently close to the true value θ^*. Note that consistency implies asymptotic unbiasedness.

Example. Recall that a multivariate Gaussian density. The probability density function of a multivariate Gaussian (normal) distribution for a random vector $\mathbf{x} \in \mathbb{R}^n$ is given by:

$$p(\mathbf{x}) = \frac{1}{\sqrt{(2\pi)^n |\Sigma|}} \exp\left(-\frac{1}{2}(\mathbf{x} - \boldsymbol{\mu})^\top \Sigma^{-1}(\mathbf{x} - \boldsymbol{\mu})\right),$$

where $\boldsymbol{\mu}$ is the mean vector, Σ is the covariance matrix (positive definite), $|\Sigma|$ is the determinant of Σ.

The multivariate Gaussian probability density function can be expressed in a Gibbs-like form. If we identify the quadratic term in the exponent as the energy function, $H(\mathbf{x})$, we have

$$H(\mathbf{x}) = \frac{1}{2}(\mathbf{x} - \boldsymbol{\mu})^\top \Sigma^{-1}(\mathbf{x} - \boldsymbol{\mu}),$$

and the multivariate Gaussian density becomes

$$p(\mathbf{x}) \propto \exp(-H(\mathbf{x})).$$

Here, $\beta = 1$, and $H(\mathbf{x})$ is quadratic in $\mathbf{x}$.

Consider the probability density function of the multivariate Gaussian distribution given by:

$$p(\mathbf{x}; M, \boldsymbol{\mu}) = \frac{1}{Z(M, \boldsymbol{\mu})} \exp\left(-\frac{1}{2}(\mathbf{x} - \boldsymbol{\mu})^\top M(\mathbf{x} - \boldsymbol{\mu})\right),$$

where M is a symmetric positive-definite matrix (the inverse of the covariance matrix). While the expression for $Z(M, \boldsymbol{\mu})$ is well-known, this serves as an illustration of the method. For clarity, we use $\mathbf{x}$ as the general n-dimensional vector. In this case, we define:

$$q(\mathbf{x}) = \exp\left(-\frac{1}{2}(\mathbf{x} - \boldsymbol{\mu})^\top M(\mathbf{x} - \boldsymbol{\mu})\right),$$

and obtain

$$\psi(\mathbf{x}; M, \boldsymbol{\mu}) = -M(\mathbf{x} - \boldsymbol{\mu}),$$

with

$$\partial_i \psi(\mathbf{x}; M, \boldsymbol{\mu}) = -m_{ii}.$$

Thus, the objective function $\tilde{J}(M, \boldsymbol{\mu})$ becomes:

$$\tilde{J}(M, \boldsymbol{\mu}) = \frac{1}{T}\sum_{t=1}^{T}\left[\sum_{i} -m_{ii} + \frac{1}{2}(\mathbf{x}^{(t)} - \boldsymbol{\mu})^{\top} MM(\mathbf{x}^{(t)} - \boldsymbol{\mu})\right].$$

To minimize this function with respect to $\boldsymbol{\mu}$, we compute the gradient

$$\nabla_{\boldsymbol{\mu}} \tilde{J} = MM\boldsymbol{\mu} - MM\frac{1}{T}\sum_{t=1}^{T}\mathbf{x}^{(t)}.$$

This is zero if and only if $\boldsymbol{\mu}$ is the sample mean:

$$\boldsymbol{\mu} = \frac{1}{T}\sum_{t=1}^{T}\mathbf{x}^{(t)}.$$

This is indeed a minimum, as the matrix MM, which defines the quadratic form, is positive definite.

Next, we compute the gradient with respect to M:

$$\nabla_M \tilde{J} = -I + M\frac{1}{2T}\sum_{t=1}^{T}(\mathbf{x}^{(t)} - \boldsymbol{\mu})(\mathbf{x}^{(t)} - \boldsymbol{\mu})^{\top} + \frac{1}{2T}\left[\sum_{t=1}^{T}(\mathbf{x}^{(t)} - \boldsymbol{\mu})(\mathbf{x}^{(t)} - \boldsymbol{\mu})^{\top}\right] M.$$

This gradient is zero if and only if M is the inverse of the sample covariance matrix:

$$\frac{1}{T}\sum_{t=1}^{T}(\mathbf{x}^{(t)} - \boldsymbol{\mu})(\mathbf{x}^{(t)} - \boldsymbol{\mu})^{\top}.$$

Thus, this yields the score-matching estimate.

5.5 Conclusion

In this chapter, we have explored how Gibbs measures originating from statistical physics can be effectively leveraged for clustering and unsupervised learning tasks. Beginning with the foundational concept of clustering in machine learning, we examined how probabilistic models based on Gibbs distributions provide a principled framework for modeling uncertainty, structure, and diversity in data partitions. These models allow for both theoretical rigor and practical flexibility, enabling interpretations grounded in energy-based formulations.

We then considered the critical issue of fairness in clustering, highlighting how Gibbs-based frameworks can incorporate fairness constraints through prior or posterior modifications, especially in the Bayesian setting. The notion of generalized Bayesian fair clustering extends this further by balancing statistical fit with fairness-aware regularization, opening pathways to ethically grounded unsupervised learning.

The challenges of clustering high-dimensional data were addressed using Bayesian and variational techniques informed by Gibbs sampling and posterior inference, showing their strength in scaling and adapting to modern data regimes. Likewise, we discussed the role of Gibbs measures in broader unsupervised learning, including their theoretical properties such as stability and generalization bounds, which bridge learning theory and statistical physics.

Finally, we turned to generative models, where Gibbs distributions serve as a foundational tool in modeling complex data manifolds and in training models via energy-based and entropy-regularized objectives.

Together, these sections underscore a central theme: Gibbs measures offer a rich and unifying language for unsupervised learning, blending probabilistic reasoning with thermodynamic intuition. Looking ahead, key research directions include the development of efficient inference algorithms for large-scale Gibbs-based models, deeper integration with fairness and interpretability, and the use of these methods in dynamic or non-Euclidean data domains.

In the domain of unsupervised learning, clustering techniques remain fundamental yet widely utilized tools in machine learning. These methods aim to partition a set of objects such that similar objects belong to the same group or cluster. A precise formulation of the clustering problem requires defining two key components: a similarity measure to quantify relationships between objects and a cost function to evaluate the quality of the partition. Consequently, clustering naturally becomes an optimization problem, often classified as NP-hard [139].

A wide variety of clustering algorithms have been proposed to address this challenge. Classic approaches include expectation-maximization (EM) methods, such as k-means and k-centers [95], which iteratively refine clusters by assigning points to the nearest cluster centers (expectation step) and then updating these centers based on the mean of the assigned points (optimization step). These methods are computationally efficient but sensitive to initial conditions and often require a pre-specified number of clusters, denoted as k.

Beyond these, hierarchical clustering methods, including percolation-like algorithms, aim to build hierarchies of clusters without requiring prior knowledge of the number of clusters. Another powerful class of methods, known as spectral clustering [8], [285], leverages the spectral properties of matrices such as the affinity matrix or diffusion operators. These approaches interpret clusters as separated components of a diffusion process, providing a mathematical foundation for clustering based on graph theory. Notably, some spectral clustering techniques utilize advanced tools like the non-backtracking matrix, which arises from the stability of belief propagation algorithms [228].

From the perspective of statistical physics, clustering can often be reframed as finding the ground state of a q-states Potts model [324]. This formulation is particularly useful when the cost function is analogous to an energy function in physics. Monte Carlo methods [389], a staple of computational physics, are employed to approximate solutions, making the connection between statistical mechanics and clustering more explicit. For certain cases, this approach is complemented by Bethe-Peierls approximations, whose algorithmic counterpart is the belief propagation algorithm—a subject explored in depth in subsequent chapters.

An important alternative to traditional center-based clustering is the Affinity Propagation (AP) algorithm. This method identifies clusters by selecting a subset of data points, called exemplars, which best represent the other points in their clusters. AP solves the clustering problem as an instance of the min-sum algorithm [135], iteratively refining the assignment of data points to exemplars based on pairwise similarities. Compared to k-means and k-centers, AP is more robust, but its computational complexity is quadratic, which can limit its scalability. A notable limitation of AP is its assumption that clusters have a roughly spherical shape. This constraint has been relaxed in later works, such as [240], [242], where soft constraints replace hard self-exemplar assignments, thereby broadening the algorithm's applicability.

A critical aspect of clustering algorithms is determining the optimal number of clusters. Many approaches, such as those based on EM [134] or information-theoretic principles [350], address this by incorporating specific parametrizations of the cluster model. However, these methods rely on precise model assumptions, which can be restrictive. Alternative strategies have emerged, such as similarity-based statistical methods [108], which are computationally intensive but provide a model-independent framework.

For example, such strategies have been combined with AP [379], albeit at the cost of increased quadratic complexity.

Determining the number of clusters in a dataset is inherently ambiguous and often considered an ill-posed problem. This ambiguity stems from the lack of a natural scale for comparing solutions obtained for different values of k. More broadly, the question relates to the consistency of clustering algorithms when applied to a given dataset. Stability-based approaches to consistency have gained prominence; these methods assess clustering reliability by comparing results across independent samples of the data, computing stability measures [108]. Parameters are chosen to optimize a stability criterion, although this does not always correspond to the best clustering from a theoretical perspective.

The question of scale invariance is central to clustering theory and remains a subject of ongoing research. Cost functions and quality measures must exhibit scale invariance [2], [217]. ensuring that clustering results are independent of arbitrary changes in measurement units or scaling factors. This requirement parallels challenges in statistical physics, where models near phase transition points are expected to exhibit insensitivity to irrelevant microscopic details while capturing large-scale behavior. Such principles guide the development of clustering algorithms that are robust, consistent, and theoretically grounded.

Chapter 6

Gibbs Measures for Structured Prediction and Graphical Models

This chapter explores the interplay between Gibbs measures, graphical models, and structured prediction in machine learning. It begins by introducing key definitions and techniques related to graphical models, illustrating how Gibbs measures underpin their probabilistic structure. The discussion then turns to structured prediction, presenting foundational concepts and algorithms such as the structured perceptron. The chapter concludes with practical applications, including lattice-based models and multivariate discrete data modeling via the Potts-Ising framework, demonstrating the effectiveness of Gibbs-based methods in capturing complex dependencies in structured data.

6.1 Graphical models and Gibbs measures

6.1.1 *Definitions and methods*

A graphical model (see [201, 219, 378] and the references therein) is represented by a graph $G = (V, E)$, where, V is a set of nodes (or vertices), each representing a random variable. E is a set of edges, which encode the dependencies between the variables.

To define a graphical model, we associate a random variable X_s with each vertex $s \in V$, where V is the set of vertices in the graph. The random variable X_s takes values in some state space $\mathcal{X}_s$, which can vary depending on the specific application. The state space $\mathcal{X}_s$ may be either continuous (e.g., $\mathcal{X}_s = \mathbb{R}$, the real numbers) or discrete (e.g., $\mathcal{X}_s = \{0, 1, \ldots, r-1\}$, a finite set of integers).

We use lowercase letters (e.g., $x_s \in \mathcal{X}_s$) to denote particular values of the random variable X_s. The notation $\{X_s = x_s\}$ then refers to the event that the random variable X_s takes the specific value $x_s \in \mathcal{X}_s$. This is a

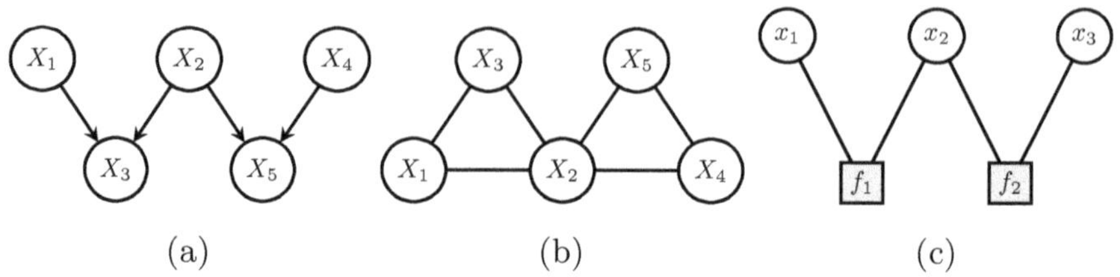

Fig. 6.1: Comparison of graphical model representations: (a) directed acyclic graph, (b) undirected graph, and (c) factor graph representation.

typical way to describe an outcome of the random variable in probability theory.

For any subset $A \subseteq V$ of the vertex set, we define the subvector X_A as the collection of random variables X_s for all vertices $s \in A$. Formally, $X_A := (X_s)_{s \in A}$, which is simply the vector of random variables corresponding to the vertices in A. The vector of particular values taken by these random variables is denoted by x_A, where $x_A := (x_s)_{s \in A}$, which corresponds to the vector of values $x_s \in \mathcal{X}_s$ for each $s \in A$. This notation is used when referring to the specific realization of the random variables associated with the set A.

Additionally, for a set $A \subseteq V$, we define the Cartesian product of the state spaces $\mathcal{X}_s$ for each $s \in A$ as $\bigotimes_{s \in A} \mathcal{X}_s$, which is the set of all possible tuples $(x_s)_{s \in A}$ where $x_s \in \mathcal{X}_s$ for each $s \in A$. This Cartesian product represents the space of all possible joint values that the random variables X_s for $s \in A$ can take. The types of edges differ between the two main graphical models.

Directed graph (Bayesian network). An edge $(X_i \to X_j)$ indicates that X_j is conditionally dependent on X_i given its parents. Following [378], suppose that $G = (V, E)$ is a directed acyclic graph (DAG), which is a graph where every edge has a direction, and it contains no directed cycles. That is, there is no way to start at a vertex, follow the directed edges, and return to the same vertex.

For any such DAG, we define a partial order on the vertex set V using the concept of ancestry. Specifically, we say that a vertex s is an ancestor of vertex u, denoted by $s \preceq u$, if there exists a directed path from s to u. A directed path is a sequence of vertices $s = t_0, t_1, t_2, \ldots, t_k, u$ such that for each pair of consecutive vertices t_i and t_{i+1}, there is a directed edge from t_i to t_{i+1}. The partial order is induced by the directed paths in the DAG.

Given a DAG G, for each vertex $s \in V$, let $\pi(s)$ denote the parent set of s, which consists of all the vertices that have directed edges pointing to s. In this context, we define $p_s(x_s \mid x_{\pi(s)})$ as a nonnegative function over the random variables x_s and $x_{\pi(s)}$, where x_s corresponds to the variable associated with vertex s, and $x_{\pi(s)}$ is the vector of variables corresponding to the parent set $\pi(s)$. This function is normalized such that

$$\int p_s(x_s \mid x_{\pi(s)})\, dx_s = 1$$

for all possible values of $x_{\pi(s)}$. The normalization condition ensures that $p_s(x_s \mid x_{\pi(s)})$ represents a valid conditional probability distribution.

A directed graphical model is defined by a collection of such probability distributions (either probability densities or mass functions) associated with the vertices of the DAG. These distributions factorize according to the following rule:

$$p(x_1, x_2, \ldots, x_m) = \prod_{s \in V} p_s(x_s \mid x_{\pi(s)}), \tag{6.1}$$

where $p(x_1, x_2, \ldots, x_m)$ is the joint probability distribution over all variables $x_1, x_2, \ldots, x_m$ corresponding to the vertices $1, 2, \ldots, m \in V$. The factorization implies that the joint distribution can be decomposed into a product of conditional distributions, where each conditional distribution depends on the variable corresponding to a vertex and its parent set in the DAG.

For example, in the case of Ising model (i.e., a system of random variables with binary values $\{-1, 1\}$), where each variable is influenced by its neighbors, one can specify the conditional probability $p_s(x_s \mid x_{\pi(s)})$ for each vertex s, as follows. The joint distribution of the random variables (for Ising model) can be expressed as a product of conditional probabilities, and each conditional probability can be written in terms of the interaction between neighboring vertices. Therefore, the conditional probability $p_s(x_s \mid x_{\pi(s)})$ has the following form (with respect to Gibbs measure of the Ising model):

$$p_s(x_s \mid x_{\pi(s)}) = \frac{1}{Z_s} \exp\left(\theta_s x_s + \sum_{t \in \pi(s)} J_{st} x_s x_t\right),$$

where J_{st} is the interaction strength between s and its parent $t \in \pi(s)$, θ_s is the external bias[1] at vertex s, and Z_s is the normalization constant to ensure that the conditional probability is normalized: $\sum_{x_s} p_s(x_s \mid x_{\pi(s)}) = 1$.

[1]The term $\theta_s x_s$ represents the bias at vertex s, which determines how likely x_s is to take specific values, independent of the other variables.

Note that the factorization (6.1) is consistent with the interpretation of $p_s(x_s \mid x_{\pi(s)})$ as the conditional probability of $X_s = x_s$ given $X_{\pi(s)} = x_{\pi(s)}$ for the distribution $p(\cdot)$ defined by the factorization. This consistency follows by an inductive argument, which uses the normalization condition on $p_s(x_s \mid x_{\pi(s)})$ and the partial order induced by the ancestor relationship in the DAG. Specifically, the structure of the DAG ensures that the factorization respects the dependencies between the variables, as defined by the ancestry relation.

Undirected graph (Markov random field). An edge (X_i, X_j) indicates that X_i and X_j are conditionally dependent. In the undirected case, the probability distribution in a graphical model factorizes according to functions that are defined on the cliques of the graph. Recall that a clique C is a fully connected subset of the vertex set V, meaning that for any two vertices $s, t \in C$, there is an edge between them, i.e., $(s,t) \in E$, where E is the set of edges in the graph.

Let us now associate a compatibility function ψ_C with each clique C, where ψ_C is a function defined on the Cartesian product of the state spaces of the random variables corresponding to the vertices in the clique. Specifically, ψ_C is defined as

$$\psi_C : \bigotimes_{s \in C} \mathcal{X}_s \to \mathbb{R}_+,$$

where $\bigotimes_{s\in C} \mathcal{X}_s$ denotes the Cartesian product of the state spaces of the random variables X_s corresponding to the vertices in the clique C, and $\mathbb{R}_+$ denotes the set of non-negative real numbers.

This compatibility function ψ_C represents a local quantity that depends only on the values x_C of the random variables X_s for the vertices $s \in C$ in the clique.

Given this definition, an undirected graphical model , also known as a Markov random field (MRF) or a Gibbs distribution, is a collection of probability distributions that factorizes as follows [378]:

$$p(x_1, x_2, \ldots, x_m) = \frac{1}{Z} \prod_{C \in \mathcal{C}} \psi_C(x_C).$$

The set $\mathcal{C}$ is typically taken to be the set of maximal cliques of the graph i.e., cliques that are not properly contained within any other clique. This condition can be imposed without loss of generality because any representation based on non-maximal cliques can always be converted to one based on maximal cliques. This is done by redefining the compatibility function on a maximal clique to be the product of the compatibility functions over the subsets of that clique.

However, it is important to note that there may be computational benefits to using a non-maximal clique representation, particularly in situations where specific factorization properties of the model can be exploited by algorithms. For this reason, we do not necessarily restrict the set of cliques $\mathcal{C}$ to just maximal cliques but instead allow it to contain all cliques in the graph. This flexibility allows for a finer-grained specification of the factorization properties, as will be discussed in the context of factor graphs below.

It is also crucial to understand that, in a general undirected graph, the compatibility functions ψ_C do not necessarily have any direct or obvious relationship to marginal or conditional distributions defined over the cliques of the graph. This is in contrast to the directed factorization (as seen in equation (6.1), where the factors correspond to conditional probabilities defined over the child-parent sets. In an undirected graphical model, the factorization is based on local dependencies within the cliques, and these dependencies may not directly correspond to marginal or conditional relationships between the variables in the traditional sense.

In undirected case, the graph captures the conditional independence relationships among variables: two variables are conditionally independent if they are not connected by an edge in the graph (see [232, 300, 349]).

Factor graphs. For large graphs, the factorization properties of a graphical model, whether undirected or directed, can be difficult to visualize from the usual graph depictions. The factor graph formalism provides an alternative graphical representation that emphasizes the factorization of the distribution [229, 254, 378]. This alternative representation is particularly useful when the primary focus is on understanding the factorization structure of the model.

Let F represent an index set for the collection of factors that define the distribution of the graphical model. In the undirected case, the set F indexes the collection $\mathcal{C}$ of cliques in the graph. In the directed case, F indexes the set of parent-child neighborhoods (i.e., sets of parent vertices and their corresponding child vertices). To facilitate this representation, we consider a bipartite graph $G' = (V, F, E')$, where V is the original set of vertices, F is the set of factors, and E' is a new set of edges connecting vertices $s \in V$ to factors $a \in F$.

In particular, an edge $(s, a) \in E'$ if and only if the random variable x_s associated with vertex s participates in the factor indexed by $a \in F$.

This representation explicitly shows the relationships between vertices and the factors that involve them.

The factorization of the joint probability distribution corresponding to the factor graph has the general form:

$$p(x_1, x_2, \ldots, x_m) = \frac{1}{Z} \prod_{a \in F} f_a(x_{\partial a}), \tag{6.2}$$

where each factor function $f_a(x_{\partial a})$ is a nonnegative function over the subset of variables $x_{\partial a}$ that participate in factor a, and Z is the normalization constant:

$$Z = \sum_{x_1, x_2, \ldots, x_m} \prod_{a \in F} f_a(x_{\partial a}). \tag{6.3}$$

The factor graph representation is particularly useful in undirected models when $\mathcal{C}$ consists of more than just the maximal cliques. In fact, when cliques are non-maximal, the compatibility functions for these cliques do not have an explicit representation in the usual undirected graphical model. However, the factor graph makes these relationships explicit by showing the factors associated with each clique, even those that are not maximal. This ability to capture non-maximal cliques is one of the primary advantages of factor graphs over traditional graph representations.

A core concept in graphical models is conditional independence. Conditional independence between two random variables means that, once we know the value of a third variable, the first two variables become independent. This property mainly holds for random fields indexed by vertices of a tree-like graph and it allowed to introduce a notion of splitting Gibbs measure (see [323], [324] and the references therein).

Example: Interpreting ∂a in a Simple Factor Graph. To make the notation $x_{\partial a}$ more concrete, consider a simple factor graph defined over variables x_1, x_2, x_3 with the following factorization:

$$p(x_1, x_2, x_3) = \frac{1}{Z} f_1(x_1, x_2) \cdot f_2(x_2, x_3),$$

where Z is the normalizing constant, the factor nodes are f_1 and f_2, and there are edges from x_1 and x_2 to f_1, and from x_2 and x_3 to f_2. Therefore, $\partial f_1 = \{x_1, x_2\}$ and $\partial f_2 = \{x_2, x_3\}$, so the factor functions are compactly expressed as $f_1(x_{\partial f_1}) = f_1(x_1, x_2)$ and $f_2(x_{\partial f_2}) = f_2(x_2, x_3)$. See the rightmost graph in Figure 6.1 for the visualization of this very example.

This interpretation aligns with the standard usage of the symbol ∂a in the graphical models literature (e.g., [229]), where it denotes the neighborhood of the factor node a in the bipartite factor graph. The notation emphasizes the role of variable-factor connections in determining the factorization structure of the joint distribution.

D-separation (Bayesian networks). In Bayesian networks, D-separation is a criterion used to determine whether two variables are conditionally independent given a set of observed variables. A path between two variables is blocked by a third variable if the third variable either:

- Is a collider (a node where two arrows meet) and is unobserved.
- Is not a collider, and either the node or its descendants are observed.

D-separation is crucial for determining which variables in a model can be ignored (i.e., are independent) when inferring the state of other variables [300].

6.1.2 *Examples*

In this section, following [201] we present examples of graphical models in which exact inference is generally infeasible. Our first example involves a diagnostic system where a fixed graphical model is used to answer queries. The remaining examples involve estimation problems where a graphical model is fit to data and subsequently used for prediction or diagnosis.

6.1.2.1 *The QMR-DT database*

The quick medical reference-decision theoretic (QMR-DT) model is a probabilistic graphical model designed for medical diagnosis, combining Bayesian networks with decision theory. Let us give a structured overview of its mathematical foundations and properties (see [196], [201], [340]):

Bayesian network structure: Constructing a bipartite graph where diseases (latent variables) and symptoms (observed variables) form a two-layer directed acyclic graph (DAG). In this structure, diseases serve as parent nodes, while symptoms are child nodes. The symptoms are conditionally independent given the diseases, which simplifies the joint probability distributions.

Parameterization: The Noisy-OR model: Efficiently models the probability of a symptom s_j being caused by multiple diseases $D_1, ..., D_n$:

$$P(s_j = 1 \mid D_1, ..., D_n) = 1 - \prod_{i:D_i=1} (1 - p_{ij})$$

where p_{ij} is the probability that disease D_i causes symptom s_j.

Each disease D_i has a prior probability $P(D_i)$, typically derived from epidemiological data.

Inference: Compute posterior probabilities $P(\text{Diseases} \mid \text{Symptoms})$. Approximate methods include: Belief propagation (exploits conditional independence), variational inference (Approximates posterior distributions),. Monte Carlo methods (Gibbs sampling for intractable cases).

Mathematical efficiency:

- *Noisy-OR*: Reduces parameters from 2^n to $O(n)$ per symptom, enabling scalability.

- *Modularity*: Symptoms and diseases can be added without restructuring the entire model.

- Quantifies diagnostic uncertainty via posterior probabilities.

Thus QMR-DT combines Bayesian networks (noisy-OR parameterization) with decision theory to model medical diagnosis. Its bipartite structure and efficient parameterization enable scalable probabilistic reasoning by Gibbs distributions, though independence assumptions and inference complexity remain challenges.

6.1.2.2 *Neural networks as graphical models*

Neural networks are layered graphs with a nonlinear "activation" function at each node. Let us consider activation functions that are bounded between zero and one, such as those obtained from the logistic function $f(z) = \frac{1}{1+e^{-z}}$. We can treat such a neural network as a graphical model by associating a binary variable s_i with each node and interpreting the activation of the node as the probability that the associated binary variable takes one of its two values. For example, using the logistic function, we write:

$$P(s_i = 1 \mid S_{\mathcal{I}(i)}) = \frac{1}{1 + \exp\left(-\sum_{j\in\mathcal{I}(i)} \mu_{ij} s_j - \mu_{i0}\right)}$$

where μ_{ij} are the parameters associated with edges between parent nodes j and node i, and μ_{i0} is the "bias" parameter associated with node i.

The sigmoid belief network (SBN), introduced by Radford M. Neal in the early 1990s (see [282]), is a directed probabilistic graphical model that uses Bernoulli variables and sigmoid activation functions to model hierarchical dependencies. Namely, SBNs are layered networks where nodes represent binary random variables (Bernoulli units). Each layer $\mathbf{h}^{(l)}$ (latent) is conditionally dependent on the layer above $\mathbf{h}^{(l+1)}$, with directed edges encoding dependencies. The visible layer $\mathbf{v}$ is at the bottom.

For example, in case of a two-layer SBN, $P(\mathbf{v}, \mathbf{h}) = P(\mathbf{v} \mid \mathbf{h})P(\mathbf{h})$.

The probability of a unit $h_i^{(l)}$ being active given its parents $\mathbf{h}^{(l+1)}$ is modeled via a sigmoid:

$$P\left(h_i^{(l)} = 1 \mid \mathbf{h}^{(l+1)}\right) = \sigma\left(\mathbf{w}_i^{(l+1\to l)} \cdot \mathbf{h}^{(l+1)} + b_i^{(l)}\right),$$

where $\sigma(z) = \frac{1}{1+e^{-z}}$ is the logistic function, $\mathbf{w}_i^{(l+1\to l)}$ are weights, and $b_i^{(l)}$ is the bias.

The joint probability distribution over all layers is:

$$P(\mathbf{v}, \mathbf{h}^{(1)}, \ldots, \mathbf{h}^{(L)}) = P(\mathbf{v} \mid \mathbf{h}^{(1)}) \prod_{l=1}^{L-1} P(\mathbf{h}^{(l)} \mid \mathbf{h}^{(l+1)}) \cdot P(\mathbf{h}^{(L)}),$$

where $P(\mathbf{h}^{(L)})$ is the prior (e.g., a factorial Bernoulli distribution) for the top layer.

Computing $P(\mathbf{h} \mid \mathbf{v})$ is intractable due to explaining-away effects (dependencies between parents given a child). Used methods include Gibbs sampling, i.e., iteratively samples layers from $P(h_i^{(l)} \mid \text{Markov blanket})$; Variational inference, i.e., approximates the posterior with a simpler distribution (e.g., mean-field).

Learning aims include maximum likelihood estimation (MLE), i.e., Parameters $\theta = \{\mathbf{W}, \mathbf{b}\}$ are learned by maximizing the log-likelihood $\log P(\mathbf{v})$; Gradient ascent, i.e., the gradient involves expectations over the posterior:

$$\nabla_\theta \log P(\mathbf{v}) = \mathbb{E}_{P(\mathbf{h}|\mathbf{v})} \left[\nabla_\theta \log P(\mathbf{v}, \mathbf{h})\right].$$

and use sampling (e.g., Gibbs) to approximate intractable expectations.

Key properties:

- Data is generated top-down: sample $\mathbf{h}^{(L)} \to \mathbf{h}^{(L-1)} \to \cdots \to \mathbf{v}$.
- Observing a child (e.g., a symptom) induces dependencies between its parents (e.g., diseases).

6.1.2.3 *Boltzmann machines*

Following [3], we define a Boltzmann machine (BM), which is an undirected graphical model with nodes representing stochastic binary units (variables) and edges representing symmetric interactions. Units are divided into visible units (v), Observed during training, and hidden units (h) which stands for latent variables capturing dependencies among visible units.

The graph allows arbitrary connections between units (except self-loops), making it a fully connected model. This structure enables BMs to model complex, higher-order interactions.

The energy function of a joint configuration (v, h) is defined as:

$$E(v,h) = -\sum_i a_i v_i - \sum_j b_j h_j - \sum_{i,j} v_i h_j w_{ij} - \sum_{i<k} v_i v_k w_{ik} - \sum_{j<l} h_j h_l w_{jl},$$

where $v_i, h_j \in \{0, 1\}$ are states of visible/hidden units, a_i, b_j are biases, w_{ij} are symmetric weights ($w_{ij} = w_{ji}$).

In vector form, for a combined state vector $s = (v, h)$, the energy simplifies to:

$$E(s) = -\frac{1}{2} s^T W s - b^T s,$$

where W is a symmetric weight matrix with zero diagonal (no self-connections).

To study the model the main probability distribution, i.e., the probability of a configuration (v, h) is the Gibbs measure:

$$P(v,h) = \frac{\exp(-E(v,h))}{Z}, \quad Z = \sum_{v',h'} \exp(-E(v', h')),$$

where Z is a normalization constant summing over all possible configurations. The exponential term assigns higher probability to low-energy states. Computing Z is intractable for large BMs due to the exponential growth of terms but sampling or approximation techniques can be used as discussed in previous chapters.

Learning aims to maximize the log-likelihood of observed data. The gradient of the log-likelihood with respect to a weight w_{ij} is:

$$\frac{\partial \log P(v)}{\partial w_{ij}} = \langle s_i s_j \rangle_{\text{data}} - \langle s_i s_j \rangle_{\text{model}}.$$

Note that BMs are a subclass of MRFs. The Hammersley-Clifford theorem establishes that any positive probability distribution satisfying the Markov properties can be factorized into Gibbs measures. BMs explicitly define these factors via their energy function.

6.1.2.4 *A simpler model*

Thus BMs are energy-based undirected graphical models that learn data distributions by minimizing energy. Their mathematical foundation combines graph theory (MRFs), statistical physics (Gibbs distributions), and approximate inference (MCMC). The probability distribution over configurations is given by the Gibbs distribution:

$$P(\mathbf{s}) = \frac{1}{Z} \exp\left(-\frac{E(\mathbf{s})}{T}\right) \tag{6.4}$$

where $Z = \sum_{\mathbf{s}} \exp(-E(\mathbf{s})/T)$ is the partition function and T is the temperature parameter. While powerful, computational limitations led to simplifications like RBMs, which remain influential in deep learning.

To address computational challenges following [3], consider the machine which is composed of primitive computing elements called units that are connected to each other by bidirectional links. A unit is always in one of two states, *on* or *off*, and it adopts these states as a probabilistic function of the states of its neighboring units and the real valued weights on its links to them. The state vector $\mathbf{s} = (s_1, s_2, \ldots, s_n)$ where $s_i \in \{0, 1\}$ represents the configuration of all units. A unit being on or off interpreted that the system currently accepts or rejects some elemental hypothesis about the domain.

The probability that unit i is in state $s_i = 1$ given the states of all other units is:

$$P(s_i = 1|\mathbf{s}_{-i}) = \sigma\left(\sum_{j \neq i} w_{ij} s_j - \theta_i\right) \tag{6.5}$$

where $\sigma(x) = \frac{1}{1+e^{-x}}$ is the sigmoid function.

In this case, consider a simpler form of the energy defined as

$$E = -\sum_{i<j} w_{ij} s_i s_j + \sum_i \theta_i s_i \tag{6.6}$$

where w_{ij} is the strength of connection between units i and j, s_i is 1 if unit i is on and 0 otherwise, and θ_i is a threshold.

To minimize the energy, a straightforward algorithm for finding a combination of truth values that represents a local minimum involves switching each hypothesis to the state that minimizes total energy, based on the current states of the other hypotheses. The gradient of energy with respect to unit i is:

$$\frac{\partial E}{\partial s_i} = -\sum_{j \neq i} w_{ij} s_j + \theta_i \tag{6.7}$$

If hardware units make decisions asynchronously and transmission delays are negligible, the system will eventually settle into a local energy minimum [3].

Since the connections are symmetric ($w_{ij} = w_{ji}$), the difference in energy between the entire system with the kth hypothesis rejected and the system with the kth hypothesis accepted can be determined locally by the kth unit. This difference, known as the "energy gap", given by

$$\Delta E_k = \sum_i w_{ki} s_i - \theta_k \tag{6.8}$$

The net input to unit k is defined as $h_k := \Delta E_k$. Identical to the energy gap, determines whether switching the unit's state reduces energy of the system.

To minimize the energy contribution of a unit, the rule is to adopt the "on" state if the total input from the other units and from outside the system exceeds its threshold. This is the standard rule for binary threshold units:

$$s_k^{(t+1)} = \begin{cases} 1 & \text{if } h_k > 0 \\ 0 & \text{if } h_k < 0 \\ s_k^{(t)} & \text{if } h_k = 0 \end{cases} \tag{6.9}$$

This deterministic rule, a simplification of the probabilistic update rule $P(s_i = 1|\mathbf{s}_{-i})$ at low temperature, ensures units choose states to minimize energy. The update rule ensures that the energy decreases (or remains constant) with each update:

$$E(\mathbf{s}^{(t+1)}) \leq E(\mathbf{s}^{(t)}), \tag{6.10}$$

where $\mathbf{s}^{(t)}$ represents the state vector at time t.

The threshold terms can be removed from Eqs. (6.6) and (6.8) by recognizing that the effect of θ_i on the global energy, or on the energy gap of an individual unit, is equivalent to the effect of a link with strength $-\theta_i$ between unit i and a special unit that is always in the "on" state by definition. This "true unit" does not need to have a physical existence, but it simplifies the calculations by allowing the threshold of a unit to be treated in the same way as the links. The value $-\theta_i$ is referred to as the bias of unit i.

We distinguish between two phases: in the clamped phase, the visible units are fixed to observed data values, allowing the model to sample hidden states conditioned on those inputs. In contrast, during free-run equilibrium, all units are allowed to update freely, and the system evolves toward its stationary distribution, representing the model's generative behavior. When the network is in free-run equilibrium, the probability distribution over the visible units is given by

$$P'(V_\alpha) = \sum_\beta P'(V_\alpha \wedge H_\beta) = \frac{\sum_\beta e^{-E_{\alpha\beta}/T}}{\sum_{\lambda\mu} e^{-E_{\lambda\mu}/T}}, \tag{6.11}$$

where V_α is a vector of states of the visible units, H_s is a vector of states of the hidden units, and $E_{\alpha\beta}$ is the energy of the system in state $V_\alpha \wedge H_\beta$

$$E_{\alpha\beta} = -\sum_{i<j} w_{ij} s_i^{\alpha\beta} s_j^{\alpha\beta}$$

Hence,

$$\frac{\partial e^{-E_{\alpha\beta}/T}}{\partial w_{ij}} = \frac{1}{T} s_i^{\alpha\beta} s_j^{\alpha\beta} e^{-E_{\alpha\beta}/T}.$$

By differentiating (6.11) we get

$$\begin{aligned}
\frac{\partial P'(V_\alpha)}{\partial w_{ij}} &= \frac{\frac{1}{T}\sum_\beta e^{-E_{\alpha\beta}/T} s_i^{\alpha\beta} s_j^{\alpha\beta}}{\sum_{\alpha\beta} e^{-E_{\alpha\beta}/T}} - \frac{\sum_\beta e^{-E_{\alpha\beta}/T} \frac{1}{T} \sum_{\lambda\mu} e^{-E_{\lambda\mu}/T} s_i{}^{\lambda\mu} s_j^{\lambda\mu}}{\left(\sum_{\lambda\mu} e^{-E_{\lambda\mu/T}}\right)^2} \\
&= \frac{1}{T}\left[\sum_\beta P'(V_\alpha \wedge H_\beta) s_i^{\alpha\beta} s_j^{\alpha\beta} \right. \\
&\quad \left. - P'(V_\alpha) \sum_{\lambda\mu} P'(V_\lambda \wedge H_\mu) s_i^{\lambda\mu} s_j^{\lambda\mu}\right].
\end{aligned}$$

This derivative is used to compute the gradient of the G-measure:

$$G = \sum_\alpha P(V_\alpha) \ln \frac{P(V_\alpha)}{P'(V_\alpha)}$$

where $P(V_\alpha)$ is the clamped probability distribution over the visible units and is independent of w_{ij}. So

$$\begin{aligned}
\frac{\partial G}{\partial w_{ij}} &= -\sum_\alpha \frac{P(V_\alpha)}{P'(V_\alpha)} \frac{\partial P'(V_\alpha)}{\partial w_{ij}} \\
&= -\frac{1}{T} \sum_\alpha \frac{P(V_\alpha)}{P'(V_\alpha)} \left[\sum_a P'(V_\alpha \wedge H_\alpha) s_i^{\alpha\beta} s_j^{\alpha\beta} \right. \\
&\quad \left. - -P'(V_\alpha) \sum_{\lambda\mu} P'(V_\lambda \wedge H_\mu) s_i^{\lambda\mu} s_j^{\lambda\mu}\right].
\end{aligned}$$

Consequently,

$$\begin{aligned}
P(V_\alpha \wedge H_\beta) &= P(H_\beta \mid V_\alpha) P(V_\alpha), \\
P'(V_\alpha \wedge H_\beta) &= P'(H_\beta \mid V_\alpha) P'(V_\alpha),
\end{aligned}$$

and

$$P'(H_\beta \mid V_\alpha) = P(H_\beta \mid V_\alpha) \tag{6.12}$$

Equation (6.12) holds because, in equilibrium, the probability of a hidden state given a visible state must be identical, regardless of whether the visible units were clamped in that state or arrived there through free-running. Hence,

$$P'(V_\alpha \wedge H_\beta)\frac{P(V_\alpha)}{P'(V_\alpha)} = P(V_\alpha \wedge H_\beta),$$

and

$$\sum_\alpha P(V_\alpha) = 1$$

Therefore,

$$\frac{\partial G}{\partial w_{ij}} = -\frac{1}{T}\left[p_{ij} - p'_{ij}\right]$$

where

$$p_{ij} := \sum_{\alpha\beta} P(V_\alpha \wedge H_\beta)\, s_i^{\alpha\beta} s_j^{\alpha\beta}, \quad p'_{ij} := \sum_{\lambda\mu} P'(V_\lambda \wedge H_\mu)\, s_i^{\lambda\mu} s_j^{\lambda\mu}.$$

The Boltzmann machine learning algorithm can also be viewed as an input-output model. The visible units are split into two sets: an input set I and an output set O. An environment defines a set of conditional probabilities of the form $P(O_\beta \mid I_\alpha)$. In the "training" phase, both the input and output units are clamped by the environment, and the parameters p_{ij} are estimated. During the "testing" phase, the input units are clamped while the output and hidden units are allowed to free-run, and the parameters p'_{ij} are estimated. The relevant G measure in this context is

$$G = \sum_{\alpha\beta} P(I_\alpha \wedge O_\beta) \ln \frac{P(O_\beta \mid I_\alpha)}{P'(O_\beta \mid I_\alpha)}$$

Similar mathematics apply in this formulation and $\partial G/\partial w_{ij}$, is the same as before.

6.1.2.5 *Gibbs measures in hidden Markov models*

The Gibbs measure formulation of Hidden Markov Models provides a statistical physics perspective on sequential modeling [65,219,232]. For an HMM with hidden states $\{X_t\}_{t=1}^T$ and observations $\{Y_t\}_{t=1}^T$, the joint distribution can be expressed in the canonical Gibbs form

$$P(X,Y) = \frac{1}{Z}\exp\left(-H(X,Y)\right), \tag{6.13}$$

where the Hamiltonian is $H(X,Y) = -\log \pi_{X_1} - \sum_{t=2}^T \log A_{X_{t-1},X_t} - \sum_{t=1}^T \log B_{X_t,Y_t}$ and the partition function $Z = \sum_X \exp(-H(X,Y))$ equals the marginal likelihood $P(Y)$.

This energy-based representation naturally connects to statistical physics concepts. The partition function Z corresponds to the normalization constant in statistical mechanics, while the free energy $F = -\log Z$ provides a thermodynamic interpretation of the model evidence. The forward algorithm computes Z through the recursion $\alpha_t(i) = \sum_j \alpha_{t-1}(j) A_{ji} B_{ik}$ with $Z = \sum_i \alpha_T(i)$.

The Gibbs formulation facilitates extensions to undirected graphical models through the Hammersley–Clifford theorem, which establishes the equivalence between Markov random fields and Gibbs measures. In conditional random fields, the directed transitions of HMMs are replaced by symmetric potential functions, maintaining the exponential family structure while accommodating more flexible dependency patterns.

Temperature scaling provides a natural mechanism for controlling the sharpness of the distribution. The tempered version $P_\beta(X, Y) = \frac{1}{Z_\beta} \exp(-\beta H(X, Y))$ interpolates between uniform ($\beta \to 0$) and concentrated ($\beta \to \infty$) distributions, enabling systematic exploration in sampling and optimization algorithms.

The energy function decomposition into local terms naturally supports message-passing algorithms. The forward-backward algorithm can be interpreted as computing marginal probabilities through local energy minimization, connecting dynamic programming solutions to variational principles in statistical physics.

6.2 Structured prediction

6.2.1 *Definitions and Gibbs formulation*

Structured prediction addresses machine learning tasks where outputs consist of complex interdependent structures—sequences, trees, or graphs—rather than independent scalar values. The fundamental challenge lies in modeling dependencies among output variables, making Gibbs measures a natural framework for these problems.

Let $\mathbf{x}$ denote input data and $\mathbf{y}$ represent the structured output. The goal is learning a mapping $f : \mathbf{x} \to \mathbf{y}$ that captures structural dependencies. Gibbs measures provide an elegant solution by modeling the conditional distribution:

$$P(\mathbf{y} \mid \mathbf{x}) = \frac{1}{Z(\mathbf{x})} \exp\left(-E(\mathbf{y}, \mathbf{x})\right), \tag{6.14}$$

where $E(\mathbf{y}, \mathbf{x})$ is the energy function encoding compatibility between input and output structures, and $Z(\mathbf{x}) = \sum_{\mathbf{y}} \exp(-E(\mathbf{y}, \mathbf{x}))$ is the partition function.

The energy function typically decomposes into local interactions:

$$E(\mathbf{y}, \mathbf{x}) = \sum_{c \in \mathcal{C}} \psi_c(\mathbf{y}_c, \mathbf{x}), \tag{6.15}$$

where $\mathcal{C}$ represents cliques of interacting variables, $\mathbf{y}_c$ denotes variables in clique c, and ψ_c captures local compatibility.

This Gibbs formulation naturally handles three key aspects of structured prediction:

- **Dependency modeling**: The energy function explicitly represents interactions between output variables
- **Global consistency**: The partition function ensures valid probability distributions over structured outputs
- **Flexible parameterization**: Local potentials ψ_c can encode domain-specific constraints and preferences

Example: Sequence Labeling as Gibbs Measures

Consider part-of-speech tagging where we assign grammatical tags to words in a sentence. The structured output $\mathbf{y} = (y_1, y_2, \ldots, y_n)$ represents the tag sequence, with dependencies between adjacent tags naturally modeled through a Gibbs measure:

$$P(\mathbf{y} \mid \mathbf{x}) = \frac{1}{Z(\mathbf{x})} \exp\left(\sum_{i=1}^{n} \phi(x_i, y_i) + \sum_{i=1}^{n-1} \psi(y_i, y_{i+1}) \right), \tag{6.16}$$

where $\phi(x_i, y_i)$ captures word-tag compatibility and $\psi(y_i, y_{i+1})$ models tag transitions. This formulation ensures that tag sequences respect both local word evidence and global sequential constraints.

6.2.2 *Inference in Gibbs measures*

Structured prediction requires computing the most probable configuration (MAP inference):

$$\hat{\mathbf{y}} = \arg\max_{\mathbf{y}} P(\mathbf{y} \mid \mathbf{x}) = \arg\min_{\mathbf{y}} E(\mathbf{y}, \mathbf{x}). \tag{6.17}$$

The exponential size of the structured output space makes exact inference intractable in general, motivating approximate methods.

Belief propagation (BP) provides an efficient framework for inference in Gibbs measures defined on graphical models. For tree-structured dependencies, BP yields exact marginals and MAP configurations. On general graphs with cycles, loopy belief propagation provides effective approximations.

Consider a factor graph representation where variables $\mathbf{y}$ interact through factors corresponding to energy terms ψ_c. BP iteratively passes messages between variable and factor nodes:

$$m_{i\to c}(y_i) = \prod_{d\in \mathrm{ne}(i)\setminus c} m_{d\to i}(y_i) \tag{6.18}$$

$$m_{c\to i}(y_i) = \min_{y_c\setminus y_i} \left[\psi_c(y_c) + \sum_{j\in \mathrm{ne}(c)\setminus i} m_{j\to c}(y_j)\right] \tag{6.19}$$

where messages are computed in the log-domain for numerical stability. Upon convergence, the MAP configuration is recovered from the final beliefs.

For sequence labeling, this reduces to the Viterbi algorithm, which efficiently computes:

$$\hat{\mathbf{y}} = \arg\max_{\mathbf{y}} P(\mathbf{y} \mid \mathbf{x}) \tag{6.20}$$

in $O(n\cdot|Y|^2)$ time, where n is sequence length and $|Y|$ is the tag vocabulary size.

6.2.3 *Learning Gibbs measures*

Parameter learning in structured prediction involves estimating the potential functions ψ_c from training data $\{(\mathbf{x}^{(i)}, \mathbf{y}^{(i)})\}_{i=1}^{N}$.

The usual log-likelihood of the data under a Gibbs measure is:

$$\mathcal{L}(\theta) = \sum_{i=1}^{N} \left[-E(\mathbf{y}^{(i)}, \mathbf{x}^{(i)}; \theta) - \log Z(\mathbf{x}^{(i)}; \theta)\right] \tag{6.21}$$

The gradient with respect to parameters θ involves both observed and model expectations:

$$\frac{\partial \mathcal{L}}{\partial \theta} = \sum_{i=1}^{N} \left[\mathbb{E}_{P(\mathbf{y}|\mathbf{x}^{(i)})}[\phi_\theta(\mathbf{x}^{(i)}, \mathbf{y})] - \phi_\theta(\mathbf{x}^{(i)}, \mathbf{y}^{(i)})\right] \tag{6.22}$$

Computing model expectations requires inference over the Gibbs measure, often approximated using belief propagation or sampling methods.

The structured perceptron avoids computing the partition function by using a margin-based approach. Given feature representations $\phi(\mathbf{x}, \mathbf{y})$ and parameters $\mathbf{w}$, the algorithm iteratively updates:

$$\mathbf{w} \leftarrow \mathbf{w} + \eta \left[\phi(\mathbf{x}^{(i)}, \mathbf{y}^{(i)}) - \phi(\mathbf{x}^{(i)}, \hat{\mathbf{y}}^{(i)})\right] \tag{6.23}$$

where $\hat{\mathbf{y}}^{(i)} = \arg\max_{\mathbf{y}} \mathbf{w}^T \phi(\mathbf{x}^{(i)}, \mathbf{y})$ is the current prediction. This approach is computationally efficient but lacks the probabilistic interpretation of Gibbs measures.

6.3 Practical considerations and applications

6.3.1 *A lattice-based model*

In this subsection, following [102], we give a lattice-based model, which represent the classical approach to using Gibbs measures in image analysis. Consider a lattice spin system (an image lattice) defined on a finite volume $S \subset \mathbb{Z}^2$ with $|S| = m$, and a spin space Λ.

Each random variable X_i, where $i \in S$, takes values from the spin space Λ. The configuration space Ω is then defined as $\Omega = \Lambda^m$, and any configuration $X \in \Omega = \Lambda^m$, where $X = \{X_i, i \in S\}$ and $X_i \in \Lambda$, corresponds to a specific image.

In practice, various types of spin spaces are used. For example, $\Lambda = \{0, \ldots, 255\}$ (the gray level space) is commonly used for image restoration or texture analysis, while $\Lambda = \{1, \ldots, M\}$ is used for semantic labeling or image segmentation with M classes.

An observable image (or observation) is obtained as a configuration $Y \in \Omega$, which typically differs from the original image $X \in \Omega$. We refer to X as the true or ideal configuration, which represents the underlying scene observed through a sensor. X is the unknown that can be interpreted as a version of Y, cleaned of artifacts (such as noise, blurring, etc.) introduced during the acquisition process. In some applications, X may also serve as an initial interpretation of the data Y, such as a segmented image or a description of the objects in the scene.

The goal is to find the solution X from a set of images that are consistent with the observed data Y. A good estimate should not only fit the data but also satisfy criteria reflecting prior knowledge about the solution, such as regularity, smoothness, and other constraints.

Let us consider finite systems of partially overlapping discs $\{d_{x_1}, \ldots, d_{x_k}\}$ of the same radius r with a hard core distance $\epsilon < r$ between any two elements, lying in a bounded domain $V \subset \mathbb{R}^2$. Let

$$\gamma = \{x_i\} \in \Gamma_d(V), \quad x_i \in V \subset \mathbb{R}^2$$

be a configuration of the centers of discs and $\Gamma_d(V)$ denotes the configuration space of the discs centers in V. Notice, the set $\Gamma_d(V)$ can be decomposed into strata:

$$\Gamma_d(V) = \bigcup_{n=0}^{N} \Gamma_d(V, n)$$

where each stratum $\Gamma_d(V, n)$ is the set of configurations containing n discs, $\Gamma_d(V, 0) = \{\emptyset\}$, and N is the maximal number of discs in V.

Consider a digital image $Y = (y_s)_{s \in S}$ defined on a finite subset of the lattice $S \subset \mathbb{Z}^2$. Next, consider configurations of disc centers $\gamma = \{x_i\} \in \Gamma_d(V)$. To prevent two discs from being centered at the same pixel, we introduce a hard core distance ϵ, which is set to the data resolution: $\epsilon = 1$ pixel.

To define the energy H, we first incorporate some prior knowledge. We want to minimize the overlap between objects. However, rather than strictly forbidding overlap, we aim for a more flexible model with respect to both the data and object shape, penalizing overlapping objects instead. The pairwise interaction is defined as follows:

$$H_2(x_i, x_j) = \max\left(0, 1 - \frac{\|x_j - x_i\|}{2r}\right), \quad \forall \{x_i, x_j\} \in \gamma^2 : \|x_j - x_i\| \geqslant 1$$

where $\|.\|$ is the Euclidean norm and r is the radius of the underlying disc.

A data term is introduced for each object to align the disc configuration with the data. We assume that there is an object, modeled by a disc centered at pixel s, in the image if the gray level values of the pixels within the projection of the disc onto the lattice differ significantly from those of the pixels in the surrounding neighborhood. To quantify this difference, we compute the Bhattacharya distance between the corresponding distributions.

Let $D_1(s)$ denote the projection of the disc with radius r centered at s onto the lattice S, and $D_2(s)$ represent the surrounding crown.

$$D_1(s) = \{t \in S : \|t - s\| \leqslant r\} \text{ and } D_2(s) = \{t \in S : r < \|t - s\| \leqslant r + 1\}$$

Recall that by $|A|$ we denote the number of elements of a set A.

We consider the mean and the variance of the data of these two subsets:

$$\mu_1(s) = \frac{1}{|D_1(s)|} \sum_{t \in D_1(s)} y_t \quad \text{and} \quad \mu_2(s) = \frac{1}{|D_2(s)|} \sum_{t \in D_2(s)} y_t,$$

$$\sigma_1^2(s) = \frac{1}{|D_1(s)|} \sum_{t \in D_1(s)} y_t^2 - \mu_1(s)^2,$$

$$\sigma_2^2(s) = \frac{1}{|D_2(s)|} \sum_{t \in D_2(s)} y_t^2 - \mu_2(s)^2.$$

Assuming Gaussian distributions, the Bhattacharya distance between the distributions in $D_1(s)$ and in $D_2(s)$ is then given by:

$$B(s) = \frac{1}{4}\left(\mu_1(s) - \mu_2(s)\right)^2 \sqrt{\sigma_1^2(s) + \sigma_2^2(s)} - \frac{1}{2}\log\frac{2\sigma_1(s)\sigma_2(s)}{\sigma_1^2(s) + \sigma_2^2(s)}$$

From this distance between the two distributions, a first order energy term is defined by

$$\forall x_i \in \gamma, H_1(x_i) = \begin{cases} 1 - \frac{B(i)}{T} & \text{if} \quad B(i) < T \\ \exp\left\{-\frac{B(i)-T}{3B(i)}\right\} - 1 & \text{if} \quad B(i) \geqslant T \end{cases}$$

where i is the closest point to x_i on the lattice, and T is a threshold parameter. Finally, under the hard core constraint, the global energy is written as follows:

$$H(\gamma) = \alpha \sum_{x_i \in \gamma} H_1(x_i) + \sum_{\{x_i, x_j\} \in \gamma^2, x_i \neq x_j} H_2(x_i, x_j),$$

where α is a weighting parameter between the data term and the prior knowledge.

The algorithm simulating the process is defined as follows:

- Computation of the data term: For each site $s \in S$ compute $H_1(s)$ from the data

- Computation of the birth map: To speed up the process, we consider a non-homogeneous birth rate to favor birth where the data term is low (where the data tends to define an object):

$$\forall s \in S, b(s) = 1 + 9\frac{\max_{t\in S} H_1(t) - H_1(s)}{\max_{t\in S} H_1(t) - \min_{t\in S} H_1(t)}.$$

The normalized birth rate is then given by:

$$\forall s \in S, B(s) = \frac{b(s)}{\sum_{t\in S} b(s)}.$$

This non-homogeneous birth rate refers to a non-homogeneous reference Poisson measure. It has no impact on the convergence to the global minima of the energy function but does have an impact on the speed of convergence in practice by favoring birth in relevant locations.

- *Main program*: initialize the inverse temperature parameter $\beta = \beta_0$ and the discretization step $\delta = \delta_0$, alternate birth and death steps

- *Birth step*: for each $s \in S$, if $x_s = 0$ (no point in s) add a point in $s\,(x_s = 1)$ with probability $\delta B(s)$ (note that the hard core constraint with $\epsilon = 1$ pixel is satisfied).

- *Death step*: consider the configuration of points $x = \{s \in S : x_s = 1\}$ and sort it from the highest to the lowest value of $H_1(s)$. For each point taken in this order, compute the death rate as follows:

$$d_x(s) = \frac{\delta a_x(s)}{1 + \delta a_x(s)}$$

where

$$a_x(s) = \exp\left\{-\beta\left(H\left(x / \{x_s\}\right) - H(x)\right)\right\}$$

and kill $s\,(x_s = 0)$ with probability $d(s)$.

- *Convergence test*: if the process has not converged, decrease the temperature and the discretization step by a given factor and go back to the birth step. The convergence criterion is usually given in practice by a fixed number of iterations.

In [102] the following results are obtained:

The first application focuses on tree crown extraction from aerial imagery. In case of 50 cm resolution images of poplars, with an example of the results shown in Figure 6.2. The outcomes are generally satisfactory, though a few false positives can be seen along the plantation's border, primarily due to shadows.

The second application addresses the issue of flamingo population counting. To better capture the flamingo's shape. The results are shown in Figure 6.3 (left), which displays the initial image of the flamingo colony, and in Figure 6.3 (right), a zoomed-in section of the image with the detected birds. The flamingos were accurately identified. For other results on this direction see [98]–[102].

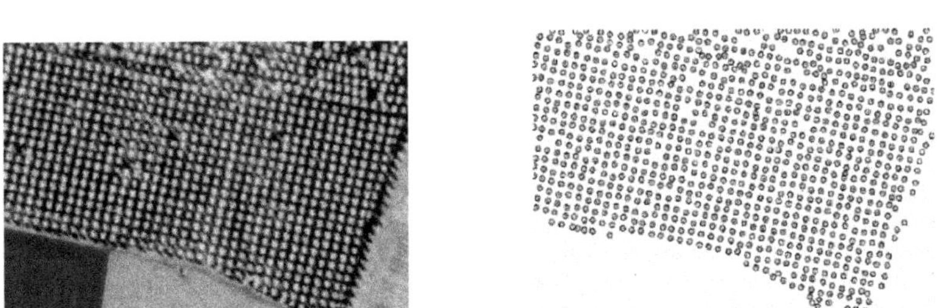

Fig. 6.2: The result on a poplar plantation (left), right: detected trees.

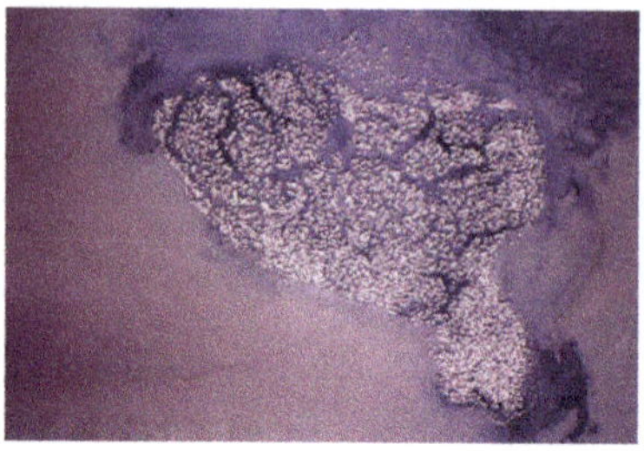 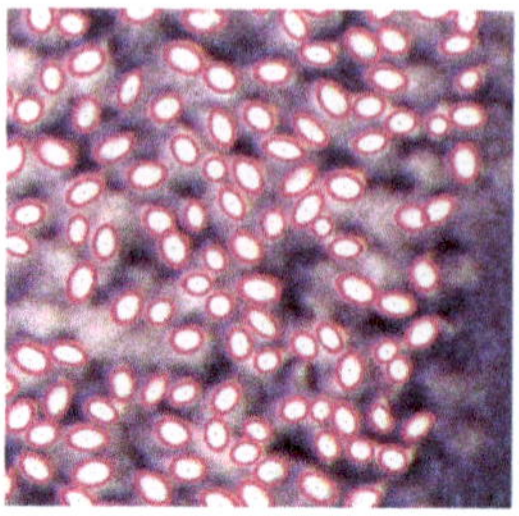

Fig. 6.3: Bird population (left). Extract of the detected birds from the image shown in the left.

6.3.2 *Discrete multivariate date: The Potts-Ising model*

The Potts model provides a flexible framework for modeling discrete multivariate data where each coordinate is categorical. While the full Potts model requires $O(d^2K^2)$ parameters, we present the Potts-Ising model (POIS) which reduces this to $O(d^2 + Kd)$ parameters by incorporating a special structure.

This model is particularly suitable for discrete data with a special level (typically "0") representing absence or non-participation, where dependencies are captured through the presence or absence of this level.

The Potts-Ising model. Consider a discrete random vector $\mathbf{z} = (z_i)_{i=1}^d$, where z_i represents the value of the i-th coordinate and takes values from the set $\mathbb{Z}_K^d := \{0, 1, \ldots, K\}^d$. In other words, each coordinate z_i can independently take values from the set $\{0, 1, \ldots, K\}$. The general Potts model defines the Gibbs distribution of $\mathbf{z}$ in the form:

$$p(\mathbf{z}) \propto \exp\left(\sum_{i=1}^{d}\sum_{k=0}^{K}\theta_{ik}z_{ik} + \sum_{\substack{i,j=1\\ i<j}}^{d}\sum_{k,l=0}^{K}\gamma_{ij,kl}z_{ik}z_{jl}\right), \quad \mathbf{z} \in \mathbb{Z}_K^d, \tag{6.24}$$

where $\mathbf{z} = (z_1, z_2, \ldots, z_d)$ is a random vector of dimension d, with $z_i \in \{0, \ldots, K\}$ for each i, z_{ik} is an indicator variable defined as $z_{ik} = 1\{z_i = k\}$, which equals 1 if $z_i = k$ and 0 otherwise, for each $k \in \{0, \ldots, K\}$, θ_{ik} represents the "bias" or "weight" associated with the event that z_i equals k, controlling the tendency of the i-th coordinate to take the value k, and $\gamma_{ij,kl}$ represents the interaction parameter between the coordinates z_i and z_j when $z_i = k$ and $z_j = l$, capturing the pairwise interaction between the i-th and j-th coordinates of the vector.

The double sum in the first term accounts for the contribution of each coordinate z_i to the energy of the system, weighted by its corresponding bias θ_{ik}. The second double sum represents pairwise interactions between coordinates, with $\gamma_{ij,kl}$ modulating the strength of the interaction depending on the specific values k and l taken by z_i and z_j, respectively.

Note that the constraint $i < j$ in the interaction sum ensures each pairwise interaction is counted only once, and the indicator variables z_{ik} are mutually exclusive for each coordinate i.

In some models, one might enforce a constraint on the interaction terms, such as setting $\gamma_{ij,kl} = 0$ for pairs of coordinates (i, j) that are not connected by an edge in some predefined edge set $E \subset [d]^2$. However, for simplicity, we do not explicitly impose this constraint in the formulation above, but it can be included if the model requires it.

This model is very flexible in capturing the multivariate dependence structure among the coordinates of the vector $\mathbf{z}$, as it allows both individual biases and pairwise interactions between any two coordinates. However, this flexibility comes at a cost:

(1) The number of parameters in the model grows rapidly with both d (the dimensionality of $\mathbf{z}$) and K (the number of possible values each coordinate can take). Specifically, the total number of parameters is of the order $O(d^2K^2)$, which can become computationally expensive when d or K is large.
(2) From a statistical perspective, accurately estimating the parameters of this model typically requires a large sample size. From a computational perspective, fitting the model can be slow and may suffer from numerical instability, especially for small sample sizes or high-dimensional settings.

Simplification: Consider a restriction of (6.24) that preserves much of the flexibility of modeling pairwise interactions among the variables, but significantly reduces the model complexity. The approach which we discuss here is suitable for sparse random vectors where level 0 has a special meaning. Instead of modeling the interactions among all levels, we simply model the interaction between level 0 and not 0. To be more precise, let

$$\sigma_i := \sigma_i(z_i) := \begin{cases} 1, & z_i = 0 \\ -1, & z_i \neq 0 \end{cases}, \tag{6.25}$$

and consider the model

$$p(z) \propto \exp\left(\sum_{i,k} \theta_{ik} z_{ik} + \sum_{(i,j):i<j} \gamma_{ij}\sigma_i\sigma_j\right). \tag{6.26}$$

which we refer to as the Potts-Ising (POIS) model. The rationale behind the naming is that the interaction term in (6.26) is similar to an Ising model which is the special case of the Potts model for $K = 1$. We, however, note that (6.26) is much richer than an Ising model because it allows for modeling arbitrary marginal distributions for the coordinates via parameters $\theta_{ik}, i \in [d], k \in [K]$. We will show that we can achieve this extra flexibility over the Ising model, at virtually no extra computational cost. Model (6.26) is a special case of the general Potts model with the following restriction on the parameters:

$$\gamma_{ij,kl} := \begin{cases} \gamma_{ij} & k = l = 0, \text{ or } k \neq 0, l \neq 0 \\ -\gamma_{ij} & k = 0, l \neq 0 \text{ or } k \neq 0, l = 0 \end{cases}$$

Through empirical validation in Section 6.3.3 on synthetic data with $d = 5$ and $K = 3$, we show that POIS retains much of the Potts model's statistical flexibility, achieving comparable sparsity with only 25 parameters compared to 105 for Potts.

6.3.3 *Numerical simulations*

To compare the three models (Ising, Potts, POIS), the experiment is conducted on a five-dimensional system ($d = 5$), with the Potts and POIS models supporting three discrete levels ($K = 3$), while the Ising model employs binary variables ($\{-1, +1\}$). The Ising model is parameterized with external fields h_i ranging from -0.2 to 0.2, introducing slight biases, and a sparse interaction matrix J_{ij} with non-zero entries of 0.3 and 0.4 to reflect moderate pairwise dependencies. The Potts model incorporates marginal parameters set to $\theta_{i,0} = 1.0$ to favor level 0 and $\theta_{i,k} = -0.5$ for other levels, promoting sparsity, with interactions $\gamma_{ij,kk} = J_{ij}$ for same-level pairs. The POIS model adopts the same marginal parameters as the Potts model but simplifies interactions to $\gamma_{ij} = J_{ij}$, focusing on level 0 versus non-0 states. Synthetic data is generated via Gibbs sampling, producing 1000 samples per model following a 100-iteration burn-in period to ensure convergence.

Evaluation centers on four key metrics to highlight the models' trade-offs. Model complexity is assessed by counting parameters: the Ising model requires $d + d(d-1)/2$, the Potts model demands $dK + d(d-1)K^2/2$, and the POIS model uses $dK + d(d-1)/2$, reflecting its reduced interaction complexity. Empirical marginal distributions are computed as average state probabilities, emphasizing the "zero" state (level 0 for Potts/POIS, $+1$ for Ising) to evaluate sparsity modeling. Pairwise correlations, derived

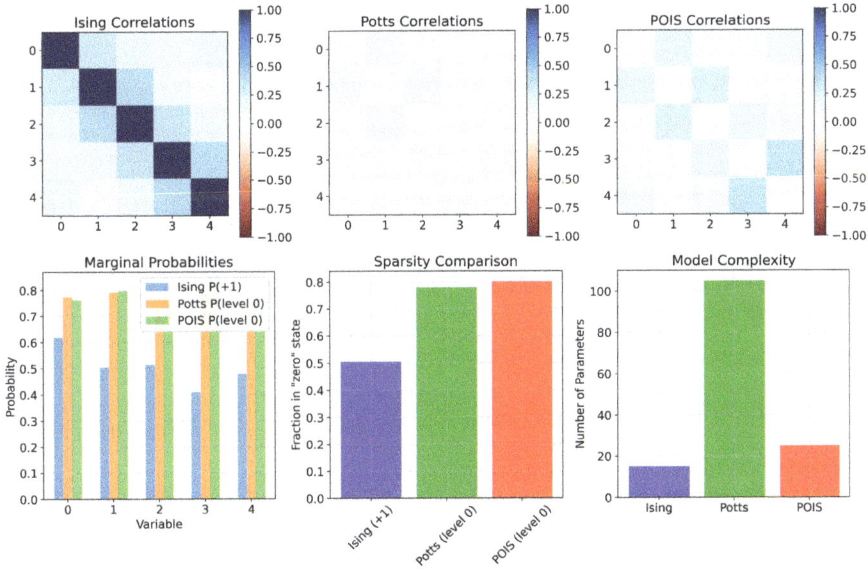

Fig. 6.4: Comparison of Ising, Potts, and Potts-Ising models. Relatively close alignment of POIS and Potts supports the claim that POIS preserves much of the Potts model's statistical flexibility for sparse scenarios.

as Pearson coefficients, use level 0 indicators for Potts and POIS to enable direct comparison with the Ising model. Sparsity is quantified as the fraction of variables in the "zero" state, testing each model's ability to capture dominant states.

The results in Figure 6.4 show that POIS level 0 marginals closely match Potts, with both achieving high sparsity compared to Ising. This alignment validates POIS as an efficient alternative, capturing Potts' sparsity with only 24% of its parameters.

6.3.4 *Model fitting*

To see model fitting, consider a random sample $\{z^t\}_{t=1}^n$ of size n where each z^t is an i.i.d. draw with the same distribution as that of z given in (6.24).

Let z_i^t be the ith coordinate of z^t and write $z_i^* := (z_i^t, t \in [n])$. For the vector z, let $z_{-i} = (z_j, j \neq i)$ and similarly for $z_{-i}^* = (z_j^t, t \in [n], j \neq i)$ and z_{-i}^t. We collect the interaction parameters from (6.26) in the $d \times d$ matrix $\Gamma = (\gamma_{ij})$ and the marginal parameters in the $d \times K$ matrix $B = (\beta_{ik})$. An effective approach for fitting graphical models of the form (6.24) is via the so-called neighborhood regression [271]: One separately fits the conditional distributions $p\left(z_i^* \mid z_{-i}^*\right)$ for all $i \in [d]$. Since these conditional densities factorize over the sample, i.e., $p\left(z_i^* \mid z_{-i}^*\right) = \prod_t p\left(z_i^t \mid z_{-i}^t\right)$, let us focus on the generic version based on z, i.e.,

$$p\left(z_i \mid z_{-i}\right) \propto \exp\left(\sum_k \theta_{ik} z_{ik} + \sum_{j:j\neq i} \gamma_{ij}\sigma_i\sigma_j\right)$$

Let us write $q_{ik} := \exp\left(\theta_{ik}\right), \gamma_{ii} = 0, \gamma_{i*} = \left(\gamma_{ij}, i = 1, \ldots, d\right)$ and $\sigma = \left(\sigma_j, j \in [d]\right)$. Then,

$$p\left(z_i = k \mid z_{-i}\right) \propto \begin{cases} q_{i0} e^{\langle \gamma_{i*}, \sigma \rangle}, & z_i = 0 \\ q_{ik} e^{-\langle \gamma_{i*}, \sigma \rangle}, & z_i = k \neq 0 \end{cases}$$

where $\langle \gamma_{i*}, \sigma \rangle = \sum_{j\neq i} \gamma_{ij}\sigma_j$ is the usual Euclidean inner product between γ_{i*} and σ. Letting $\beta_{ik} := q_{ik}/q_{i0}$, and some algebra gives

$$p\left(z_i \mid z_{-i}\right) = \frac{\prod_{k\neq 0}\left[\beta_{ik} e^{-2\langle \gamma_{i*}, \sigma \rangle}\right]^{z_{ik}}}{1 + \beta_{i\oplus} e^{-2\langle \gamma_{i*}, \sigma \rangle}}$$

where $\beta_{i\oplus} := \sum_{k'\neq 0} \beta_{ik'}$. The notation $\oplus$ means that we are summing over all values of the index except 0 . This notation is helpful because of the special role level 0 plays in the model. Later we use $+$ in place of an index to mean summing over all values of that index (without exception). Since $\beta_{i0} = 1$, in these notations, we have $\beta_{i\oplus} = \beta_{i+} - 1$.

6.3.4.1 *Reduction to logistic regression*

Consider the sample of size n, then with $z_{ik}^t = 1\left\{z_i^t = k\right\}$, we have

$$p\left(z_i^* \mid z_{-i}^*\right) := \prod_{t=1}^n p\left(z_i^t \mid z_{-i}^t\right) = \prod_{t=1}^n \frac{\prod_{k\neq 0}\left[\beta_{ik} e^{-2\langle \gamma_{i*}, \sigma_t \rangle}\right]^{z_{ik}^t}}{1 + \beta_{i\oplus} e^{-2\langle \gamma_{i*}, \sigma_t \rangle}}$$

Here $\sigma_t = \left(\sigma_{tj}, j \in [n]\right)$ where $\sigma_{tj} = \sigma\left(z_{tj}\right)$ and $\sigma(\cdot)$ is defined in (6.25). The conditional loglikelihood of the model, $\ell_i\left(\beta_i, \gamma_{i*}\right) := \log p\left(z_i^* \mid z_{-i}^*\right)$, is

$$\ell_i\left(\beta_i, \gamma_{i*}\right) = \sum_{k\neq 0} z_{ik}^+ \log \beta_{ik} - \sum_t \left[2 z_{i\oplus}^t \langle \gamma_{i*}, \sigma_t \rangle + \log\left(1 + \beta_{i\oplus} e^{-2\langle \gamma_{i*}, \sigma_t \rangle}\right)\right]$$

where $z_{i\oplus}^t = \sum_{k\neq 0} z_{ik}^t$ and $z_{ik}^+ = \sum_{t=1}^n z_{ik}^t$, using the summation convention discussed earlier. To estimate the parameters, it is common to maximize a penalized version of the conditional log-likelihood. For simplicity, let's assume there is no additional penalty at this point.

Given $\beta_{i\oplus}$, the tasks of estimating β_{i*} and γ_{i*} become independent. One approach is to solve the problem with the added constraint $\beta_{i\oplus} = e^u$, where $u \in \mathbb{R}$, and then jointly optimize over both u and γ_{i*}:

$$\max_{\beta_i > 0, \gamma_{i*}} \ell_i\left(\beta_i, \gamma_{i*}\right) = \max_{u, \gamma_{i*}} \max_{\beta_i > 0: \sum_{k \neq 0} \beta_{ik} = e^u} \ell_i\left(\beta_i, \gamma_{i*}\right).$$

The solution of the inner optimization problem over β is simply

$$\widehat{\beta}_{ik}(u) = e^u \frac{z_{ik}^+}{z_{i\oplus}^+}, \quad k \neq 0 \tag{6.27}$$

where $z_{i\oplus}^+ = \sum_{k \neq 0} z_{ik}^+ = \sum_{t=1}^n \sum_{k \neq 0} z_{ik}^t = n - z_{i0}^+$, where the last equation follows since, by definition, $\sum_{k=0}^K z_{ik}^t = 1$ for all $t \in [n]$. Consequently, we get:

$$\begin{aligned}
&\max_{\beta > 0, \gamma_{i*}} \ell_i\left(\beta, \gamma_{i*}\right) \\
&= C + \max_{u, \gamma_{i*}} \left\{ z_{i\oplus}^+ u - \sum_t \left[2 z_{i\oplus}^t \left\langle \gamma_{i*}, \sigma_t \right\rangle + \log\left(1 + e^{u - 2\langle \gamma_{i*}, \sigma_t \rangle}\right)\right]\right\} \\
&= C - \min_{u, \gamma_{i*}} \sum_t \left[z_{i\oplus}^t \left(2 \left\langle \gamma_{i*}, \sigma_t \right\rangle - u\right) + \log\left(1 + e^{u - 2\langle \gamma_{i*}, \sigma_t \rangle}\right)\right]
\end{aligned}$$

where $C = \sum_{k \neq 0} z_{ik}^+ \log\left(z_{ik}^+ / z_{i\oplus}^+\right)$. The above is a convex problem jointly in u and γ_{i*}. To simplify, let $\widetilde{x} = \left(-2\gamma_{ij}, j \neq i\right), \widetilde{a}_t = \left(\sigma_{tj}, j \neq i\right)$, and

$$x = \begin{pmatrix} u \\ \tilde{x} \end{pmatrix}, \quad a_t = \begin{pmatrix} 1 \\ \tilde{a}_t \end{pmatrix}, \quad b_t = z_{i\oplus}^t$$

so that $u - 2\left\langle \gamma_{i*}, \sigma_t \right\rangle = \left\langle x, a_t \right\rangle$. Note that $b_t = 1 - z_{i0}^t$. The optimization can be written as

$$\min_{x \in \mathbb{R}^d} \sum_t \left[-b_t \left\langle x, a_t \right\rangle + \log\left(1 + e^{\langle x, a_t \rangle}\right)\right]$$

which is exactly the optimization problem for computing the MLE in a logistic regression model, based on the data $\left(b_t, a_t\right), t \in [n]$.

The parameter u in equation (6.27) represents the intercept in this logistic regression model. Once the logistic regression is fitted, we obtain the estimate $\widehat{\gamma}_{i*}$ for γ_{i*} and $\widehat{u}$ for u. Using $\widehat{u}$ in equation (6.27), we can then compute the estimates $\widehat{\beta}_{i*}(\widehat{u})$ for the marginal parameters. An alternative method for optimizing the conditional log-likelihood, utilizing coordinate descent, is outlined in the Supplement. However, the global approach described above performs better in practice.

Since all parameters can be estimated by performing neighborhood regressions $p\left(z_i^* \mid z_{-i}^*\right)$ for each $i \in [d]$, the problem of fitting equation (6.26) reduces to conducting d parallel logistic regressions. The i-th regression estimates the i-th rows of Γ and B. Because Γ is symmetric, we obtain two estimates for each row/column. This situation is common in neighborhood regression, and there are various methods for combining these estimates. In this case, we simply average them.

Regularization: Since the conditional neighborhood regression reduces to logistic regression, any regularization technique applicable to logistic regression can be directly applied to this model. Specifically, one can minimize $(\beta_i, \gamma_{i*}) \mapsto -\log \ell_i\left(\beta_i, \gamma_{i*}\right) + \rho_{\lambda_i}\left(\gamma_{i*}\right)$, where ρ_λ represents the penalty function, and λ is a regularization parameter. This approach adds $\rho_{\lambda_i}\left(\gamma_{i*}\right)$ to the logistic regression objective. One can also interpret this penalty as introducing a prior on γ_{i*}, specifically one that is proportional to $e^{-\rho_\lambda(\cdot)}$, in a Bayesian framework.

6.4 Conclusion

In conclusion, Gibbs measures provide a powerful framework for structured prediction tasks and graphical models, offering a principled approach to modeling complex dependencies and relationships among variables. This chapter underscores the significance of Gibbs measures in advancing machine learning techniques in structured prediction and graphical modeling domains, paving the way for innovative research and practical applications in diverse fields (see e.g. [41], [219], [278], [378]).

In this chapter, starting with graphical models, we introduced the formal connection between Gibbs distributions and Markov random fields, which allows for compact representation and efficient inference in high-dimensional joint distributions. Examples such as the Ising and Potts models illustrate how local interactions can model global phenomena through energy-based formulations.

The second part of the chapter focused on structured prediction, where the output variables are interdependent and must be predicted jointly. We introduced core concepts and practical methods, including the structured perceptron and energy-based structured prediction models. These approaches leverage Gibbs distributions to score structured outputs, allowing learning algorithms to balance fit and complexity in richly structured spaces.

Finally, we highlighted practical applications, particularly lattice-based models and discrete multivariate data. The Potts-Ising framework offers a powerful approach for modeling spatial interactions in image analysis, computational biology, and networked systems, emphasizing the versatility and interpretability of Gibbs-based models.

Recent works for further reading are [170], [221], [293], [368], [378].

Chapter 7

Advanced Topics in Gibbs Measures and Their Implications in Deep Learning

This chapter synthesizes advanced concepts from statistical mechanics—particularly Gibbs measures—to elucidate theoretical and practical aspects of deep learning. We begin by establishing the foundational role of Gibbs distributions in modeling neural network dynamics. Subsequently, we explore the statistical mechanics of deep learning, addressing: Learning in neural networks via thermodynamic analogies; Kernel renormalization and signal propagation in the infinite-width limit; Entropy-driven analyses of model capacity and generalization.

Moreover, we formalize deep learning optimization through the lens of Boltzmann machines, linking energy minimization to gradient-based training. Then we contextualize Gibbs measures in contemporary models, covering: Bayesian learning via stochastic gradient Langevin dynamics; Attention mechanisms, reinterpreting queries/keys as measure-theoretic operations; Dense associative memories, analyzed through Gibbsian energy landscapes.

Collectively, this chapter demonstrates how Gibbs measures offer a unifying framework for analyzing convergence, generalization, and representational power in deep learning, bridging statistical mechanics with modern architectures.

7.1 Introduction

Deep learning, a subset of machine learning, employs multilayered artificial neural networks trained on large-scale datasets to address complex information processing challenges [42]. Over the past decade, it has become the dominant approach in machine learning, demonstrating remarkable performance across various domains such as computer vision, automatic speech recognition, and natural language processing [162, 234]. Its widespread adoption has extended into numerous sectors, including healthcare, manufacturing, finance, commerce, and scientific research, where it supports applications ranging from medical diagnostics to predictive maintenance and autonomous systems [124, 202].

Current phase in the evolution of neural networks began in the early 2010s (see [41, 42]), marking a significant turning point in the field of artificial intelligence. This period saw the emergence and rapid advancement of techniques that made it possible to train neural networks with a large number of layers—networks that came to be known as deep neural networks. Prior to this era, training deep architectures was a major challenge due to modeling issues like vanishing gradients and lack of powerful computational resources (e.g. GPUs), which hindered performance and scalability.

Several key breakthroughs helped overcome these obstacles. These included the development of more effective training algorithms, such as improved versions of backpropagation, the availability of large datasets for supervised learning (like ImageNet), and the use of high-performance hardware, particularly GPUs, which drastically accelerated training times. As a result, researchers were able to build and optimize much deeper and more complex models than ever before.

More recently, the advent of Transformer-based architectures [374] has enabled the development of large language models (LLMs), marking a significant milestone in artificial intelligence. These models, often comprising hundreds of billions to over a trillion parameters, have demonstrated unprecedented generalization capabilities across a broad spectrum of linguistic and cognitive tasks [59]. As a result, they are increasingly viewed as early indicators of artificial general intelligence (AGI), catalyzing transformative shifts in technology and society [51].

The goals of deep learning theory aim to explain several fundamental aspects which include:

1. *Generalization in over-parameterized models*: Explain why models with vastly more parameters than training samples generalize well, avoiding overfitting. Investigate the role of implicit regularization (e.g., gradient descent dynamics, architectural choices) and data structure [32], [284], [399].

2. *Optimization dynamics*: Understand how non-convex loss landscapes are efficiently navigated by gradient-based methods. Analyze why local minima or saddle points do not hinder convergence [83], [91], [244].

3. *Architecture design principles*: Identify why specific architectures (CNNs, Transformers) excel at tasks like vision or language. Explore inductive biases and connectivity patterns [235], [374].

4. *Interpretability and feature learning*: Decipher how networks represent hierarchical features and make decisions [398].

5. *Scalability and emergent behaviors*: Characterize benefits of scaling model size, data, and compute [206], [381].

6. *Robustness and security*: Diagnose vulnerabilities and establish theoretical guarantees for robustness [162], [163], [261].

7. *Statistical principles in high dimensions*: Address how deep learning circumvents the curse of dimensionality [29], [105], [154].

8. *Connections to neuroscience*: Bridge artificial and biological neural networks (see [171], [392] and references therein).

9. *Theoretical frameworks and tools*: Develop mathematical tools to analyze network behavior [90], [369].

10. *Transfer and Meta-learning*: Explain adaptation to new tasks with limited data [66], [127].

11. *Learning dynamics*: Model the evolution of parameters during training [333].

12. *Ethical and societal implications*: Provide insights into fairness, bias, and societal impact [62], [270].

For deeper exploration, follow citation trails within these papers or refer to textbooks [162], [337].

7.2 Statistical mechanics of deep learning

Today, there is a growing body of research exploring the interplay between statistical mechanics and deep learning, aiming to better understand the theoretical foundations and behavior of complex learning systems. This interdisciplinary approach has led to the development of novel analytical tools for characterizing learning dynamics, generalization, and loss landscapes in neural networks. Notable early and recent contributions have applied concepts such as energy landscapes, phase transitions, and entropy-based measures to study neural network behavior (see [17], [21], [26] and references therein).

For instance, statistical mechanics has been instrumental in analyzing over parameterization, double descent, and the role of noise in stochastic gradient descent. Works such as [269] and [83] have shown that the loss surfaces of deep networks exhibit properties analogous to spin glass models, while [336] and [121] provide foundational insights into the learning dynamics from a thermodynamic perspective. This line of research continues to bridge the gap between machine learning and physics, fostering deeper theoretical insights and practical improvements in model design and training algorithms.

7.2.1 *Statistical mechanics of learning with neural networks*

In this subsection, we follow [387] (see also references therein). Consider the function $H(x;\theta)$ implemented by a neural network parametrized by θ, where x denotes the input arguments. Occasionally, a specific attention in deep learning theory is given to the analysis of the distribution of the network's outputs $H(x;\theta)$ at the start of training. A theoretical appeal of this scenario is that each parameter in the parameter set/vector θ is initialized independently at random.

A Gibbs distribution (measure) is defined over outputs of neural networks with quenched parameters and annealed inputs. Focusing on the Hamiltonian

$$H_{\text{tot}}(x;\theta) = \frac{1}{2\beta}\|x\|^2 + H(x;\theta), \tag{7.1}$$

one can study the structure of network inputs x that minimize H_{tot} as network width tends to infinity.

Using replica methods, one derives saddle-point equations for the disorder-averaged free energy and present numerical evidence for full replica symmetry breaking (RSB) in shallow networks with sinusoidal activations versus replica symmetry in ReLU[1]/tanh networks. For deep networks, the problem can be reduced to solving coupled ODEs and demonstrate dominance of replica-symmetric solutions. These approaches provide insights into adversarial examples, feature visualization, and the statistical mechanics of neural networks.

The Gibbs measure $\exp(-\beta H_{\text{tot}}(x;\theta))$ for neural networks is characterized by the Hamiltonian (7.1), in which θ represents quenched network parameters and $H(x;\theta)$ is the network output.

A notable question in this setting is: *What is the distribution of inputs x with anomalously large negative outputs $H(x;\theta)$?*

On one hand, this question already contains the phenomenon called "adversarial examples". Adversarial examples are the perturbed inputs $x+\delta x$, where x is a sample from the training or test set. The distinguishing

[1] Rectified Linear Unit (ReLU), given by $\text{ReLU}(x) := \max(0, x)$, is one of the most commonly used activation functions in neural networks, especially deep learning models.

property of an adversarial input is that even though δx is an absolutely small perturbation, the network output $H(x + \delta x; \theta)$ differs significantly from $H(x; \theta)$.

On the other hand, it helps us provide a proof of concept for applying the replica method from statistical mechanics to analyze neural networks with quenched weights.

7.2.1.1 *The replica trick and multilayer perceptrons*

To define Hamiltonian H as function of the inputs x, first define the pre-activations recursively as follows:

$$z_i^{(1)}(x) \equiv \sum_{j=1}^{N^{(0)}} W_{ij}^{(1)} x_j, \quad \text{for } 1 \leq i \leq N^{(1)},$$

$$z_i^{(\ell+1)}(x) \equiv \sum_{j=1}^{N^{(\ell)}} W_{ij}^{(\ell+1)} \sigma\left(z_j^{(\ell)}\right), \quad \text{for } 1 \leq \ell \leq L-1,\ 1 \leq i \leq N^{(\ell+1)},$$

$$H(x) \equiv \sum_{i=1}^{N^{(L)}} W_i^{(L+1)} \sigma\left(z_i^{(L)}\right).$$

Hence, the function is computed layer by layer. The ℓ-th layer has width $N^{(\ell)}$ and is calculated in terms of the previous layer. We consider the limit where all $N^{(\ell)}$ are proportional to a large parameter N that tends to infinity:

$$N^{(\ell)} = a^{(\ell)} N, \quad \ell = 1, \ldots, L, \quad N \to \infty$$

We refer to $\sigma(z_i^{(\ell)})$ as the activation of the i-th neuron in layer ℓ, and $z_i^{(\ell)}$ as the pre-activation. The weight matrix $W^{(\ell)}$ has dimensions $N^{(\ell)} \times N^{(\ell-1)}$, with entries drawn i.i.d. from a centered Gaussian distribution with variance $1/N^{(\ell-1)}$, except for $W^{(L+1)}$, whose entries have variance 1.

This convention ensures that the scale of pre-activations $z_i^{(\ell)}$ is independent of N, but the overall Hamiltonian $H(x)$ will scale extensively with N.

Single-system partition function and three types of averages: For a given realization of the network weights $W_{ij}^{(\ell)}$, the partition function of a single system is:

$$Z_\beta = \int \exp\left(-\beta H(x) - \frac{1}{2}\|x\|^2\right) \frac{d^{N^{(0)}} x}{\sqrt{2\pi}^{N^{(0)}}}.$$

This can also be interpreted as the expected value of $\exp(-\beta H(x))$ when x is drawn from a multivariate Gaussian. Since the neural network weights are random, Z_β itself is a random variable.

At this point, it's helpful to clarify the three types of averages:

- **Disorder averages:** These are averages over the random weights of the network, denoted by $\mathbb{E}$. This includes quantities like the average partition function of a given architecture, the average free energy, or the kernel (covariance between outputs for two fixed inputs). The term "disorder average" comes from condensed matter physics, where randomness in a system (e.g., impurities in a material) leads to randomness in the Hamiltonian.
- **Thermodynamic averages:** These are averages over the thermodynamic ensemble, i.e., with respect to the Boltzmann distribution[2]:
$$p(x) \sim \frac{1}{\sqrt{2\pi}^{N^{(0)}}} \exp\left(-\frac{1}{2}\|x\|^2 - \beta H(x)\right)$$
For a fixed neural network, we can talk about the average value of x or $H(x)$. One can also combine these with disorder averages to compute expressions like $\mathbb{E}\left[\int dx\, p(x)\cdot\right]$.
- **Gaussian averages:** These refer to averages over the input distribution when x is drawn from a Gaussian:
$$\frac{1}{\sqrt{2\pi}^{N^{(0)}}} \exp\left(-\frac{1}{2}\|x\|^2\right)$$

These averages are primarily mathematical tools, but they serve an important purpose: the partition function, which normalizes thermodynamic averages, is itself the average of $e^{-\beta H(x)}$ under the Gaussian input distribution. We often consider the *combined* Gaussian and disorder average, since the partition function can be viewed as the disorder average of the Gaussian average of $e^{-\beta H(x)}$.

We are particularly interested in computing $\mathbb{E}[Z_\beta^n]$, the average of the *replicated partition function* over the random weights. More crucially, we aim to compute $\mathbb{E}[\log Z_\beta]$, since many thermodynamic observables such as the disorder-averaged energy can be derived from it. For example, the disorder average of the thermodynamic average of the energy $H(x)$ is:

$$\mathbb{E}_{\text{disorder}}[H(x)] = -\frac{\partial}{\partial\beta}\mathbb{E}[\log Z_\beta] \propto \mathbb{E}\left[\int H(x) e^{-\beta H(x) - \frac{1}{2}\|x\|^2} d^{N^{(0)}}x\right]$$

[2]Boltzmann distribution is a special case of the Gibbs measure applied to classical particles.

Other observables, beyond just energy, are similarly best understood via $\mathbb{E}[\log Z_\beta]$. In some contexts, it's tempting to use the approximation:

$$\mathbb{E}[\log Z] \approx \log \mathbb{E}[Z]$$

However, for the types of Hamiltonians we study, this approximation is highly inaccurate. To evaluate $\mathbb{E}[\log Z_\beta]$, we employ a method known as the *replica trick*, which we now introduce.

7.2.1.2 *The replica trick*

The replica trick, a technique from the statistical mechanics of systems with random (quenched) disorder, is based on two core observations.

(1) The first one is a simple application of the identity $\log Z_\beta = \lim_{n\to 0} \frac{Z_\beta^n - 1}{n}$, namely,

$$\mathbb{E}[\log Z_\beta] = \lim_{n\to 0} \frac{\partial}{\partial n}\mathbb{E}[Z_\beta^n]. \tag{7.2}$$

(2) For positive integers n:

$$\mathbb{E}\left(Z_\beta^n\right) = \int \mathbb{E}\left(\exp\left(-\beta\sum_{a=1}^{n} H\left(x_a\right)\right)\right)\exp\left(-\frac{1}{2}\sum_{a=1}^{n} x_a^2\right)\frac{d^{nN^{(0)}}x_a}{\sqrt{2\pi}^{nN^{(0)}}} \tag{7.3}$$

where each x_a is an independent copy (or *replica*) of the input, conditioned on the same realization of the weights.

In other words, the nth power of the partition function corresponds to a "supersystem" of n identical copies (or *replicas*). The insight of the replica trick is that we can compute $\mathbb{E}[Z_\beta^n]$ for integer n, and then analytically continue the expression to $n \to 0$ to recover $\mathbb{E}[\log Z_\beta]$.

Let us illustrate this with a simple example that introduces two key ideas: the usefulness of *overlap matrices* and the power of the *saddle-point approximation*.

Consider the Hamiltonian:

$$H(x) = \sum_{i=1}^{N} h_i x_i,$$

where the h_i are i.i.d. standard normal random variables. This is a special case of the MLP Hamiltonian with no hidden layers and linear activations.

The partition function is:

$$Z_\beta = \int \exp\left(-\beta h \cdot x - \frac{1}{2}\|x\|^2\right) \frac{d^N x}{\sqrt{2\pi}^N} = \exp\left(\frac{\beta^2}{2} h^2\right).$$

Taking expectations, we observe

$$\begin{aligned} \log \mathbb{E}[Z_\beta] &= -\frac{N}{2}\log(1-\beta^2), \\ \mathbb{E}[\log Z_\beta] &= \frac{N\beta^2}{2}. \end{aligned} \tag{7.4}$$

This shows that the left hand sides of Equation (7.4) cannot be used interchangeably. Now fix $n \geq 1$ and compute the replicated partition function:

$$Z_\beta^n = \int \exp\left(-\beta \sum_{a=1}^n h \cdot x_a - \frac{1}{2}\sum_{a=1}^n \|x_a\|^2\right) \frac{d^{nN} x_a}{\sqrt{2\pi}^{nN}}$$

Integrating over h introduces correlations between replicas:

$$\mathbb{E}[Z_\beta^n] = \int \exp\left(-\frac{N\beta^2}{2}\sum_{a,b=1}^n Q_{ab} - \frac{1}{2}\sum_{a=1}^n \|x_a\|^2\right) \prod_{a=1}^n \frac{d^N x_a}{\sqrt{2\pi}^N},$$

where $Q_{ab} = \frac{1}{N} x_a \cdot x_b$ is the *overlap matrix*, encoding the pairwise similarities between replicas.

To simplify further, we perform a change of variables via the standard identity:

$$f(u) = \frac{2\pi}{N}\int D\Sigma DQ \exp(N\Sigma(Q-u)) f(Q),$$

this is shorthand representation of a measure[3] over the matrix variables Σ and Q. Here f is an essentially arbitrary function f, with the contour for the Σ integral is parallel to the imaginary axis. In terms of the saddle point approximation we will perform later, this actually means we will be minimizing with respect to Σ, not maximizing. This allows us to write

$$\mathbb{E}\left(Z_\beta^n\right) = \int D\Sigma DQ \exp\left(-\frac{N\beta^2}{2}\sum_{a,b=1}^N Q_{ab}\right.$$

$$\left. + \frac{1}{2}\sum_{ab} \Sigma_{ab}\left(NQ_{ab} - x_a \cdot x_b\right) - \sum_{a=1}^n \|x_a\|^2/2\right) \frac{d^{nN} x}{\sqrt{2\pi}^{nN}},$$

[3]That mean integration over the space of matrices Σ and Q, respectively. These matrices are typically symmetric $n \times n$ matrices, and the measure is over all their independent components. For example, if Q is a symmetric $n \times n$ matrix, DQ represents $\prod_{a \leq b} dQ_{ab}$, i.e., a product of differentials over each independent component of Q.

where we drop some factors for $\sqrt{2\pi/N}^{n^2}$ which will go to 1 anyways. We will adopt the notation from physics literature, where Σ is considered a self-energy. However, we note that in neural network literature this quantity is sometimes written $\hat{Q}$, where it is considered a dual kernel. At this point, the x integral is purely Gaussian and we have

$$\mathbb{E}\left(Z_\beta^n\right) = \int D\Sigma DQ \exp(S[\Sigma, Q])$$

$$S = -\frac{N\beta^2}{2} \sum_{a,b=1}^{N} Q_{ab} + \frac{N}{2} \sum_{ab} \Sigma_{ab} Q_{ab} - \frac{N}{2} \log\det\left(\Sigma + \delta_{ab}\right)$$

Note that the quantity in the exponent of the last formula is proportional to N. For large N, one can use steepest descent (a generalization of Laplace's method) to argue that the integral should be approximated by $\exp\left(S\left(\Sigma^*, Q^*\right)\right)$, where Σ^*, Q^* is a stationary point of S. To find the stationary point we take derivatives to obtain

$$\frac{\partial S}{\partial \Sigma_{ab}} = -\frac{N}{2}\left(\Sigma + \delta_{ab}\right)_{ab}^{-1} + \frac{N}{2} Q_{ab}, \qquad \frac{\partial S}{\partial Q_{ab}} = \frac{N}{2}\Sigma_{ab} - \frac{N\beta^2}{2}.$$

Setting these both to zero we get

$$Q_{ab}^* = \left(\delta_{ab} - \beta^2\right)^{-1}$$

In other words, we need to invert the matrix which is equal to the identity minus β^2 times the all 1's matrix. This matrix has one eigenvector which is proportional to $(1, 1, 1, \ldots, 1)$, whose eigenvalues is $1 - n\beta^2$. The other $n-1$ eigenvalues are degenerate, and equal to 1 . Therefore

$$Q_{ab}^* = \delta_{ab} + \frac{1}{n}\left(\frac{1}{1 - n\beta^2} - 1\right)$$

Plugging this back into the action and using the steepest descent approximation gives

$$\begin{aligned}\log \mathbb{E}(Z_\beta^n) = &-\frac{nN\beta^2}{2} \cdot \frac{1}{1 - n\beta^2} + \frac{nN}{2} \\ &- \frac{nN}{2}\left(1 + \frac{1}{n} \cdot \frac{1}{1 - n\beta^2}\right) - \frac{N}{2}\log(1 - n\beta^2).\end{aligned}$$

This simplifies to

$$\log \mathbb{E}(Z^n) = -\frac{N}{2}\log(1 - n\beta^2).$$

The strategy for this simple problem was to rewrite the partition function in terms of an $n \times n$ overlap matrix Q and a Lagrange multiplier Σ, obtain a large action proportional to the system size, solve the stationary point of this action, and then take the $n \to 0$ limit. This same basic strategy will be used to compute $\mathbb{E}(Z_\beta^n)$ and $\mathbb{E}[\log Z_\beta]$ for the multilayer perceptron in the next sections.

The replicated action: Now derive a path integral expression (i.e., an explicit action) for the average value of Z_β^n:

$$\mathbb{E}\left(Z_\beta^n\right) = \int \mathbb{E}\left(\exp\left(-\beta \sum_{a=1}^{n} H\left(x_a\right)\right)\right) \exp\left(-\frac{1}{2}\sum_{a=1}^{n} \|x_a\|^2\right) \frac{d^{nN^{(0)}} x_a}{\sqrt{2\pi}^{nN^{(0)}}}. \tag{7.5}$$

To do this, we define an $n \times n$ overlap matrix $Q^{(\ell)}$ in each layer:

$$Q_{ab}^{(0)} = \frac{1}{N^{(0)}} \sum_{i=1}^{N^{(0)}} x_{ia} x_{ib}, \quad Q_{ab}^{(\ell)} = \frac{1}{N^{(\ell)}} \sum_{i=1}^{N^{(\ell)}} \sigma\left(z_{ia}^{(\ell)}\right) \sigma\left(z_{ib}^{(\ell)}\right) \qquad \text{for } 1 \le \ell \le L$$

The integral in (7.5) requires to compute the Laplace transform of replica summed Hamiltonian $\sum_a H\left(x_a\right)$. The strategy for performing this computation starts by noticing that while $\sum_a H\left(x_a\right)$ is a complicated function of x_a, it is in fact a simple function of the overlap matrix $Q^{(L)}$. Then one needs to compute the distribution of $Q^{(L)}$ as a function of x_a. This is again a complicated distribution. However, one can introduce dual overlap matrices $\Sigma^{(\ell)}$ to efficiently compute the distribution of $Q(\ell)$ given $Q^{(\ell-1)}$ for all $\ell = L, \ldots, 1$. Finally, the distribution of $Q^{(1)}$ given x_a is simple and we will obtain an explicit action for $\mathbb{E}\left(Z_\beta^n\right)$.

To implement this plan, the key insight is that conditional on a given $Q_{ab}^{(\ell)}$, the preactivations of the next layer $z_{ia}^{(\ell+1)}$ are jointly Gaussian with mean 0 and covariance

$$\operatorname{Cov}\left(z_{ia}^{(\ell+1)}, z_{jb}^{(\ell+1)}\right) = \delta_{ij} Q_{ab}^{(\ell)}.$$

Hence,

$$\mathbb{P}\left(Q^{(\ell+1)} \mid Q^{(\ell)}\right)$$

$$= \int \exp\left(-\frac{1}{2}\sum_{ab} \Sigma_{ab}^{(\ell+1)} \left(N^{(\ell+1)} Q_{ab}^{(\ell+1)} - \sum_i \sigma\left(z_{ia}^{(\ell+1)}\right) \sigma\left(z_{ib}^{(\ell+1)}\right)\right)\right) \times$$

$$D^{N^{(\ell+1)}} z^{(\ell+1)} D\Sigma_{ab}^{(\ell+1)},$$

where the integral is over the space of Lagrange multiplier $n \times n$ matrices $\Sigma_{ab}^{(\ell+1)}$ and the measure $D^{N^{(\ell+1)}} z^{(\ell+1)}$ is shorthand for the Gaussian integrand

$$D^{N^{(\ell+1)}} z^{(\ell+1)} = \frac{d^{nN^{(\ell+1)}} z_{ia}^{(\ell+1)}}{\sqrt{\det 2\pi Q^{(\ell)}}^{N^{(\ell)}}} \exp\left(-\frac{1}{2}\sum_{i}\sum_{ab} z_{ia}^{(\ell+1)} \left(Q_{ab}^{(\ell)}\right)^{-1} z_{ib}^{(\ell+1)}\right).$$

To simply the notation, let us define

$$\begin{aligned}\mathcal{Z}_{\Sigma_{ab};Q_{ab}} &\equiv \int \exp\left(-\frac{1}{2}\sum_{ab} z_a \left(Q_{ab}\right)^{-1} z_b + \frac{1}{2}\Sigma_{ab}\sigma\left(z_a\right)\sigma\left(z_b\right)\right) \frac{d^n z}{\sqrt{\det 2\pi Q}} \\ &= \int \exp\left(\frac{1}{2}\sum_{ab}\Sigma_{ab}\sigma\left(z_a\right)\sigma\left(z_b\right)\right) Dz \qquad (7.6)\end{aligned}$$

to be the expected value of $\exp\left(\frac{1}{2}\Sigma_{ab}\sigma\left(z_a\right)\sigma\left(z_b\right)\right)$ where z is a Gaussian with covariance Q. Now we can write

$$\mathbb{P}\left(Q^{(\ell+1)} \mid Q^{(\ell)}\right)$$

$$= \int D\Sigma_{ab}^{(\ell+1)} \exp\left(-\frac{1}{2}N^{(\ell+1)}\sum_{ab}\Sigma_{ab}^{(\ell+1)}Q_{ab}^{(\ell+1)} + N^{(\ell+1)}\log\mathcal{Z}_{\Sigma_{ab}^{(\ell+1)};Q^{(\ell)}}\right)$$

Similarly to derivation of the distribution of $Q^{(\ell+1)}$ conditioned on $Q^{(\ell)}$, we can also derive the distribution of H conditioned on $Q^{(L)}$. If we condition on a given value of $Q_{ab}^{(L)}$, the n variables $H\left(x_{ia}\right)$ are joint Gaussian with covariance $N^{(L)}Q_{ab}^{(L)}$. Thus

$$\begin{aligned}\mathbb{E}\left(Z^n(b) \mid Q_{ab}^{(L)}\right) &= \mathbb{E}\left(\exp\left(-b\sum_{a} H\left(x_{ia}\right)\right) \mid Q_{ab}^{(L)}\right) \\ &= \exp\left(-\frac{1}{2}\beta^2 N^{(L)}\sum_{ab} Q_{ab}^{(L)}\right).\end{aligned}$$

We now have a way to get the distribution of H conditioned on $Q^{(L)}$ and the distribution of $Q^{(\ell+1)}$ conditioned on $Q^{(\ell)}$. All we need is an expression for the probability distribution of $Q^{(0)}$ in the Gaussian input distribution, and we can derive an expression for the disorder+Gaussian average for $\exp\left(-\beta\sum_{a=1}^{n} H\left(x_a\right)\right)$, i.e. the replicated partition function. We can write

$$\begin{aligned}p\left(Q^{(0)}\right) &= \int DxD\Sigma_{ab}^{(0)} \exp\left(-\frac{1}{2}\sum_{ab}\Sigma_{ab}^{(0)}\left(N^{(0)}Q_{ab}^{(0)} - \sum_{i} x_{ia}x_{ib}\right)\right) \\ &= \int D\Sigma_{ab}^{(0)} \exp\left(-\frac{1}{2}N^{(0)}\left(\operatorname{tr}\Sigma^{(0)}Q^{(0)} + \operatorname{tr}\log\left(I - \Sigma^{(0)}\right)\right)\right).\end{aligned}$$

We can use a saddle-point approximation for $\Sigma^{(0)}$, getting $\Sigma^{(0)} = I - Q^{(0)-1}$. Plugging this in gives

$$p\left(Q^{(0)}\right) = \exp\left(-N^{(0)}\mathrm{KL}\left(Q^{(0)} \mid I\right)\right)$$

where KL $\left(Q^{(0)} \mid I\right)$ the Kullback-Liebler (KL) divergence between a multivariate Gaussian with covariance $Q_{ab}^{(0)}$ and one with covariance δ_{ab}. This can be written

$$\mathrm{KL}\left(Q^{(0)} \mid I\right) = \frac{1}{2}\left(\operatorname{tr} Q^{(0)} - \operatorname{tr}\log Q^{(0)} - n\right)$$

Combining everything, we have the following expression for the expected value of the replicated partition function:

$$\begin{aligned}
\mathbb{E}\left(Z_\beta^n\right) &= \int D^{L+1}\Sigma D^{L+1}Q \exp(S[\Sigma, Q]) \\
S[\Sigma, Q] &= \frac{1}{2}\beta^2 N^{(L)} \sum_{ab} Q_{ab}^{(L)} + \sum_{\ell=1}^{L} N^{(\ell)}\left(\log \mathcal{Z}_{\Sigma^{(\ell)};Q^{(\ell-1)}} - \frac{1}{2}\sum_{ab}\Sigma_{ab}^{(\ell)} Q_{ab}^{(\ell)}\right) \\
&\quad - \frac{1}{2}N^{(0)}\left(\operatorname{tr}\Sigma^{(0)}Q^{(0)} + \operatorname{tr}\log\left(I - \Sigma^{(0)}\right)\right).
\end{aligned} \tag{7.7}$$

We note in passing that one might also be interested in the extreme value statistics of the squared magnitude of layer L rather than the statistics of a linear combination. Since the squared magnitude of replica a is just $N^{(L)}Q_{aa}^{(L)}$, the action is the same with the first term replaced with $\beta N^{(L)}\sum_a Q_{aa}^{(L)}$.

Saddle-point equations for the replicated action: The equations (7.7) provide an exact expression (up to the saddle point approximation for the distribution of $Q^{(0)}$) for calculating the replicated partition function $\mathbb{E}\left(Z^n\right)$ for multilayer perceptrons in terms of model hyperparameters like L and the N s. But actually evaluating this integral over $2Ln^2$ variables is infeasible (especially when n isn't even an integer). Instead, we derive explicit forms of the saddle point equations for the action S:

$$\frac{\partial S}{\partial Q^{(\ell)}} = 0, \qquad \frac{\partial S}{\partial \Sigma^{(\ell)}} = 0, \qquad \text{for all } 0 \leq \ell \leq L$$

Evaluating the derivatives of S with respect to the Q s and Σs will involve taking derivatives of $\log \mathcal{Z}_{\Sigma;Q}$. For this it is useful to define

$$\begin{aligned}
\langle \bullet \rangle_{\Sigma;Q} \equiv \frac{1}{\mathcal{Z}_{\Sigma;Q}} \int \bullet \exp\left(-\frac{1}{2}\sum_{ab} z_a \left(Q_{ab}\right)^{-1} z_b - \Sigma_{ab}\sigma\left(z_a\right)\sigma\left(z_b\right)\right) \\
\times \frac{d^n z}{\sqrt{\det 2\pi Q}}.
\end{aligned}$$

This means that $\langle \bullet \rangle_{\Sigma;Q}$ is the expectation value of $\bullet$ with respect to a probability density over n variables z_a whose joint density is proportional to the exponential in the integrand. A direct computation then gives

$$\frac{\partial \mathcal{Z}_{\Sigma;Q}}{\partial Q_{ab}} = \frac{1}{2} Q_{a\mu}^{-1} \langle z_\mu z_\nu - Q_{\mu\nu} \rangle_{\Sigma;Q} Q_{\nu b}^{-1}, \quad \frac{\partial \mathcal{Z}_{\Sigma;Q}}{\partial \Sigma_{ab}} = \frac{1}{2} \langle \sigma(z_a) \sigma(z_b) \rangle_{\Sigma;Q}.$$

After a bit of algebra, the saddle point equations take the following form:

$$Q^{(0)} = \left(I - \Sigma^{(0)} \right)^{-1}, \tag{7.8}$$

$$Q^{(\ell)} = \langle \sigma(z_a) \sigma(z_b) \rangle_{\Sigma^{(\ell)}; Q^{(\ell-1)}} \quad \text{for } 1 \leq \ell \leq L, \tag{7.9}$$

$$\Sigma^{(\ell)} = \frac{N^{(\ell-1)}}{N^{(\ell)}} \left(\frac{1}{2} Q_{a\mu}^{(\ell)-1} \left\langle z_\mu z_\nu - Q_{\mu\nu}^{(\ell)} \right\rangle_{\Sigma^{(\ell+1)}; Q^{(\ell)}} Q_{\nu b}^{(\ell)-1} \right),$$
$$\text{for } 0 \leq \ell \leq L-1, \tag{7.10}$$

$$\Sigma_{ab}^{(L)} = \beta^2. \tag{7.11}$$

Now we focus on evaluating the right-hand sides of the saddle-point equations in the limit $n \to 0$, particularly in the replica-symmetric case, where we assume the forms:

$$Q_{ab}^{(\ell)} = c^{(\ell)} \delta_{ab} + q^{(\ell)}, \quad \Sigma_{ab}^{(\ell)} = \Lambda^{(\ell)} \delta_{ab} + \Gamma^{(\ell)}.$$

From equation (7.8), we obtain:

$$c^{(0)} = \frac{1}{1 - \Lambda^{(0)}}, \qquad \frac{q^{(0)}}{c^{(0)}} = \frac{\Gamma^{(0)}}{1 - \Lambda^{(0)}}.$$

Moreover, equation (7.11) yields directly:

$$\Lambda^{(L)} = 0, \qquad \Gamma^{(L)} = \beta^2.$$

The more challenging part is evaluating the right-hand sides of equations (7.9) and (7.10), for which we refer to Appendix A in [387]. Once the equations of motion are known, solving for the quantities $c^{(\ell)}$, $q^{(\ell)}$, $\Lambda^{(\ell)}$, and $\Gamma^{(\ell)}$ becomes surprisingly straightforward using what is called the zipper method:

(1) Initialize all values of $c^{(\ell)}$, $q^{(\ell)}$, $\Lambda^{(\ell)}$, and $\Gamma^{(\ell)}$. For small or intermediate values of β, initializing to zero is often sufficient. For larger β, it may be beneficial to initialize using a solution from a smaller β and proceed recursively.
(2) Use equations (7.8) and (7.9) to solve for $c^{(\ell)}$, $q^{(\ell)}$ step-by-step, starting from $\ell = 0$ up to $\ell = L$.

(3) Use equations (7.11) and (7.10) to solve for $\Lambda^{(\ell)}$, moving backward from $\ell = L$ to $\ell = 0$.
(4) Repeat steps 2 and 3 until convergence is observed. Typically, four to five iterations suffice.

This method is guaranteed to work in the high-temperature regime (small β) and, empirically, shows good convergence even for larger β. However, one must proceed with caution: the zipper method cannot spontaneously break replica symmetry. If the initial conditions are replica-symmetric (RS), they will remain RS at every iteration, even in cases where a replica symmetry breaking (RSB) solution is more thermodynamically favorable.

To address this, one may introduce a slight RSB perturbation and monitor whether it grows or diminishes as the zipper algorithm progresses.

Throughout the work [387], the authors have assumed that the overlap matrices obey the Parisi ansatz. This ansatz implies that the dominant states of the Boltzmann distribution satisfy the ultrametric triangle inequality (see [327] and references therein for a theory on ultrametric):

$$d(A, C) \leq \max\left(d(A, B), d(B, C)\right), \quad \forall A, B, C.$$

However, this structure is inconsistent with certain key observations regarding concept embeddings in neural activation space. A classic example is the word analogy:

$$\texttt{embed}(\text{Man}) - \texttt{embed}(\text{Woman}) \approx \texttt{embed}(\text{King}) - \texttt{embed}(\text{Queen}),$$

which suggests the four associated vectors form a parallelogram in activation space [275]. Nonetheless, parallelograms violate the ultrametric triangle inequality. Specifically, for any parallelogram $ABCD$, we have:

$$AC^2 + BD^2 = 2(AB^2 + BC^2),$$

implying that:

$$\max(AC, BD)^2 \geq AB^2 + BC^2.$$

This directly contradicts the ultrametric inequality when either AC^2 or BD^2 exceeds $AB^2 + BC^2$.

While we cannot exhaustively search the space of $n \times n$ matrices to confirm the dominance of the Parisi ansatz, several lines of evidence support its validity. For instance, a generalization of the replicon analysis to layered overlap structures reveals no instability corresponding to parallelogram-like configurations in activation space.

Deep linear networks: Here we solve equations (7.8)–(7.11) in case of deep linear networks in which the activation function is $\sigma(t) = t$.

Note that when the activation function σ is the identity, the energy function $H(x)$ becomes linear in x. Consequently, the function

$$\frac{1}{2}\|x\|^2 + \beta H(x)$$

has only one local minimum. This ensures replica symmetry for the saddle points of the replicated action introduced in equation (7.7). Thus, we may write the overlap matrix as:

$$Q_{ab}^{(\ell)} = c^{(\ell)}\delta_{ab} + q^{(\ell)}. \tag{7.12}$$

Moreover, for linear networks, the partition function expression in equation (7.6) simplifies to:

$$Z^{-2}_{\Sigma^{(\ell)};Q^{(\ell-1)}} = \det\left(I - \Sigma^{(\ell)}Q^{(\ell-1)}\right).$$

Substituting this into equation (7.10) allows us to exactly solve for the equation of motion for $\Sigma^{(\ell)}$:

$$\Sigma^{(\ell)} = \left(Q^{(\ell)}\right)^{-1} - \left(Q^{(\ell-1)}\right)^{-1}. \tag{7.13}$$

Plugging equation (7.13) back into equation (7.7), the action can be rewritten solely in terms of the overlap matrices Q:

$$S = \frac{1}{2}\beta^2 N^{(L)} \sum_{ab} Q_{ab}^{(L)} - \sum_{i=0}^{L} N^{(i)}\,\mathrm{KL}(Q^{(i)} \,|\, Q^{(i-1)}),$$

where we define $Q^{(-1)} \equiv I$. Using the replica-symmetric form from equation (7.12), the action becomes:

$$\frac{S}{n} = \frac{1}{2}\beta^2 N^{(L)} c^{(L)}$$

$$-\frac{1}{2}\sum_{i=0}^{L} N^{(i)}\left[\frac{c^{(i)}}{c^{(i-1)}} - \log\left(\frac{c^{(i)}}{c^{(i-1)}}\right) + \left(\frac{q^{(i)}}{c^{(i)}} - \frac{q^{(i-1)}}{c^{(i-1)}}\right)\cdot\left(\frac{c^{(i)}}{c^{(i-1)}} - 1\right)\right], \tag{7.14}$$

with the initial conditions $c^{(-1)} \equiv 1$, $q^{(-1)} \equiv 0$.

Recomputing the equations of motion for $q^{(i)}$ from this action shows that they are only satisfied when:

$$\frac{c^{(i)}}{c^{(i-1)}} = 1 \quad \text{for all } i,$$

which implies $c^{(i)} = 1$ for all i. That is, the thermodynamically random component of the activations (parameterized by the $c^{(i)}$) does not change with depth.

Differentiating with respect to $c^{(i)}$ and substituting $c^{(i)} = 1$, we obtain for $0 \leq i < L$:

$$N^{(i)}\left(q^{(i)} - q^{(i-1)}\right) = N^{(i+1)}\left(q^{(i+1)} - q^{(i)}\right),$$

and at the final layer $i = L$:

$$\beta^2 = q^{(L)} - q^{(L-1)}.$$

From these relations, we deduce a recursive formula:

$$q^{(i)} - q^{(i-1)} = \frac{N^{(L)}}{N^{(i)}}\beta^2,$$

with $q^{(-1)} = 0$. Solving this recurrence yields:

$$c^{(i)} = 1, \qquad q^{(i)} = \beta^2 \sum_{j=0}^{i} \frac{N^{(L)}}{N^{(j)}}, \tag{7.15}$$

for all $0 \leq i \leq L$.

Let us interpret equation (7.15) in words following [387]. As we progress deeper into the network, the random part of the activation vectors (captured by $c^{(i)}$) remains unchanged, but the deterministic part (encoded in $q^{(i)}$) increases monotonically. This leads to the angle between activation vectors of different replicas, given by

$$\arccos\left(\frac{q^{(i)}}{c^{(i)} + q^{(i)}}\right),$$

gradually becoming smaller, reflecting the increasing alignment between replicas.

7.2.2 *The infinite width limit and backpropagating kernel renormalization*

The paper [246] (see also [26]) presents a theoretical framework for analyzing deep linear neural networks (DLNNs) using tools from statistical mechanics. The authors introduce the backpropagating kernel renormalization (BPKR), a method to compute equilibrium properties of weight distributions induced by supervised learning. Despite focusing on linear networks, they provide insights into nonlinear phenomena in deep learning, such as generalization in overparameterized regimes and the role of depth.

Bayesian learning and statistical mechanics: In the statistical mechanics view of learning, a neural network's weights $\Theta \in \mathbb{R}^M$ are treated as microscopic degrees of freedom, where M denotes the total number of parameters. The training dataset $\mathcal{D} = \{(x^\mu, y^\mu)\}_{\mu=1}^P$, consisting of P input-output pairs with $x^\mu \in \mathbb{R}^d$ and $y^\mu \in \mathbb{R}$, plays the role of quenched disorder.

Assuming a Gaussian likelihood:

$$P(y^\mu|x^\mu, \Theta) \propto \exp\left(-\frac{\beta}{2}[f(x^\mu, \Theta) - y^\mu]^2\right),$$

and a Gaussian prior over weights:

$$P(\Theta) \propto \exp\left(-\frac{1}{2\sigma^2}|\Theta|^2\right),$$

the posterior distribution over parameters becomes the Gibbs distribution:

$$P(\Theta|\mathcal{D}) = Z^{-1}\exp\left(-\frac{\beta}{2}\sum_{\mu=1}^{P}[f(x^\mu, \Theta) - y^\mu]^2 - \frac{1}{2\sigma^2}|\Theta|^2\right), \tag{7.16}$$

where $f(x^\mu, \Theta)$ is the output of the neural network for input x^μ, $\beta = 1/T$ is the inverse temperature controlling the stochasticity (or noise level), σ^2 is the prior variance on weights, $Z = \int d\Theta \, \exp(-\beta L(\Theta|\mathcal{D}) - \frac{1}{2\sigma^2}|\Theta|^2)$ is the partition function, and $L(\Theta|\mathcal{D}) = \frac{1}{2}\sum_{\mu=1}^{P}[f(x^\mu, \Theta) - y^\mu]^2$ is the empirical risk (negative log-likelihood).

The generating functional, central to path integral and statistical field theory methods, is given by:

$$Z(l) = \int d\Theta \exp\left(-\beta L(\Theta|\mathcal{D}) - \frac{1}{2\sigma^2}|\Theta|^2 - il\frac{1}{\sqrt{N}}\sum_{i=1}^{N} a_i \, x_i^L(W, x)\right), \tag{7.17}$$

where $l \in \mathbb{R}$ is an auxiliary source parameter, $a_i \in \mathbb{R}$ are fixed probe vectors, $x_i^L(W, x)$ denotes the ith component of the network output at layer L with weights W, and N is the width (number of neurons) in each hidden layer, see Fig. 7.1.

This functional enables computation of output statistics such as moments and correlations via functional derivatives with respect to l.

Infinite width limit. As the width $N \to \infty$, randomly initialized feedforward neural networks converge in distribution to a Gaussian Process (GP) [237, 282]. For a deep network with L layers, consider the layer-wise outputs $x^l(x) \in \mathbb{R}^N$ defined recursively by:

$$x^l(x) = \phi\left(\frac{1}{\sqrt{N}}W^l x^{l-1}(x) + b^l\right), \quad l = 1, \ldots, L,$$

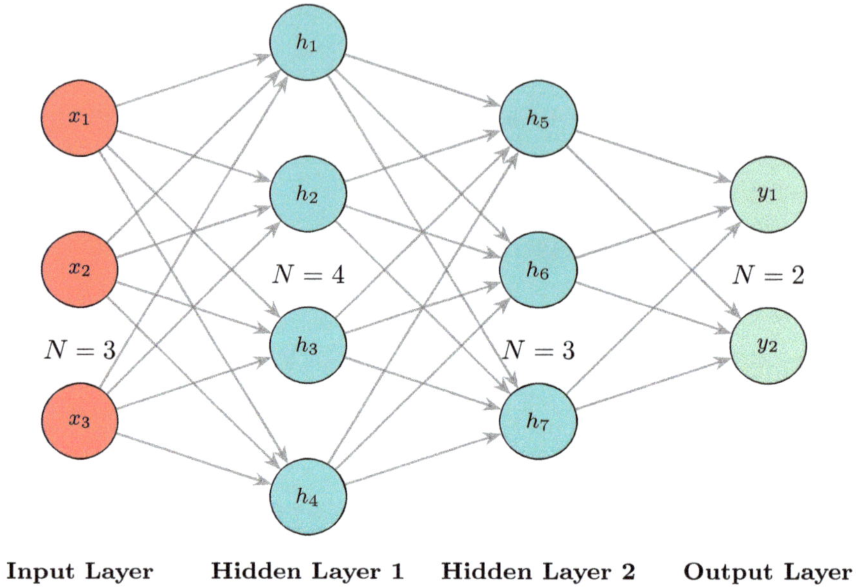

Fig. 7.1: Simple feedforward neural network with fully connected layers of width N.

where $W^l \in \mathbb{R}^{N\times N}$ is the weight matrix at layer l with entries $W^l_{ij} \sim \mathcal{N}(0, \sigma_w^2)$, $b^l \sim \mathcal{N}(0, \sigma_b^2 I)$ is the bias vector at layer l, $\phi : \mathbb{R} \to \mathbb{R}$ is the activation function (e.g., ReLU, tanh), and $x^0(x) = x$ is the input vector.

The kernel of the induced Gaussian process at layer L is:

$$K_L(x,y) = \sigma_w^2 \, \mathbb{E}_{w^L}\left[\phi\left(\frac{1}{\sqrt{N}} w^L \cdot x^{L-1}(x)\right)\phi\left(\frac{1}{\sqrt{N}} w^L \cdot x^{L-1}(y)\right)\right], \tag{7.18}$$

where the expectation is taken over random weights $w^L \sim \mathcal{N}(0, I_N)$, and $x^{L-1}(\cdot)$ is the penultimate layer output.

This kernel evolves recursively via:

$$K_l(x,y) = \mathbb{E}_{(z_1,z_2)\sim\mathcal{N}(0,\Sigma_{l-1})}[\phi(z_1)\phi(z_2)], \quad \Sigma_{l-1} = \begin{pmatrix} K_{l-1}(x,x) & K_{l-1}(x,y) \\ K_{l-1}(y,x) & K_{l-1}(y,y) \end{pmatrix}, \tag{7.19}$$

starting from the initial kernel $K_0(x,y) = \frac{1}{d} x^\top y$ (assuming normalized inputs). This recursion converges to a fixed-point kernel $K_L(x,y)$, which defines the GP prior over function space in the infinite-width limit.

The connection to the statistical mechanics framework becomes clear when we recognize that the kernel $K_L(x,y)$ determines the covariance structure of the Gaussian process, which in turn governs the partition function through the likelihood term. The generating functional $Z(l)$ can be evaluated exactly in this limit, yielding insights into the learning dynamics.

Deep linear networks. Deep linear networks provide a particularly tractable case within the statistical mechanics framework because the absence of nonlinearities allows for exact analytical solutions. In these networks, the activation functions are identity maps, so the output at layer l is given by:

$$x^l = W^l x^{l-1}, \quad \text{for } l = 1, \dots, L,$$

where $W^l \in \mathbb{R}^{N \times N}$ are the weight matrices, and $x^0 = x \in \mathbb{R}^d$ is the input. The overall network function is thus a composition of linear maps, resulting in:

$$f(x) = W^L W^{L-1} \cdots W^1 x = \mathcal{W}x, \quad \text{with } \mathcal{W} := \prod_{l=1}^{L} W^l.$$

Despite the absence of nonlinearities, these models exhibit rich behavior due to the multiplicative structure of weight matrices. Crucially, the kernel recursion simplifies dramatically: for linear activations $\phi(z) = z$, the kernel evolution becomes $K_l(x, y) = \sigma_w^2 K_{l-1}(x, y)$, leading to $K_L(x, y) = \sigma_w^{2L} K_0(x, y)$.

In the statistical mechanics framework, the partition function quantifies the volume of weight configurations that fit the training data under a given prior and likelihood. For a dataset $\mathcal{D} = \{(x^\mu, y^\mu)\}_{\mu=1}^P$, and under a Gaussian prior and Gaussian likelihood, the partition function can be written as:

$$Z = \int \left[\prod_{l=1}^{L} dW^l\right] \exp\left(-\frac{\beta}{2}\sum_{\mu=1}^{P} |f(x^\mu) - y^\mu|^2 - \sum_{l=1}^{L} \frac{1}{2\sigma^2}\|W^l\|_F^2\right).$$

To compute this analytically, particularly in the thermodynamic limit where width $N \to \infty$, one can integrate out the top-layer weights using the generating functional approach. The key technique is to exploit the Gaussian structure of both the prior and likelihood, which makes the high-dimensional integrals tractable through standard Gaussian integration formulas. After integrating out the top layer and introducing a Fourier representation for the delta function enforcing the data constraints, we obtain a reduced partition function at the $(L-1)$-th layer:

$$Z_{L-1} = \int dt \exp\left(it^T Y + NG(t) - \frac{T}{2} t^T t\right), \tag{7.20}$$

where $t \in \mathbb{R}^P$ is an auxiliary integration variable (dual to the data constraints), $Y \in \mathbb{R}^P$ is the vector of labels y^μ, $G(t)$ encodes the contribution of the network kernel, and $T = \beta^{-1}$ is the effective temperature.

The function $G(t)$ is given by:

$$G(t) = -\frac{1}{2}\log\left(1 + \frac{\sigma^2}{N}t^T K_{L-1} t\right), \tag{7.21}$$

where $K_{L-1} \in \mathbb{R}^{P\times P}$ is the empirical kernel matrix of the network up to layer $L-1$, defined by:

$$[K_{L-1}]_{\mu\nu} = x^{L-1}(x^\mu)^T x^{L-1}(x^\nu).$$

This analytical tractability allows for a renormalization group (RG) interpretation of the kernel dynamics: the transformation $K_l \mapsto K_{l+1}$ across layers mimics RG flow in statistical physics, describing how data representations evolve in depth. The backpropagating kernel renormalization (BPKR) method leverages this connection, treating each layer integration as a renormalization step that progressively integrates out degrees of freedom while preserving the essential statistical properties of the learning problem. This perspective unifies the infinite width limit, the kernel evolution, and the statistical mechanics framework into a coherent analytical approach.

7.2.3 *Gibbs measures in neural network training and generalization*

7.2.3.1 *The Gibbs posterior in neural networks*

In the Bayesian framework for neural networks, the posterior distribution over parameters Θ given training data $\mathcal{D} = \{(x_i, y_i)\}_{i=1}^P$ takes the form of a Gibbs measure. Starting from Bayes' theorem:

$$P(\Theta|\mathcal{D}) = \frac{P(\mathcal{D}|\Theta)P(\Theta)}{P(\mathcal{D})} = \frac{1}{Z}\exp\left(-\beta L(\Theta)\right),$$

where $L(\Theta) = -\log P(\mathcal{D}|\Theta) - \log P(\Theta)$ is the negative log-posterior (loss function plus regularization), $\beta = 1/T$ is the inverse temperature, and $Z = \int \exp(-\beta L(\Theta))d\Theta$ is the partition function.

This canonical Gibbs form reveals that neural network training fundamentally involves sampling from or finding modes of a Gibbs measure defined by the loss landscape [260, 282].

The temperature parameter T controls the concentration of the posterior. At $T \to 0$ (cold regime), the posterior concentrates on global minima, corresponding to maximum a posteriori (MAP) estimation. At $T \to \infty$ (hot regime), the posterior approaches the prior, ignoring the data. At finite T, the posterior balances data fit with prior beliefs and regularization.

The partition function Z encodes the marginal likelihood (model evidence):

$$Z = P(\mathcal{D}) = \int P(\mathcal{D}|\Theta)P(\Theta)d\Theta,$$

which is crucial for Bayesian model selection but computationally intractable for neural networks [141].

7.2.3.2 *SGD as Gibbs sampling*

Stochastic gradient descent (SGD) can be viewed as an approximate sampling algorithm for the Gibbs posterior. In the continuous-time limit, SGD follows the Langevin dynamics:

$$d\Theta_t = -\nabla_\Theta L(\Theta_t)dt + \sqrt{2T_{\text{eff}}}dW_t,$$

where dW_t is a Wiener process (Brownian motion) and T_{eff} is the effective temperature arising from the learning rate η, minibatch noise, and discrete time effects. Higher learning rates increase effective temperature, while stochastic gradients introduce thermal fluctuations, and finite step size creates additional noise. The effective temperature can be approximated using N, the dataset size, B, the minibatch size, and Σ_{grad}, the gradient covariance matrix [262].

The Fokker-Planck equation governing the evolution of the parameter distribution $P(\Theta, t)$ is:

$$\frac{\partial P(\Theta, t)}{\partial t} = \nabla_\Theta \cdot \left(\nabla_\Theta L(\Theta)P(\Theta, t) + T_{\text{eff}}\nabla_\Theta P(\Theta, t)\right),$$

which has the Gibbs measure as its unique equilibrium distribution:

$$P_{\text{eq}}(\Theta) = \frac{1}{Z}\exp\left(-\frac{L(\Theta)}{T_{\text{eff}}}\right).$$

This connection explains why SGD can find good solutions: it naturally explores the loss landscape according to the Gibbs measure, spending more time in low-loss regions while occasionally escaping local minima through thermal fluctuations [78].

7.2.3.3 *Generalization through Gibbs Entropy*

The generalization capability of neural networks is intimately connected to the entropy of the Gibbs posterior. The entropy quantifies the effective volume of parameter space consistent with the training data:

$$S = -\int P(\Theta|\mathcal{D}) \log P(\Theta|\mathcal{D}) d\Theta = \langle L(\Theta) \rangle_{\text{Gibbs}}/T + \log Z,$$

where $\langle \cdot \rangle_{\text{Gibbs}}$ denotes expectation under the Gibbs posterior.

Flat minima and high entropy: Flat minima in the loss landscape correspond to high-entropy regions of the Gibbs measure. The local entropy around a minimum Θ^* can be approximated using the Hessian $H = \nabla^2 L(\Theta^*)$:

$$S_{\text{local}} \approx \frac{1}{2} \log \det(2\pi T H^{-1}) = \frac{1}{2} \sum_{i=1}^{M} \log \left(\frac{2\pi T}{\lambda_i} \right),$$

where λ_i are the eigenvalues of the Hessian and M is the number of parameters.

Smaller eigenvalues (flatter directions) contribute more to the entropy, explaining why flat minima are associated with better generalization [189, 212].

Double descent as phase transitions: The double descent phenomenon in overparameterized models can be understood through phase transitions in the Gibbs measure [32, 279]. In the underparameterized regime ($M < P$), the Gibbs measure has low entropy, concentrated around a few solutions, where limited model capacity prevents overfitting. At the interpolation threshold ($M \approx P$), this critical phase sees the measure become concentrated on sharp minima that perfectly fit the training data, leading to poor generalization. In the overparameterized regime ($M \gg P$), the measure develops high entropy with many interpolating solutions, where implicit regularization from SGD dynamics guides the system toward flatter, more generalizable solutions.

The free energy $F = -T \log Z$ exhibits characteristic scaling behavior at these transitions, providing theoretical insight into the double descent curve.

7.2.3.4 *Temperature effects in deep learning*

Temperature plays a crucial role in various aspects of deep learning beyond basic training dynamics.

Temperature scaling for calibration: Modern neural networks often produce overconfident predictions. Temperature scaling [167] applies a learned temperature parameter T_{cal} to the logits before the final softmax:

$$P(y = k|x) = \frac{\exp(z_k/T_{\text{cal}})}{\sum_{j=1}^{C} \exp(z_j/T_{\text{cal}})},$$

where z_k are the network logits and C is the number of classes. This transforms the output distribution to better match the true posterior probabilities.

Annealing schedules: Temperature annealing in training can improve convergence and generalization through several approaches: simulated annealing [50, 198] gradually decreases temperature during training using $T(t) = T_0/\log(1+t)$ for instance. Cyclical annealing [140] employs periodic temperature cycles to help escape local minima, and tempered transitions use parallel chains at different temperatures for better mixing.

Ensemble methods as Gibbs sampling: Deep ensembles can be viewed as approximate samples from the Gibbs posterior [132]. Training multiple networks with different random initializations and data orderings produces a collection $\{\Theta_1, \ldots, \Theta_K\}$ that approximates sampling from $P(\Theta|\mathcal{D})$.

The ensemble prediction:

$$P_{\text{ensemble}}(y|x) = \frac{1}{K} \sum_{k=1}^{K} P(y|x, \Theta_k)$$

approximates the Bayesian predictive distribution:

$$P(y|x, \mathcal{D}) = \int P(y|x, \Theta) P(\Theta|\mathcal{D}) d\Theta.$$

This connection explains why ensembles often outperform single models: they better capture the uncertainty encoded in the Gibbs posterior, leading to more robust and calibrated predictions [231].

The Gibbs measure framework thus provides a unifying perspective on neural network training, generalization, and uncertainty quantification, connecting seemingly disparate phenomena through the lens of statistical mechanics.

7.3 Gibbs measures in contemporary deep learning models

7.3.1 *Attention mechanisms*

Attention mechanisms [20] have revolutionized machine learning by enabling dynamic focus on relevant input features, particularly in sequential data tasks when employed within Transformer models [374]. At the core of these models is the dot-product self-attention mechanism, which learns self-alignment between tokens in an input sequence by estimating the relative importance of each token with respect to all others. The mechanism then transforms each token into a weighted average of the feature representations of the other tokens with weights proportional to the learned importance scores. The relative importance scores capture contextual information among tokens and are key to the success of the transformer architecture.

The critical significance of understanding and controlling how Transformer works makes it vital to develop approachable theoretical frameworks both to explain currently available model variants as well as shed light on potential improvements that could be gained through such proof-friendly frameworks. Consequently, recent research has attempted to study Transformers, especially the self-attention mechanism, through variety of theoretical viewpoints such as kernel smoothing [371], dynamical systems [256], variational methods [286], measure theory [376], and recently, interacting particle systems [150] to name but a few. Diversity of such perspectives reduces the friction in engagement of researchers from various backgrounds in studying Transformers.

This subsection provides a rigorous theoretical framework, including definitions, propositions, theorems, open problems, and key references.

Transformers rely on multiple layers of self-attention to capture dependencies between tokens in an input sequence. Each Transformer layer consists of a self-attention block followed by feedforward layers and residual connections. The self-attention mechanism is responsible for dynamically aggregating contextual information by attending to relevant tokens, making it a crucial component for learning expressive representations. In this section, we formally define the self-attention computation and its role within the network.

Definition 27 (Attention function). *An attention function maps a query* $\mathbf{q} \in \mathbb{R}^{d_q}$ *and a set of key-value pairs* $\{(\mathbf{k}_i, \mathbf{v}_i)\}_{i=1}^{n}$ *(where* $\mathbf{k}_i \in \mathbb{R}^{d_k}, \mathbf{v}_i \in \mathbb{R}^{d_v}$*) to an output* $\mathbf{o} \in \mathbb{R}^{d_v}$*. The output is a weighted sum of values, with*

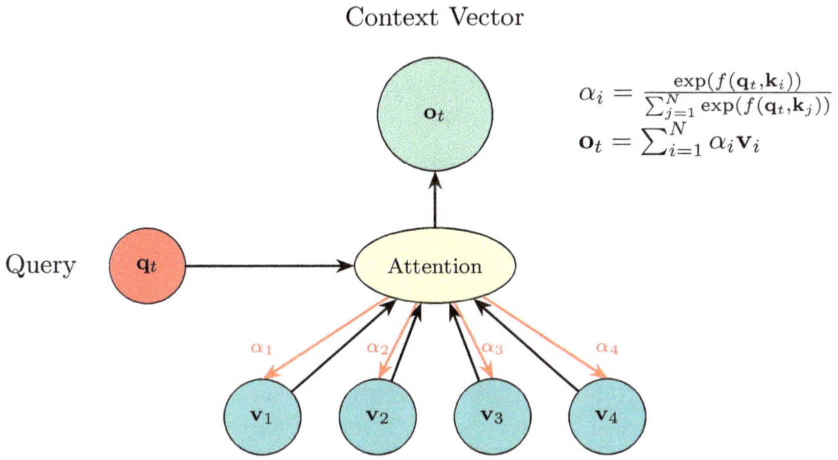

Fig. 7.2: Attention mechanism as in Definition 27.

weights determined by compatibility between $\mathbf{q}$ *and* $\mathbf{k}_i$*:*

$$\mathbf{o} = \sum_{i=1}^{n} \alpha_i \mathbf{v}_i, \quad \alpha_i = \mathit{softmax}(f(\mathbf{q}, \mathbf{k}_i)),$$

where $f : \mathbb{R}^{d_q} \times \mathbb{R}^{d_k} \to \mathbb{R}$ *is a compatibility function, see Fig. 7.2.*

Example: (Scaled dot-product attention [374]) Let $d_k = d_v = d$ and $f(q, k) = q^\top k/\sqrt{d}$ in the definition above. Then in practice, given an input sequence $X = [x_1, \ldots, x_N]^\top \in \mathbb{R}^{N\times d}$ of N feature vectors, the self-attention mechanism transforms the input to $U := [u_1, \ldots, u_N]^\top \in \mathbb{R}^{N\times d}$ as follows:

$$\begin{aligned} u_i &= \sum_{j=1}^{N} \operatorname{softmax}\left(\frac{(W_K x_i)^\top (W_Q x_j)}{\sqrt{d}}\right) W_V x_j \\ &= \sum_{j=1}^{N} \operatorname{softmax}\left(\frac{q_i^\top k_j}{\sqrt{d}}\right) v_j \end{aligned} \tag{7.22}$$

for $i = 1, \ldots, N$, where $\operatorname{softmax}(a_j) := \operatorname{softmax}(a)_j$ for $a = [a_1, \ldots, a_N]$ is an abuse of notation for convenience. The vectors q_i, k_j, and v_j, $j = 1, \ldots, N$, are the query, key, and value vectors, respectively. They are computed as

$$\begin{aligned} Q &:= [q_1, \ldots, q_N]^\top = XW_Q^\top \in \mathbb{R}^{N\times d}, \\ K &:= [k_1, \ldots, k_N]^\top = XW_K^\top \in \mathbb{R}^{N\times d}, \\ V &:= [v_1, \ldots, v_N]^\top = XW_V^\top \in \mathbb{R}^{N\times d}, \end{aligned}$$

where $W_Q, W_K, W_V \in \mathbb{R}^{d\times d}$ are the corresponding learnable projection matrices. Eqn. 7.22 can then be expressed in matrix form as:

$$U = \text{softmax}\left(\frac{QK^\top}{\sqrt{d}}\right)V, \tag{7.23}$$

where the softmax function is applied row-wise to the matrix $QK^\top/\sqrt{d}$.

Since the introduction of self-attention mechanisms within Transformers in [374], there has been a significant amount of research on different attention variants [1, 33, 116, 287, 289, 395] and tailored theoretical frameworks [150, 256, 286, 371, 376].

Building upon the conceptual understanding of attention mechanisms and the scaled dot-product self-attention introduced earlier, we now explore a rigorous mathematical framework that leverages measure theory to model attention. This perspective, first posed in [376], offers a profound lens through which we can interpret attention mechanisms, particularly in the context of Gibbs measures, which are central to this book's exploration of machine learning mathematics.

7.3.2 *A measure-theoretic perspective on attention*

We now establish a rigorous mathematical foundation for attention mechanisms through the lens of measure theory and statistical mechanics. This framework reveals that attention is fundamentally a process of reweighting probability distributions, where the familiar softmax operation emerges naturally from measure-theoretic transformations.

The key insight is to view attention not merely as matrix operations on vectors, but as transformations of probability measures that govern how information flows between tokens. This perspective connects attention mechanisms to the rich theory of Boltzmann-Gibbs distributions from statistical physics, providing both theoretical depth and practical insights.

From tokens to probability measures

Before diving into the mathematical machinery, we need to bridge the gap between discrete token sequences and continuous probability measures. Given a collection of tokens $X = \{x_1, \ldots, x_T\} \subset \mathbb{R}^d$, we can naturally associate to it a probability measure through the empirical measure mapping:

$$X \mapsto m(X) := \frac{1}{|T|}\sum_{t\in T}\delta_{x_t}, \tag{7.24}$$

where δ_{x_t} denotes the Dirac measure concentrated at token x_t. This transformation allows us to treat discrete token configurations as elements of $\mathcal{P}(\mathbb{R}^d)$, the space of probability measures on $\mathbb{R}^d$, enabling the application of powerful measure-theoretic tools.

The empirical measure $m(X)$ encodes the distribution of tokens in the sequence and serves as the foundation for modeling how attention dynamically redistributes probability mass based on token interactions.

The Boltzmann-Gibbs transformation

At the heart of attention mechanisms lies a fundamental operation from statistical mechanics: the reweighting of probability distributions according to energy functions. This operation, known as the Boltzmann-Gibbs transformation, provides the mathematical foundation for the softmax operation ubiquitous in attention.

Definition 28 (Boltzmann-Gibbs transformation). *Let $g : E \to \mathbb{R}_{>0}$ be a bounded measurable function on a measurable space E. The Boltzmann-Gibbs transformation associated to g is the mapping $\Psi_g : \mathcal{P}(E) \to \mathcal{P}(E)$ defined as:*

$$\Psi_g(\nu)(\mathrm{d}x) := \frac{g(x)\nu(\mathrm{d}x)}{\nu(g)}, \tag{7.25}$$

where $\nu(g) = \int g(x)\nu(\mathrm{d}x)$ is the normalization constant ensuring the result remains a probability measure.

This transformation takes a probability measure ν and reweights it according to the function g, then normalizes to maintain unit total mass. The function g can be interpreted as an "importance weight" or, in the language of statistical physics, as exp(−energy). Points where g is large receive higher probability mass in the transformed measure.

The connection to attention becomes clear when we recognize that the softmax operation is precisely a discrete version of this transformation, where g encodes the similarity between queries and keys.

Softmatch kernels: Modeling query-key interactions

To model attention, we need to capture how queries interact with collections of keys. This is achieved through the concept of softmatch kernels, which are families of Markov kernels parametrized by probability measures.

For attention mechanisms, we define an interaction potential $G : E \times E \to \mathbb{R}^+$ of the form $G(x, y) = \exp(a(x, y))$, where $a(x, y)$ represents a similarity function (such as the scaled dot-product $a(x, y) = x^\top y/\sqrt{d}$). This potential encodes the strength of interaction between any pair of tokens.

Definition 29 (Softmatch kernel). *For an interaction potential G, we define the softmatch kernel as the family of Markov kernels $\{\Psi_G(\nu)\}_{\nu \in \mathcal{P}(E)}$ indexed by measures $\nu \in \mathcal{P}(E)$, where for any measurable set $A \subset E$:*

$$\Psi_G(\nu)(x, A) = \int_A \Psi_{G(x,\cdot)}(\nu)(\mathrm{d}y). \tag{7.26}$$

This kernel captures how a query token x "attends to" a measurable set A of key tokens, with the attention distribution determined by the Boltzmann-Gibbs transformation using the interaction potential $G(x, \cdot)$.

Recovering classical attention

To see how this abstract framework recovers the familiar attention formulas, consider a concrete scenario with a single query $q \in \mathbb{R}^d$ and a finite set of keys $K = \{k_1, \ldots, k_N\}$. The empirical measure of the key set is $m(K) = \frac{1}{N}\sum_{i=1}^N \delta_{k_i}$.

The softmatch operation between the query (represented by the Dirac measure δ_q) and the key set becomes:

$$\delta_q \Psi_G(m(K)) = \int \delta_q(\mathrm{d}q')\Psi_{G(q',\cdot)}(m(K)) \tag{7.27}$$

$$= \Psi_{G(q,\cdot)}(m(K)) \tag{7.28}$$

$$= \sum_{s=1}^{N} \frac{G(q, k_s)}{\sum_{r=1}^N G(q, k_r)} \delta_{k_s}. \tag{7.29}$$

When we substitute $G(q, k_s) = \exp(q^\top k_s/\sqrt{d})$, we recover exactly the softmax attention weights:

$$\alpha_s = \frac{\exp(q^\top k_s/\sqrt{d})}{\sum_{r=1}^N \exp(q^\top k_r/\sqrt{d})}. \tag{7.30}$$

This demonstrates that the classical softmax attention mechanism is precisely the discrete realization of the Boltzmann-Gibbs transformation applied to empirical measures.

Multi-query attention: Leveraging linearity

The measure-theoretic framework naturally extends to multiple queries through the linearity of integration. Consider a set of queries $Q = \{q_1, \ldots, q_M\}$ with empirical measure $m(Q) = \frac{1}{M}\sum_{t=1}^M \delta_{q_t}$.

The interaction between the query set and key set is captured by:

$$m(Q)\Psi_G(m(K)) = \frac{1}{M}\sum_{t=1}^{M}\int \delta_{q_t}(\mathrm{d}q')\Psi_{G(q',\cdot)}(m(K)) \tag{7.31}$$

$$= \frac{1}{M}\sum_{t=1}^{M}\Psi_{G(q_t,\cdot)}(m(K)) \tag{7.32}$$

$$= \frac{1}{M}\sum_{t=1}^{M}\sum_{s=1}^{N}\frac{G(q_t,k_s)}{\sum_{r=1}^{N}G(q_t,k_r)}\delta_{k_s}. \tag{7.33}$$

This expression represents the joint configuration of attention weights after all queries have interacted with all keys through the softmatch transformation. The resulting measure is a weighted sum of Dirac measures, where each weight corresponds to the normalized attention score between a specific query-key pair.

Theoretical implications and connections

This measure-theoretic formulation reveals several important insights:

Statistical mechanics connection: The softmax operation in attention is fundamentally a Boltzmann-Gibbs reweighting, where the similarity function $a(q, k)$ plays the role of negative energy. Higher similarity corresponds to lower energy and thus higher probability in the attention distribution.

Equilibrium interpretation: The attention weights can be viewed as an equilibrium distribution over key tokens, where the equilibrium is determined by the interaction potential G. This provides a principled way to think about attention as a process that finds stable configurations of information flow.

Measure-theoretic tools: By casting attention in terms of measure transformations, we gain access to powerful analytical tools from probability theory and functional analysis, enabling more sophisticated theoretical analysis of attention mechanisms.

The framework thus provides both a unifying theoretical perspective and practical insights into the nature of attention, revealing its deep connections to fundamental principles in statistical physics and probability theory.

7.3.3 *Dense associative memories*

Almost analogously to attention mechanisms and their interpretation as Boltzmann-Gibbs measure transformation, we shall also provide a similar view on another important learning model – Dense associative memories (DAMs).

DAMs represent a significant advancement in neural associative memory architectures [193], offering exponential storage capacity and robust retrieval properties that have found applications across machine learning and neuroscience [95, 223, 276, 307].

Classical associative memory: The Hopfield model

Associative memories are neural network models designed to store and retrieve patterns through content-addressable access. The foundational model, introduced by John Hopfield in 1982 [193], consists of N binary neurons with symmetric connections that store P binary patterns $\{\xi^1, \xi^2, \ldots, \xi^P\}$ where $\xi_i^\mu \in \{-1, +1\}$.

The classical Hopfield network uses the energy function:

$$E_{\text{Hopfield}}(x) = -\frac{1}{2} \sum_{i,j=1}^{N} J_{ij} x_i x_j, \tag{7.34}$$

where the synaptic weights are given by Hebb's rule: $J_{ij} = \frac{1}{N} \sum_{\mu=1}^{P} \xi_i^\mu \xi_j^\mu$. The fundamental limitation of classical Hopfield networks is their *linear storage capacity*—the number of storable patterns grows only linearly with the number of neurons, severely constraining their practical applicability.

Dense associative memories, introduced by Krotov and Hopfield [224], overcome the linear capacity limitation through an innovative approach: replacing the quadratic Hopfield energy with higher-order polynomial interactions. This seemingly simple modification leads to *exponential storage capacity*, where $P = \exp(\alpha N)$ patterns can be stored for some $\alpha > 0$.

The key insight is that biological neural networks exhibit rich, nonlinear dynamics that cannot be captured by simple pairwise interactions. By incorporating higher-order terms, DAMs more accurately model the complex energy landscapes of biological associative memory while achieving unprecedented computational performance [223, 224].

Mathematical formulation of dense associative memories

A dense associative memory stores P continuous-valued patterns $\Xi = \{\xi^1, \xi^2, \ldots, \xi^P\} \subset \mathbb{R}^d$ using the energy function:

$$E_{\text{DAM}}(x) = -\frac{1}{\lambda} \log\left(\sum_{\rho=1}^{P} e^{\lambda x \cdot \xi^\rho}\right) + \frac{1}{2}\|x\|^2, \tag{7.35}$$

where $x \in \mathbb{R}^d$ represents the current network state and $\lambda > 0$ is a temperature parameter.

This energy function consists of two terms:

(1) *Memory term:* $-\frac{1}{\lambda}\log\left(\sum_{\rho=1}^{P} e^{\lambda x\cdot\xi^{\rho}}\right)$ captures interactions with stored patterns
(2) *Regularization Term:* $\frac{1}{2}\|x\|^2$ prevents unbounded growth and ensures well-posed dynamics

Definition 30 (Dense associative memory network). *A dense associative memory is a dynamical system defined by the gradient flow*

$$\frac{\mathrm{d}x}{\mathrm{d}t} = -\nabla E_{DAM}(x) = \sum_{\rho=1}^{P} \alpha_\rho(x)\xi^\rho - x, \tag{7.36}$$

where the retrieval coefficients are given by

$$\alpha_\rho(x) = \frac{e^{\lambda x\cdot\xi^\rho}}{\sum_{s=1}^{P} e^{\lambda x\cdot\xi^s}}. \tag{7.37}$$

The retrieval coefficients $\alpha_\rho(x)$ represent the network's current "attention" to each stored pattern, with patterns more similar to the current state x receiving higher weight. The dynamics evolve the network state toward a weighted combination of stored patterns, with weights determined by pattern-state similarity.

Connection to modern deep learning

Recent developments have revealed deep connections between DAMs and contemporary deep learning architectures:

Attention mechanisms: The retrieval coefficients $\alpha_\rho(x)$ in DAMs are mathematically closely aligned with the attention weights in transformer networks, suggesting unified computational principles [174, 257].

Modern Hopfield networks: Generalized Hopfield models with continuous states and exponential interaction functions have been shown to be able to implement attention-like mechanisms [307].

Memory-augmented networks: DAMs provide theoretical foundations for memory modules in neural networks, enabling efficient storage and retrieval of task-relevant information [292, 307, 402].

These connections suggest that associative memory principles may be more fundamental to intelligence than previously recognized, motivating the development of unified theoretical frameworks.

7.3.4 A measure-theoretic perspective on dense associative memories

Having briefly discussed the origins and foundations of Dense Associative Memories, we now demonstrate how the same measure-theoretic framework developed for attention mechanisms naturally encompasses DAMs. This connection reveals that DAMs and attention are manifestations of the same underlying principle: the Boltzmann-Gibbs reweighting of probability measures, albeit applied to different computational contexts.

The key insight is that while attention mechanisms reweight key-value pairs based on query-key interactions, Dense Associative memories reweight stored memory patterns based on the similarity between the current network state and memorized configurations. Both processes can be understood as transformations of empirical measures through exponential tilting, connecting associative memory to the same statistical mechanical principles that govern attention.

From memory patterns to probability measures

Following similar steps as before, consider a collection of P memory patterns $\Xi = \{\xi^1, \xi^2, \ldots, \xi^P\} \subset \mathbb{R}^d$ that we wish to store in an associative memory network. Following the same measure-theoretic approach developed for attention, we associate to this pattern collection the empirical measure:

$$\mu(\Xi) := \frac{1}{P} \sum_{\rho=1}^{P} \delta_{\xi^\rho}, \tag{7.38}$$

where δ_{ξ^ρ} denotes the Dirac measure concentrated at pattern ξ^ρ. This transformation embeds the discrete memory patterns into the space of probability measures $\mathcal{P}(\mathbb{R}^d)$, enabling the application of measure-theoretic tools to analyze memory storage and retrieval.

The empirical measure $\mu(\Xi)$ encodes the distribution of stored patterns and serves as the foundation for understanding how the network dynamically retrieves memories through measure transformations.

The memory energy function as measure transform

The core of dense associative memory lies in defining an energy function that captures the interaction between the current network state $x \in \mathbb{R}^d$ and the stored patterns. This energy function emerges naturally from our measure-theoretic framework.

For a network state x, we define the *memory interaction potential* $G_\lambda : \mathbb{R}^d \times \mathbb{R}^d \to \mathbb{R}_{>0}$ as:

$$G_\lambda(x, \xi) := \exp(\lambda x \cdot \xi), \tag{7.39}$$

where $\lambda > 0$ is a temperature parameter controlling the sharpness of interactions. This potential quantifies the *resonance* between the current state x and a stored pattern ξ.

The total memory energy function can then be expressed as a logarithmic moment generating function of the pattern measure:

$$\Phi_\lambda(x; \mu(\Xi)) := \frac{1}{\lambda} \log \int G_\lambda(x, \xi)\, \mu(\Xi)(\mathrm{d}\xi) = \frac{1}{\lambda} \log \left(\frac{1}{P} \sum_{\rho=1}^{P} e^{\lambda x \cdot \xi^\rho} \right). \tag{7.40}$$

This function represents the logarithmic partition function of the memory patterns weighted by their interaction with the current state x.

Dense associative memory as Gibbs measure

The complete energy function of a dense associative memory network combines the memory interaction term with a quadratic regularization:

$$E_{\mathrm{DAM}}(x) := -\Phi_\lambda(x; \mu(\Xi)) + \frac{1}{2}\|x\|^2 = -\frac{1}{\lambda} \log \left(\sum_{\rho=1}^{P} e^{\lambda x \cdot \xi^\rho} \right) + \frac{1}{2}\|x\|^2. \tag{7.41}$$

The equilibrium distribution of the network is given by the Gibbs measure:

$$\pi_{\mathrm{DAM}}(\mathrm{d}x) := \frac{1}{Z} \exp(-E_{\mathrm{DAM}}(x))\, \mathrm{d}x, \tag{7.42}$$

where Z is the normalization constant. Substituting the energy function:

$$\pi_{\mathrm{DAM}}(\mathrm{d}x) = \frac{1}{Z} \left(\frac{1}{P} \sum_{\rho=1}^{P} e^{\lambda x \cdot \xi^\rho} \right)^{1/\lambda} \exp\left(-\frac{1}{2}\|x\|^2 \right) \mathrm{d}x. \tag{7.43}$$

This is precisely a *mixture of Gaussian measures* weighted by the Boltzmann-Gibbs transformation of the pattern empirical measure. Each stored pattern ξ^ρ contributes a Gaussian component centered at $\lambda \xi^\rho$ with variance 1.

Retrieval dynamics and measure concentration

Memory retrieval in DAMs corresponds to the dynamics of gradient descent on the energy function:

$$\frac{\mathrm{d}x}{\mathrm{d}t} = -\nabla E_{\mathrm{DAM}}(x) = \sum_{\rho=1}^{P} \alpha_\rho(x)\xi^\rho - x, \tag{7.44}$$

where the *retrieval weights* are given by:

$$\alpha_\rho(x) = \frac{e^{\lambda x\cdot\xi^\rho}}{\sum_{s=1}^{P} e^{\lambda x\cdot\xi^s}}. \tag{7.45}$$

These weights are precisely the result of applying the Boltzmann-Gibbs transformation $\Psi_{G_\lambda(x,\cdot)}$ to the pattern empirical measure $\mu(\Xi)$:

$$\Psi_{G_\lambda(x,\cdot)}(\mu(\Xi)) = \sum_{\rho=1}^{P} \alpha_\rho(x)\delta_{\xi^\rho}. \tag{7.46}$$

The retrieval dynamics can thus be interpreted as the network state x evolving toward the barycenter of the transformed measure, weighted by the current attention to each stored pattern.

Continuous pattern distributions

The measure-theoretic framework naturally extends to continuous pattern distributions. Instead of discrete empirical measures, consider patterns drawn from a continuous distribution $\nu \in \mathcal{P}(\mathbb{R}^d)$. The memory energy function becomes:

$$\Phi_\lambda(x;\nu) = \frac{1}{\lambda}\log\int e^{\lambda x\cdot\xi}\,\nu(\mathrm{d}\xi), \tag{7.47}$$

which is the logarithmic moment generating function of ν evaluated at λx.

This formulation connects DAMs to the theory of *exponential families* in statistics, where the sufficient statistics are precisely the inner products $x\cdot\xi$ between network states and pattern configurations.

7.4 A deep learning approach to Boltzmann optimization problems

Mathematics plays a foundational role in analyzing and interpreting deep learning systems as has been discussed until now in this book. Conversely, deep learning has also emerged as a valuable tool for addressing certain mathematical problems, offering computationally feasible solutions to otherwise intractable tasks. This section is based on a recent paper [218], where deep learning models are applied to solve Boltzmann optimization problems, leveraging the Gibbs measure to connect the system energy to probabilistic outcomes.

We begin by introducing the reverse Ising problem, which involves solving for the parameters of an Ising system. This problem is intrinsically tied to optimizing the Gibbs measure. The work focuses on optimizing Ising parameters to shape this measure, ensuring desired outputs attain high probability under specified inputs.

Traditional methods for solving the reverse Ising problem rely on gradient-based optimization of the Gibbs measure, which becomes intractable for large systems due to the computational cost of estimating Z. We circumvent this by reformulating the problem into a supervised learning task. By generating training data that maps input spin configurations to optimal Ising parameters, we train neural networks to predict these parameters directly. This approach replaces slow gradient-based updates with efficient forward passes through a learned model.

The core contribution of [218] is a novel method for optimizing Ising parameters to maximize the likelihood of target outputs under the Gibbs measure. By avoiding explicit computation of the partition function Z, this machine learning models enable scalable solutions for complex systems. This advancement unlocks the potential for designing next-generation Ising-based hardware capable of solving real-world optimization problems.

Recall that the Ising model's equilibrium behavior is characterized by the Gibbs measure. For a system with spins $s = \{s_i\} \in \{\pm 1\}^N$, the Hamiltonian defines the energy landscape:

$$H(s) = \sum_{i=1}^{N} h_i s_i + \sum_{\langle i,j \rangle} J_{i,j} s_i s_j, \tag{7.48}$$

where h_i is external field acting on spin i; $J_{i,j}$ is interaction coefficient between neighboring spins i and j; the second sum over all neighboring pairs (i, j), each counted once. $s_i, s_j \in \{-1, +1\}$. The Gibbs measure quantifies the probability of each state. The ground state, the configuration minimizing $H(s)$, dominates the measure at low temperatures but competes with thermally excited states at finite temperatures.

The reverse problem involves tuning parameters $\{h_i, J_{i,j}\}$ such that the Gibbs measure assigns high probability to predefined output spins when input spins are fixed. This requires solving a high-dimensional optimization over the Gibbs measure, traditionally approached via gradient ascent on the log-likelihood. However, computing gradients involving Z is prohibitive for large N. The method sidesteps this by training models to predict parameters that implicitly maximize the Gibbs probability of desired states.

Performance and implications: These models outperform traditional gradient-based methods in both speed and scalability, as they eliminate reliance on Z. By learning the mapping from inputs to optimal parameters, we enable rapid design of Ising systems for complex tasks. This framework generalizes to systems with non-uniform temperatures or external fields, broadening the applicability of Ising machines.

Reverse Ising optimization:

We are given a desired spin state

$$s = (s_1, \ldots, s_n, s_{n+1}, \ldots, s_N) \in \{\pm 1\}^N,$$

and we want to determine Hamiltonian coefficients $h_i \in \mathbb{R}$, $J_{i,j} \in \mathbb{R}$ such that this state becomes the ground state (i.e., the state of minimum energy) with high probability. Specifically, we require:

$$H(s_1, \ldots, s_n, s_{n+1}, \ldots, s_N) < H(s_1, \ldots, s_n, t_1, \ldots, t_m),$$

for all $t_k \in \{\pm 1\}$, $1 \leq k \leq m$, where $t_k \neq s_{n+k}$ for at least one k.

Since Ising systems are non-deterministic, we maximize the likelihood of the state s by minimizing its energy:

$$\min_{\psi} H_{\psi}(s),$$

where

$$\psi = \{h_1, \ldots, h_N, J_{1,2}, \ldots, J_{N-1,N}\}$$

is the vector of Hamiltonian coefficients, and H_{ψ} is the Hamiltonian defined by these coefficients.

Example:

Let $s = (-1, 1, -1)$, where the third spin is variable. We want to find Hamiltonian coefficients such that $(-1, 1, -1)$ has the lowest energy among possible configurations:

$$\min_{\psi} H_{\psi}(-1, 1, -1) \quad \text{s.t.} \quad H_{\psi}(-1, 1, -1) < H_{\psi}(-1, 1, 1)$$

Generalization to higher dimensions:

Let $s = (u, v) \in \{\pm 1\}^N$, where $u \in \{\pm 1\}^n$ and $v \in \{\pm 1\}^m$. The optimization problem becomes:

$$\min_{\psi} H_{\psi}(u, v) \quad \text{s.t.} \quad H_{\psi}(u, v) < H_{\psi}(u, t), \quad \forall t \neq v$$

Sometimes, the system of inequalities $\{H_\psi(u, v) < H_\psi(u, t)\}_{t \neq v}$ is infeasible. To resolve this, auxiliary spins are introduced.

Let α be the number of auxiliary spins and $a \in \{\pm 1\}^\alpha$ their values. The total number of spins becomes $N = n + m + \alpha$.

We now minimize over both the Hamiltonian coefficients and the auxiliary spins:

$$\min_{\psi} \min_{a \in \{\pm 1\}^\alpha} H_\psi(u, v, a)$$

subject to:

$$H_\psi(u, v, a) < H_\psi(u, t, a), \quad \forall t \neq v,$$

$$H_\psi(u, v, a) < H_\psi(u, t, a_1, \ldots, a_\alpha), \quad \text{for all } t \neq v$$

Reverse Ising problem (Multiple states):

In the generalized reverse Ising problem, we are given a set of target states:

$$\mathcal{S} = \{s^{(i)}\}_{i=1}^{\ell}, \quad s^{(i)} \in \{\pm 1\}^N$$

We aim to find coefficients $\psi = \{h_1, \ldots, h_N, J_{1,2}, \ldots, J_{N-1,N}\}$ such that each desired state $s^{(i)}$ has lower energy than all other possible spin configurations.

Multiple desired states with shared spins.

In the reverse Ising problem, we are given a set of desired states $\mathcal{S} = \{s^{(i)}\}_{i=1}^{\ell}$, where each state is of the form:

$$s^{(i)} = \left(s_1^{(i)}, \ldots, s_n^{(i)}, s_{n+1}^{(i)}, \ldots, s_{n+m}^{(i)}\right),$$

and is augmented with a corresponding auxiliary spin configuration:

$$a^{(i)} = \left(a_1^{(i)}, \ldots, a_\alpha^{(i)}\right).$$

We aim to find Hamiltonian coefficients ψ such that each desired state has the lowest energy among all other states with the same first n spins. That is, for each $1 \leq i \leq \ell$:

$$H_\psi\left(s_1^{(i)}, \ldots, s_n^{(i)}, s_{n+1}^{(i)}, \ldots, s_{n+m}^{(i)}, a_1^{(i)}, \ldots, a_\alpha^{(i)}\right)$$

$$< H_\psi\left(s_1^{(i)}, \ldots, s_n^{(i)}, t_1, \ldots, t_m, a_1^{(i)}, \ldots, a_\alpha^{(i)}\right),$$

for all $t_k \in \{\pm 1\}$, where $t_k \neq s_{n+k}^{(i)}$ for at least one k.

We note that for each desired spin state $s^{(i)} = (s_1^{(i)}, \ldots, s_{n+m}^{(i)})$, the auxiliary spin configuration $a^{(i)} = (a_1^{(i)}, \ldots, a_\alpha^{(i)})$ can differ - that is, each state can have its own independent auxiliary spin vector.

Recall that the Ising model is a nondeterministic system. The probability of the system settling into a particular state (including the ground state) is a function of that state's energy. As a result, rather than focusing solely on minimizing the energy of each desired state, it is often more effective to consider the probability distribution over all states induced by the Hamiltonian, and to maximize the probability of the system settling into each desired state.

The probability of the system taking on state $\tilde{s}$ given Hamiltonian coefficients ψ is defined by Gibbs measure:

$$\mathrm{Prob}(S = \tilde{s}) := \frac{\exp(H_\psi(\tilde{s})/(\beta T))}{\sum_{s\in\{\pm 1\}^N} \exp(H_\psi(s)/(\beta T))}, \tag{7.49}$$

where β represents the Boltzmann constant and T the temperature. Here, we consider the temperature to be fixed. More specifically, we let $\beta T = 1$.

Viewing the probability of a system taking on state $\tilde{s}$ in terms of the probability of all other states, (7.49) is equivalent to:

$$\mathrm{Prob}(S = \tilde{s}) = \frac{1}{\sum\limits_{w\neq\tilde{s}} \frac{\exp(H_\psi(w))}{\exp(H_\psi(\tilde{s}))} + 1} = \frac{\exp(H_\psi(\tilde{s}))}{\exp(H_\psi(\tilde{s})) + \sum\limits_{w\neq\tilde{s}} \exp(H_\psi(w))}.$$

In light of the reverse Ising problem, we desire to maximize the probability of obtaining a set of desired states, which we denote as $\{s^{(i)} = (u^{(i)}, v^{(i)}, a^{(i)})\}$, where the auxiliary spins $a^{(i)} = (a_\alpha^{(i)})$ specific to each state are free to take on any value.

Equivalently, we want to minimize the maximum probability of obtaining a state with an incorrect second portion v over all the possible first portions u.

More specifically, let $S = \{s^{(i)} = (u^{(i)}, v^{(i)}, a^{(i)})\}_{1\leq i\leq \ell}$ represent the set of desired states, where $a^{(i)}$ is free. Let $W^{(i)} = \{(u^{(i)}, t, a^{(i)})\}_t$, where t ranges over all vectors in $\{\pm 1\}^m$ such that $t \neq v^{(i)}$. That is, $W^{(i)}$ represents the set of states where the second portion differs from $v^{(i)}$ in at least one position.

Then, the probability of not being in state $s^{(i)} = (u^{(i)}, v^{(i)}, a^{(i)})$ but rather in a state in $W^{(i)}$ is

$$\mathrm{Prob}_{\mathrm{boltz}}(s^{(i)}, W^{(i)}) := \frac{\sum\limits_{w\in W^{(i)}} \exp(H_\psi(w))}{\exp(H_\psi(s^{(i)})) + \sum\limits_{w\in W^{(i)}} \exp(H_\psi(w))}.$$

The resulting objective function of interest is as follows:

$$\rho(a^{(1)}, \ldots, a^{(\ell)}) = 1 - \min_{\psi} \max_{1 \le i \le \ell} \mathrm{Prob}_{\mathrm{boltz}}(s^{(i)}, W^{(i)}) \tag{7.50}$$

$$= 1 - \min_{\psi} \max_{1 \le i \le \ell} \frac{\sum_{w \in W^{(i)}} \exp(H_\psi(w))}{\exp(H_\psi(s^{(i)})) + \sum_{w \in W^{(i)}} \exp(H_\psi(w))}.$$

Here, ψ is the set of Hamiltonian coefficients and $a^{(i)} \in \{\pm 1\}^\alpha$ for $1 \le i \le \ell$, each i corresponding to the desired state $s^{(i)}$. Throughout the rest of the section, $(a^{(1)}, \ldots, a^{(\ell)})$ is referred to as the *auxiliary array.*

Here we demonstrate the successful application of machine learning algorithms to solve our optimization problem (7.50). The result is a computationally efficient approach for evaluating Ising system design parameters. This is a notable advancement over the prohibitively slow state-of-the-art methods for performing this evaluation.

Thus we implement a deep learning model that solves a complex optimization problem both accurately and efficiently. This result enables the study of larger and more complex Ising circuits. The sophisticated process required to generate a training dataset for this problem and the machine learning architectures used are discussed next, followed by a discussion of the performance of the ML models.

Methods: *Size of the system and the complexity of the problem.*

The size and complexity of the reverse Ising problem is determined by the number of fixed, non-fixed, and auxiliary spins of the system under consideration. For $N = n + m + \alpha$ total spins, the total number of possible states considered when computing the Boltzmann probability is 2^N, and the vector ψ of Hamiltonian coefficients has

$$\frac{1}{2}(N^2 + N + 2)$$

variables. The number of linear inequality constraints in our resulting optimization problem is

$$2^{n+\alpha}(2^m - 1).$$

Finally, the number of possible combinations of auxiliary spin values scales double exponentially as $2^{\alpha 2^n}$. This relationship is obtained as the auxiliary spins for each desired state are independent from one another and can take on any of the 2^α values. As a result, we obtain 2^α possibilities for each of the 2^n equations, giving us the relationship

$$(2^\alpha)^{2^n} = 2^{\alpha 2^n}.$$

For example, a system with 6 total spins ($n = 2, m = 2, \alpha = 2$) will have 48 inequality constraints and 256 possible auxiliary arrays. In this scenario, one auxiliary array takes on the form $(a^{(1)}, \dots, a^{(4)})$, where each vector $a^{(i)}$ can take on four different values. A problem with twice as many total spins ($n = 4, m = 4, \alpha = 4$) will have 3840 inequality constraints and $2^{64} \approx 1.8 \times 10^{19}$ possible auxiliary arrays.

If the system had no auxiliary spins, the constraints are linear with respect to the set of variables

$$\psi = (h_1, \dots, h_N, J_{1,2}, \dots, J_{N-1,N})$$

due to the linearity of the Hamiltonian (7.48). With the addition of auxiliary spins, the constraints become nonlinear and non-convex. Further, the objective function (7.50) is a nonlinear, non-differentiable function with an intractable number of local minima. This rules out the possibility of using standard simplex methods to find a global minimum. Because of this, faster methods for evaluating the Boltzmann probability for arbitrary-sized systems with large numbers of auxiliary spins are essential to optimization problems like the reverse Ising problem.

Modeling the Boltzmann probability optimization problem:

We continue to assume there are a total of $N = n + m + \alpha$ spins in the system, where α is the number of auxiliary spins, n spins are fixed, leaving m of the spins to vary. Recall our minimization-maximization programming problem:

$$\rho(a^{(1)}, \dots, a^{(\ell)}) = 1 - \min_{\psi} \max_{1 \le i \le \ell} \mathrm{Prob}_{\mathrm{boltz}}(s^{(i)}; W^{(i)}), \tag{7.51}$$

where auxiliary spin values $a^{(i)} \in \{\pm 1\}^{\alpha}$ are the input, and $s^{(i)} = (u^{(i)}, v^{(i)}, a^{(i)})$ represent the set of desired states. The set $W^{(i)} = \{(u^{(i)}, t, a^{(i)}) \,|\, t \neq v^{(i)}\}$ represents the complement of state $s^{(i)}$.

For a given problem size and a desired set of states, our goal is to create a model that provides a means of calculating ρ both quickly and accurately, within some tolerance ε.

To do this, we first need to generate training data. (7.51) presents several challenges for generating such data. Firstly, it is numerically unstable, and secondly, it is not everywhere differentiable, ruling out the use of gradient-based methods. We describe the process by which we construct a numerically stable approximation to (7.51). This development is crucial for generating high-quality training data required to train supervised machine learning models, which proved essential for optimizing Ising system design parameters.

Generating the training data:

We desire to transform (7.51) such that we can easily and reliably compute solutions. The nonlinear optimization solver which used in [218] is based on the Sequential Least Squares Programming (SLSQP) algorithm (see also references therein).

For the majority of choices for Hamiltonian coefficients $\psi = (h_1, \ldots, h_N, J_{1,2}, \ldots, J_{N-1,N})$, the maximum probability of landing in an undesired state will be extremely close to one. To assist in distinguishing these probabilities, we minimize the maximum of the logarithm of the probability of an undesired state:

$$\rho(a^{(1)}, \ldots, a^{(\ell)}) = 1 - \exp\left(\min_{\psi} \max_{1 \le i \le \ell} \log \mathrm{Prob}_{\mathrm{boltz}}(s^{(i)}; W^{(i)}) \right).$$

This amplifies small changes in probability resulting from updates to the Hamiltonian coefficients.

An additional transformation is applied to the maximum function. For a fixed set of Hamiltonian coefficients ψ, the maximum function

$$\max_i \log \mathrm{Prob}_{\mathrm{boltz}}(s^{(i)}; W^{(i)})$$

$$= \max_i \log \left(\frac{\sum_{w \in W^{(i)}} \exp(H_\psi(w))}{\exp(H_\psi(s^{(i)})) + \sum_{w \in W^{(i)}} \exp(H_\psi(w))} \right)$$

is not guaranteed to be continuously differentiable. Because SLSQP requires the gradient to be defined everywhere, we approximate the maximum function. For a sufficiently large scaling parameter λ, the following approximation holds for an arbitrary function g evaluated at points x_i:

$$\max_i g(x_i) \approx \frac{1}{\lambda} \log \left(\sum_i \exp(\lambda g(x_i)) \right).$$

Applying this approximation, we obtain a continuously differentiable version of the log-probability term:

$$\log \mathrm{Prob}_{\mathrm{boltz}}(s^{(i)}; W^{(i)}) \approx \frac{1}{\lambda} \log \left(\sum_i \exp\left(\lambda \log \mathrm{Prob}_{\mathrm{boltz}}(s^{(i)}; W^{(i)}) \right) \right).$$

This completes our transformation process. The Boltzmann probability optimization function can now be approximated in a numerically stable, continuously differentiable form as follows:

$$\rho(a^{(1)}, \ldots, a^{(\ell)}) \approx 1 - \exp\left(\min_{\psi} f(\psi) \right). \tag{7.52}$$

$$f(\psi) = \frac{1}{\lambda} \log \left(\sum_i \exp\left(\lambda \log \mathrm{Prob}_{\mathrm{boltz}}(s^{(i)}; W^{(i)}) \right) \right).$$

To speed up the work done by the SLSQP solver and to further ensure numerical stability, we also apply standard log-sum-exp transformations and vectorize $f(\psi)$. Details are found in the supplementary information.

Our training data is of the form:

$$\text{input; target} = a, \rho(a),$$

where $0 \leq \rho(a) \leq 1$ and $a = (a^{(1)}, \ldots, a^{(\ell)})$ is the auxiliary array, with $a^{(i)} \in \{\pm 1\}^{\alpha}$ being the vector of α auxiliary spins.

For a given a, $f(\psi)$ is minimized using the SLSQP solver in [218]. From this, they calculated the associated target value using (7.52).

7.5 Conclusion

In this chapter, we explored the intersection of Gibbs measures and deep learning, highlighting how foundational concepts from statistical mechanics provide powerful tools for understanding and improving modern machine learning models. Beginning with an overview of the statistical mechanics of neural networks, we examined how learning dynamics can be interpreted in terms of energy landscapes, phase transitions, and equilibrium distributions. The infinite-width limit of neural networks—approached through kernel methods and renormalization—provided insights into model behavior in high-dimensional settings and underscored the role of Gaussian processes in this regime.

We further discussed the intricate balance between entropy, model capacity, and generalization, illustrating how Gibbs measures offer a principled framework to analyze and optimize this trade-off. The formulation of learning as a Boltzmann optimization problem allowed us to reframe training objectives in terms of energy minimization, leading to connections with simulated annealing, variational inference, and probabilistic sampling.

Then, we addressed the application of Gibbs measures in contemporary Bayesian deep learning, particularly through stochastic gradient Langevin dynamics (SGLD), which operationalizes a Gibbs sampler in high-dimensional parameter spaces. These methods not only provide uncertainty quantification but also enhance robustness and generalization in complex models.

Finally, we showed that both attention mechanisms and DAMs can be understood as performing *exponential tilting* of probability measures on different manifolds:

- *Attention:* Tilts the uniform distribution over keys toward query-relevant keys, creating focused information flow.
- *DAMs:* Tilts the uniform distribution over patterns toward state-resonant patterns, enabling memory retrieval.

This unifying perspective reveals that both mechanisms implement the same fundamental computational primitive – Boltzmann-Gibbs reweighting – adapted to their respective domains. The measure-theoretic framework thus provides a common mathematical language for understanding these seemingly disparate neural computation paradigms, opening new avenues for theoretical analysis and algorithmic development in both attention-based architectures and associative memory networks.

Overall, Gibbs measures serve as a unifying language bridging physics and machine learning. Their continued integration into deep learning frameworks promises to deepen our theoretical understanding and open new avenues for algorithmic innovation, particularly in the design of scalable, probabilistic, and interpretable AI systems.

For further reading on mathematics of deep learning, see [21, 26, 92, 139, 150, 157, 227, 246].

Chapter 8

Recent Advances and Emerging Trends in the Use of Gibbs Measures in Machine Learning

This chapter explores the evolving role of Gibbs measures in machine learning, tracing their influence from foundational applications to the cutting-edge developments in contemporary artificial intelligence. We begin by examining the current state of Gibbs measures in classical machine learning contexts, including Bayesian neural networks, Markov chain Monte Carlo methods, and statistical learning theory.

The chapter then presents a comprehensive treatment of deep learning and modern neural architectures through the lens of Gibbs measures. We explore how structured Gibbs measures are driving innovations in deep learning, encompassing new training methodologies, architectural advances, and reinterpretations of diffusion models and deep belief networks within a Gibbsian framework. This section culminates with an extensive analysis of Large Language Models, examining the statistical mechanics of autoregressive language modeling, the role of temperature in sampling and generation, and how information-theoretic quantities in language modeling can be understood as thermodynamic observables.

We then delve into probabilistic programming and Gibbs sampling, highlighting algorithmic foundations, hybrid inference engines, convergence diagnostics, and applications such as topic modeling. The chapter explores emerging developments in Bayesian nonparametrics through the lens of infinite-dimensional Gibbs measures, nonparametric mixture models, and neural processes. We examine graphical models and message passing algorithms, including belief propagation, federated learning applications, and quantum enhancements.

The chapter concludes by synthesizing unifying themes across these diverse applications, identifying emerging horizons in the field, and discussing the broader societal implications of these developments.

This comprehensive survey underscores the foundational and unifying role of Gibbs measures in advancing machine learning theory and practice, from classical statistical methods to the latest developments in artificial intelligence.

8.1 The state of Gibbs measures in machine learning

8.1.1 *Bayesian neural networks*

Bayesian Neural Networks (BNNs) combine the representational power of deep learning with Bayesian principles to model uncertainty in predictions. Unlike standard neural networks, which produce point estimates of model parameters, BNNs infer a posterior distribution over weights θ given data $\mathcal{D} = \{(x_i, y_i)\}_{i=1}^N$. This yields predictive distributions rather than deterministic outputs, allowing BNNs to express epistemic uncertainty, a crucial property in safety-critical applications and scientific modeling.

The posterior over parameters is typically approximated by a tempered (Gibbs) version:

$$\pi(\theta \mid \mathcal{D}) \propto \exp\left(-\beta \sum_{i=1}^{N} \ell(y_i, f_\theta(x_i))\right) p_0(\theta),$$

where $\ell(y, f_\theta(x))$ is a loss function (e.g., negative log-likelihood), $p_0(\theta)$ is the prior over parameters, and $\beta > 0$ is the inverse temperature controlling the concentration of the posterior. When $\beta = 1$, this coincides with the standard Bayesian posterior; as $\beta \to \infty$, it approaches the maximum a posteriori (MAP) solution.

Scalable inference: Exact inference in BNNs is computationally infeasible due to the high dimensionality and non-convexity of neural network parameter spaces. Therefore, approximate methods are required. One such method is stochastic gradient Langevin dynamics (SGLD), introduced by Welling and Teh [384], which augments stochastic gradient descent (SGD) with Gaussian noise to simulate Langevin dynamics:

$$\theta_{t+1} = \theta_t - \eta_t \nabla \mathcal{L}(\theta_t) + \sqrt{2\eta_t/\beta} \cdot \xi_t, \quad \xi_t \sim \mathcal{N}(0, I),$$

where η_t is a decreasing learning rate and $\mathcal{L}(\theta)$ is the loss function over mini-batches. Under Robbins–Monro conditions (i.e., $\sum_t \eta_t = \infty$, $\sum_t \eta_t^2 < \infty$), SGLD converges to samples from the Gibbs posterior. Recent advancements, such as Hamiltonian Monte Carlo (HMC), variational inference, and Stein variational gradient descent (SVGD) offer alternative trade-offs between fidelity to the posterior and computational efficiency. For instance, a hybrid VI-SVGD approach enhances posterior approximation for complex BNN architectures [252], while HMC with discretized priors enables exact uncertainty quantification in specific settings [129].

BNNs are supported by PAC-Bayesian theory, which provides high-probability generalization bounds. For any posterior ρ over weights (possibly obtained by approximate inference) and prior π, the following inequality holds with probability at least $1 - \delta$ (see [9], [73]):

$$\mathbb{E}_\rho[\mathcal{R}] \leq \mathbb{E}_\rho[\hat{\mathcal{R}}_n] + \sqrt{\frac{D_{\mathrm{KL}}(\rho\|\pi) + \ln(1/\delta)}{2\beta n}},$$

where $\mathcal{R}$ is the true (population) risk, $\hat{\mathcal{R}}_n$ is the empirical risk, and $D_{\mathrm{KL}}(\rho\|\pi)$ is the Kullback–Leibler divergence between the posterior and prior. This result highlights the trade-off between data fitting and model complexity, and offers formal generalization guarantees even for over-parameterized models. Extensions to data-dependent priors and localized bounds (e.g., Dziugaite and Roy [114]) further adapt these results to practical deep learning scenarios, with recent refinements tightening bounds for over-parameterized BNNs [109].

Cold posterior effect: In practice, BNNs often perform best when the posterior is colder than what standard Bayesian theory prescribes—that is, when $\beta > 1$. This phenomenon, known as the cold posterior effect, has been reported in several empirical studies [385]. It challenges the conventional wisdom that $\beta = 1$ should yield optimal results under a well-specified model. Several hypotheses have been proposed to explain this, including:

- Model misspecification: The assumed likelihood may not capture real-world data noise.
- Prior mismatch: Chosen priors are often too diffuse or uninformative.
- Implicit regularization: Larger β may reduce overfitting by flattening the posterior.
- Data augmentation and optimization biases: These introduce implicit temperature scaling effects.

Recent studies suggest that colder posteriors act as implicit regularizers, improving generalization in tasks like time-series forecasting, and link the effect to distinct posterior modes under discretized priors [129]. Understanding and resolving the cold posterior effect remains an active area of theoretical and empirical research, with implications for the foundations of Bayesian deep learning.

8.1.2 *Markov chain Monte Carlo*

Recall that Markov chain Monte Carlo (MCMC) methods are a class of algorithms designed to sample from complex probability distributions, particularly those that arise as Gibbs measures in statistical mechanics and Bayesian inference. MCMC offers a practical approach to approximate expectations under Gibbs measures.

The core idea of MCMC is to construct a Markov chain whose stationary distribution is the target distribution (e.g. Gibbs measure). Over time, the samples produced by the chain approximate draws from the Gibbs measure, making MCMC a cornerstone of inference in both physics and machine learning.

Metropolis-Hastings: One of the foundational MCMC algorithms, the Metropolis-Hastings (MH) method, operates by proposing a new state x' from a proposal distribution $q(x' \mid x)$ and accepting it with probability

$$\alpha(x \to x') = \min\left(1, \frac{\pi(x')q(x \mid x')}{\pi(x)q(x' \mid x)}\right).$$

When $\pi(x) \propto e^{-\beta H(x)}$, this becomes

$$\alpha(x \to x') = \min\left(1, e^{-\beta \Delta H}\right),$$

where $\Delta H = H(x') - H(x)$. This form ensures detailed balance, a key property guaranteeing convergence to the Gibbs measure.

Gibbs Sampling: A special case of MH, Gibbs sampling avoids the need for an acceptance step by updating each variable x_i sequentially or randomly using its full conditional distribution:

$$x_i^{(t+1)} \sim \pi(x_i \mid x_{-i}),$$

where x_{-i} denotes all variables except x_i. When the conditional distributions are tractable, this method allows direct sampling and has strong theoretical ties to the DLR equations that characterize Gibbs measures on lattices and graphs.

Hamiltonian Monte Carlo (HMC): HMC improves sampling efficiency in high-dimensional settings by augmenting the state space with momentum variables and simulating Hamiltonian dynamics:

$$H(x, p) = U(x) + K(p),$$

where $U(x) = -\log \pi(x)$ and $K(p)$ is kinetic energy. Using the gradient of $U(x)$, HMC enables proposals that explore the space with reduced random walk behavior. While not rooted in classical Gibbs sampling, HMC preserves a modified Gibbs measure over phase space (x, p) and obeys a generalized detailed balance.

Recent Innovations and Variants:

Stein Variational Gradient Descent (SVGD): A deterministic, particle-based method for approximating posterior distributions. SVGD evolves particles to minimize the Stein discrepancy between the empirical particle distribution and the target, using functional gradients. Unlike classical MCMC, SVGD blends optimization and sampling [252].

Normalizing Flows: Flows are often used to approximate Gibbs posteriors in Bayesian deep learning and modern flow based generative models [247], and allow efficient likelihood evaluation [297]. They provide a general way of constructing flexible probability distributions over continuous random variables. Let $\mathbf{x}$ be a D-dimensional real vector, and suppose we would like to define a joint distribution over $\mathbf{x}$. The main idea of flow-based modeling is to express $\mathbf{x}$ as a transformation T of a real vector $\mathbf{u}$ sampled from $p_u(\mathbf{u})$:

$$\mathbf{x} = T(\mathbf{u}) \quad \text{where} \quad \mathbf{u} \sim p_u(\mathbf{u}).$$

Here $p_u(\mathbf{u})$ is *base distribution* of the flow-based model.[1]

The defining property of flow-based models is that the transformation T must be invertible and both T and T^{-1} must be differentiable. Such transformations are known as *diffeomorphisms* and require that $\mathbf{u}$ be D-dimensional as well under these conditions, the density of $\mathbf{x}$ is well-defined and can be obtained by a change of variables (see [297] and references therein):

$$p_x(\mathbf{x}) = p_u(\mathbf{u}) \left|\det J_T(\mathbf{u})\right|^{-1} \quad \text{where} \quad \mathbf{u} = T^{-1}(\mathbf{x}).$$

Equivalently, we can also write $p_x(\mathbf{x})$ in terms of the Jacobian of T^{-1}:

$$p_x(\mathbf{x}) = p_u\left(T^{-1}(\mathbf{x})\right) \left|\det J_{T^{-1}}(\mathbf{x})\right|.$$

The Jacobian $J_T(\mathbf{u})$ is the $D \times D$ matrix of all partial derivatives of T given by:

$$J_T(\mathbf{u}) = \begin{bmatrix} \frac{\partial T_1}{\partial u_1} & \cdots & \frac{\partial T_1}{\partial u_D} \\ \vdots & \ddots & \vdots \\ \frac{\partial T_D}{\partial u_1} & \cdots & \frac{\partial T_D}{\partial u_D} \end{bmatrix}.$$

[1]The transformation T and the base distribution $p_u(\mathbf{u})$ can have parameters of their own (denote them as ϕ and ψ respectively); this induces a family of distributions over $\mathbf{x}$ parameterized by $\{\phi, \psi\}$.

In practice, a flow-based model is constructed by implementing T (or T^{-1}) with a neural network and taking $p_u(\mathbf{u})$ to be a simple density such as a multivariate normal.

Intuitively, one can think of the transformation T as warping the space $\mathbb{R}^D$ in order to mold the density $p_u(\mathbf{u})$ into $p_x(\mathbf{x})$. The absolute Jacobian determinant $|\det J_T(\mathbf{u})|$ quantifies the relative change of volume of a small neighborhood around $\mathbf{u}$ due to T.

Parallel Tempering (Replica Exchange MCMC): To overcome metastability in multimodal distributions, parallel tempering runs several chains at different temperatures β_k, each sampling from $\pi_k(x) \propto e^{-\beta_k H(x)}$. Periodically, states are exchanged between chains using Metropolis-style acceptance criteria, allowing better global exploration [115].

The deep connection between MCMC and Gibbs measures lies in their shared goal: modeling equilibrium distributions of complex systems. Gibbs measures are the mathematical foundation for defining the target distribution, and MCMC provides the computational machinery to approximate it.

8.2 Deep Learning with Structured Gibbs Measures

8.2.1 *Training methodologies*

Recall that Gibbs measures incorporate domain knowledge through energy functions:

$$p(\mathbf{x};\theta) = \frac{1}{Z(\theta)} \exp\left(-E(\mathbf{x};\theta)\right)$$

Key architectures include:

Restricted Boltzmann machines (RBMs): Energy function for bipartite graph:

$$E(\mathbf{v},\mathbf{h}) = -\mathbf{v}^T\mathbf{W}\mathbf{h} - \mathbf{b}^T\mathbf{v} - \mathbf{c}^T\mathbf{h}$$

Deep Energy-based models: Multi-layer generalizations with hierarchical latent variables

Contrastive divergence: Approximates gradient using short MCMC runs [183], [186].

Persistent contrastive divergence: Maintains persistent Markov chains between updates [367].

Score matching: Minimizes Fisher divergence:

$$J(\theta) = \mathbb{E}_{p_{\text{data}}}\left[\frac{1}{2}\left\|\nabla_{\mathbf{x}} \log p(\mathbf{x};\theta)\right\|^2 + \operatorname{tr}(\nabla^2_{\mathbf{x}} \log p(\mathbf{x};\theta))\right]$$

Avoids partition function computation [194].

Some of the relatively recent architectures are as follows:

Implicit EBMs: Energy functions defined by divergence minimization [107].

Joint energy models: Hybrid generative-discriminative formulations [164].

Latent space EBMs: Gibbs priors over latent variables in a Variational Autoencoder, a generative model that learns a probabilistic mapping between observed data and a latent space [296].

8.2.2 *Diffusion models as Gibbs processes*

Diffusion models [346, 393] are a powerful class of generative models that define a forward noising process and learn to reverse it to generate samples from complex data distributions. This process is intrinsically related to the theory of Gibbs measures and stochastic differential equations.

The forward diffusion process gradually adds Gaussian noise to the data over a sequence of time steps. At each time $t \in [0, T]$, the perturbed data $\mathbf{x}_t$ is conditionally distributed as:

$$p_t(\mathbf{x}_t|\mathbf{x}_0) = \mathcal{N}(\mathbf{x}_t; \sqrt{\bar{\alpha}_t}\mathbf{x}_0, (1 - \bar{\alpha}_t)\mathbf{I}),$$

where $\bar{\alpha}_t = \prod_{s=1}^{t} \alpha_s$, with $\alpha_t \in (0, 1)$ controlling the noise schedule.

This process can be interpreted as defining a time-dependent Gibbs distribution where the "energy" is quadratic in $\mathbf{x}_t - \sqrt{\bar{\alpha}_t}\mathbf{x}_0$, and the inverse temperature scales with $1/(1-\bar{\alpha}_t)$. As $t \to T$, the distribution converges to an isotropic Gaussian, analogous to a high-temperature limit in statistical physics.

The reverse process, which the model learns, corresponds to sampling from the target distribution by reversing the noise. The learned dynamics are modeled via:

$$\mathbf{x}_{t-1} = \frac{1}{\sqrt{\alpha_t}}\left(\mathbf{x}_t - \frac{1 - \alpha_t}{\sqrt{1 - \bar{\alpha}_t}}\epsilon_\theta(\mathbf{x}_t, t)\right) + \sigma_t \mathbf{z},$$

where $\epsilon_\theta(\mathbf{x}_t, t)$ is a neural network trained to approximate the score function (gradient of the log-density), and $\mathbf{z} \sim \mathcal{N}(0, \mathbf{I})$. This is a discretization of Langevin dynamics, a sampling algorithm commonly used to generate samples from a Gibbs distribution:

$$\mathrm{d}\mathbf{x}_t = \nabla \log \pi(\mathbf{x}_t)\, \mathrm{d}t + \sqrt{2}\, \mathrm{d}\mathbf{w}_t,$$

where $\pi(\mathbf{x}) \propto e^{-H(\mathbf{x})}$ is a Gibbs measure with Hamiltonian H, and $\mathbf{w}_t$ is Brownian motion.

This connection is formalized through the Fokker–Planck equation (see [299]), which describes the time evolution of the probability density under diffusion:

$$\frac{\partial p_t(\mathbf{x})}{\partial t} = \nabla \cdot (p_t(\mathbf{x}) \nabla H(\mathbf{x})) + \Delta p_t(\mathbf{x}),$$

indicating that diffusion models evolve a distribution p_t toward the target Gibbs measure e^{-H}. In the denoising diffusion framework, the target distribution is the complex data distribution, while intermediate steps approximate intermediate Gibbs-like states.

Thus, diffusion models realize generative sampling as a sequence of time-varying Gibbs measures, where the forward process introduces noise (annealing toward high entropy), and the reverse process denoises via score-based Langevin steps, effectively cooling toward the data distribution.

This theoretical foundation also relates diffusion models to score-based generative modeling [347], non-equilibrium thermodynamics [346] all of which are generalizations of probabilistic transitions governed by Gibbs-type principles.

8.2.3 *Deep belief networks*

Recall that Deep Belief Networks (DBNs) are deep generative models that represent complex data distributions by stacking multiple layers of Restricted Boltzmann Machines (RBMs). Each RBM models a bipartite Markov Random Field between a visible layer $\mathbf{v}$ and a hidden layer $\mathbf{h}$. The learning and sampling dynamics in DBNs are intimately connected to the theory of Gibbs measures, and each RBM defines a probability distribution of the Gibbs form:

$$p(\mathbf{v}, \mathbf{h}) = \frac{1}{Z} \exp\left(-E(\mathbf{v}, \mathbf{h})\right),$$

with energy function:

$$E(\mathbf{v}, \mathbf{h}) = -\mathbf{v}^\top W \mathbf{h} - \mathbf{b}^\top \mathbf{v} - \mathbf{c}^\top \mathbf{h},$$

where W is the weight matrix and $\mathbf{b}, \mathbf{c}$ are bias vectors. This is a discrete Gibbs distribution with Hamiltonian $H(\mathbf{v}, \mathbf{h}) = E(\mathbf{v}, \mathbf{h})$. The marginal distribution over the visible units defines a nontrivial Gibbs measure on the configuration space of $\mathbf{v}$, capturing complex dependencies among observed variables.

Layer-wise Construction via Gibbs Sampling. DBNs are constructed by training RBMs in a greedy, layer-wise manner. The training procedure is as follows [184]:

1. Initialize the bottom RBM with data $\mathbf{v}^{(0)}$.

2. Sample hidden features from the conditional distribution using the sigmoid activation:

$$p(\mathbf{h}^{(0)} \mid \mathbf{v}^{(0)}) = \sigma(W^{(0)}\mathbf{v}^{(0)} + \mathbf{c}^{(0)}).$$

3. Freeze weights and treat $\mathbf{h}^{(0)}$ as data for the next RBM layer. Repeat the sampling procedure layer by layer.

4. Fine-tune the full DBN using either:

The wake-sleep algorithm [182], which alternates between inference and generative steps; Or supervised backpropagation, if labels are available.

Each layer of a DBN represents a Gibbs distribution over hidden units conditioned on the previous layer. The full model defines a hybrid generative model, where the top two layers form a generative RBM (undirected), and lower layers define a directed generative model from latent variables to data.

Each RBM in the DBN hierarchy models a Gibbs distribution where the energy function defines the local structure of interactions. The global model can be seen as an approximation to a more complex Gibbs distribution over the joint space of visible and latent variables, learned via layer-wise energy decomposition. The wake-sleep algorithm mimics the iterative refinement of conditional Gibbs distributions in alternating directions of the network.

These connections are not only conceptual but also foundational to the mathematical formulation of DBNs as energy-based models that approximate complex high-dimensional Gibbs distributions through compositional structures.

Despite their foundational role in deep generative modeling, several open problems remain in the theory and application of DBNs:

1. *Convergence Analysis of Gibbs Sampling in Deep Architectures*: While each RBM uses Gibbs sampling to approximate marginals, there is limited theoretical understanding of the convergence rate and ergodicity in deep DBNs where errors compound across layers. Backpropagation through MCMC chains requires non-trivial gradient estimators [106], [187].

2. *Global Partition Function Estimation*: The partition function Z of the full DBN is intractable due to its hierarchical structure. Approximate methods (e.g., contrastive divergence) are widely used, but the statistical and variational properties of such estimates remain under-explored.

3. *Phase Transitions and Critical Behavior*: Analogous to physical systems, DBNs may exhibit phase transitions (e.g., between ordered and disordered representations). Studying how the structure of the energy landscape changes with depth and data remains an open theoretical challenge.

4. *Expressivity vs. Learnability Trade-off*: While DBNs are universal approximators under certain conditions, it remains unclear how architectural choices (depth, width, weight sparsity) affect the learnability of high-quality Gibbs-like approximations.

5. *Thermodynamic Interpretations of Fine-tuning Algorithms*: The wake-sleep algorithm and contrastive divergence have interpretations as minimizing variational free energy. Developing a rigorous thermodynamic formulation could bridge learning algorithms with nonequilibrium statistical mechanics.

6. *Extension to Continuous or Structured Variables*: Generalizing DBNs to non-binary variables, structured data (e.g., graphs, sequences), or non-Euclidean geometries while preserving their Gibbsian semantics is an active research area.

8.3 Gibbs measures in Large Language Models

Large Language Models (LLMs) such as GPT-4, Claude 3, and Gemini 1.5 have become one of the hottest research directions in AI. They are able to compete and even outperform traditional internet search engines and auto-complete as they enjoy state-of-the-art deep learning architectures to learn from vast amount of data. This section attempts to translate some of the core components of modern LLMs into statistical mechanics terms to showcase the potential utility of the Gibbsian framework.

8.3.1 *Statistical mechanics of autoregressive language modeling*

The connection between language modeling and statistical mechanics emerges naturally when we view language as a stochastic process over discrete symbols. In this framework, text sequences can be understood as configurations of a statistical system, where each token represents a microscopic state, and the probability of a sequence corresponds to the Gibbs-Boltzmann weight of that configuration.

8.3.1.1 *Language as a Stochastic Process and Statistical Mechanics Connection*

Consider a language model operating over a vocabulary $\mathcal{V}$ of size $|\mathcal{V}| = V$. A text sequence $\mathbf{x} = (x_1, x_2, \ldots, x_T)$ of length T can be viewed as a configuration in the configuration space $\mathcal{V}^T$. The fundamental insight is that the probability distribution over sequences

$$P(\mathbf{x}) = P(x_1, x_2, \ldots, x_T)$$

can be expressed as a Gibbs measure of the form

$$P(\mathbf{x}) = \frac{1}{Z} \exp\left(-\beta H(\mathbf{x})\right)$$

where $H(\mathbf{x})$ is the energy function (Hamiltonian) of the configuration, β is the inverse temperature, and Z is the partition function ensuring normalization.

8.3.1.2 *Sequence probability distributions and the chain rule*

The chain rule of probability provides the foundation for autoregressive language modeling [30, 354, 374]:

$$P(\mathbf{x}) = P(x_1) \prod_{t=2}^{T} P(x_t | x_{<t})$$

where $x_{<t} = (x_1, \ldots, x_{t-1})$ denotes the context preceding position t. Taking the negative logarithm, we obtain an additive energy function:

$$H(\mathbf{x}) = -\log P(x_1) - \sum_{t=2}^{T} \log P(x_t | x_{<t})$$

This decomposition reveals that the sequence energy is the sum of local energy contributions, each depending on the conditional probability of a token given its context. This additive structure is characteristic of many-body systems in statistical mechanics.

In autoregressive language models, the conditional distribution $P(x_t | x_{<t})$ is parameterized by a neural network f_θ (usually Transformer models) as in Fig. 8.1. The model computes contextual representations $\mathbf{h}_t = f_\theta(x_{<t})$ and uses these to define the conditional probabilities. The energy contribution at position t becomes:

$$H_t(\mathbf{x}) = -\log P(x_t | x_{<t}) = -\log \operatorname{softmax}(\mathbf{W}\mathbf{h}_t)_{x_t}$$

where $\mathbf{W}$ is the output embedding matrix, see Fig. 8.1.

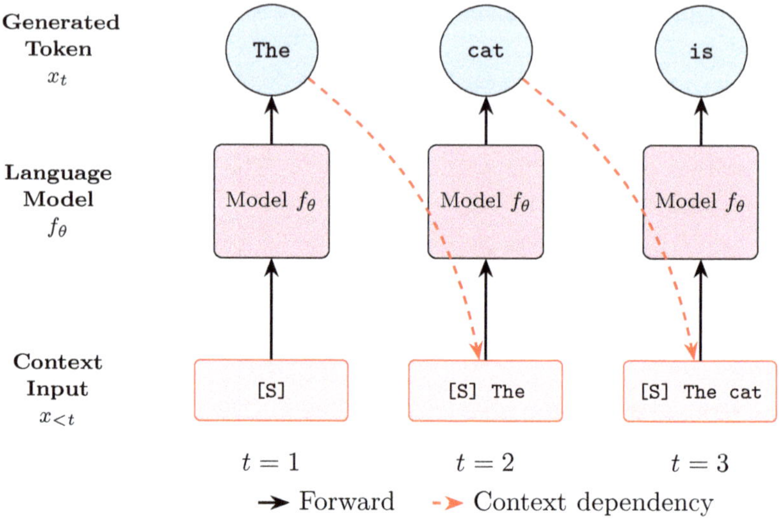

Fig. 8.1: Autoregressive text generation: each token is generated based on all previous tokens in the sequence.

8.3.1.3 *Next token prediction as sampling from Gibbs distribution over vocabulary*

For next token prediction , given a context $x_{<t}$, the model defines a Gibbs distribution over the vocabulary:

$$P(x_t = v|x_{<t}) = \frac{\exp(-\beta E_v(\mathbf{h}_t))}{\sum_{v' \in \mathcal{V}} \exp(-\beta E_{v'}(\mathbf{h}_t))},$$

where $E_v(\mathbf{h}_t) = -(\mathbf{W}\mathbf{h}_t)_v$ is the energy of token v given the context representation $\mathbf{h}_t$.

As has been mentioned many times throughout the book, the softmax function, ubiquitous in language modeling, is precisely the Gibbs-Boltzmann distribution:

$$\text{softmax}(\mathbf{z}/\tau)_i = \frac{\exp(z_i/\tau)}{\sum_j \exp(z_j/\tau)},$$

where $\mathbf{z} = \mathbf{W}\mathbf{h}_t$ are the logits and τ is the temperature parameter. This establishes the direct correspondence:

Energy: $E_i = -z_i$,

Inverse temperature: $\beta = 1/\tau$,

Partition function: $Z = \sum_j \exp(z_j/\tau)$.

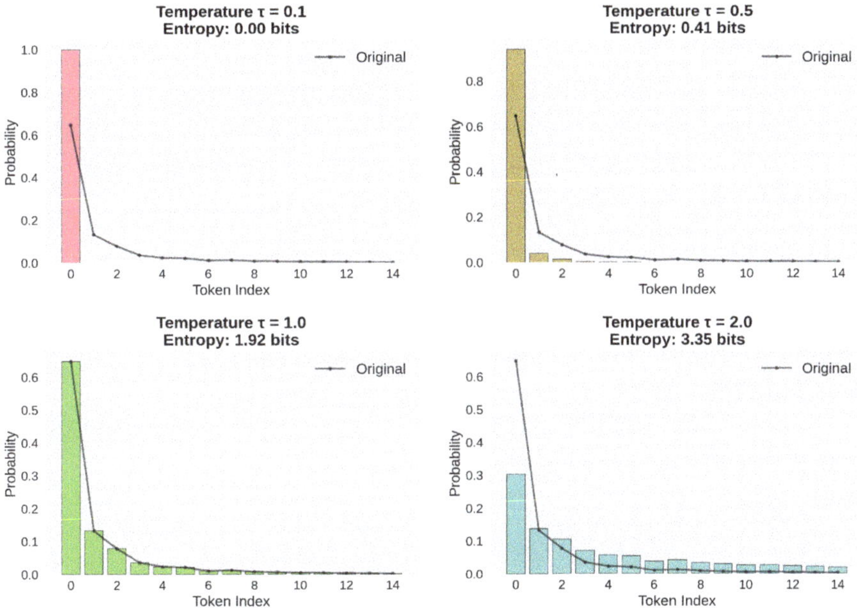

Fig. 8.2: Effect of the temperature parameter on the next token distribution over the vocabulary $\mathcal{V}$: small temperature is for sharper distributions, higher temperature for smoother/more uniform distributions.

The temperature parameter τ controls the sharpness of the distribution. As $\tau \to 0$ (high inverse temperature), the distribution becomes concentrated on the maximum logit (greedy sampling), while as $\tau \to \infty$ (low inverse temperature), the distribution approaches uniform, see Fig. 8.2. This behavior mirrors the thermodynamic limit where low temperatures favor ground states and high temperatures lead to maximum entropy configurations whose discussion follows next.

8.3.2 *Temperature and sampling in language generation*

The temperature parameter in language generation provides a direct bridge between the statistical mechanics of Gibbs measures and practical text generation algorithms. Understanding this connection illuminates the tradeoffs between creativity and coherence in generated text, while revealing the deep mathematical structure underlying modern sampling methods.

8.3.2.1 *Temperature parameter as inverse thermodynamic temperature*

The temperature parameter τ in language generation corresponds precisely to the thermodynamic temperature in statistical mechanics. Given logits $\mathbf{z} = (z_1, z_2, \ldots, z_V)$ for vocabulary $\mathcal{V}$, the temperature-scaled Gibbs distribution is:

$$P(v|x_{<t}) = \frac{\exp(z_v/\tau)}{\sum_{v' \in \mathcal{V}} \exp(z_{v'}/\tau)} = \frac{\exp(-\beta E_v)}{\sum_{v' \in \mathcal{V}} \exp(-\beta E_{v'})}$$

where $\beta = 1/\tau$ is the inverse temperature and $E_v = -z_v$ is the energy associated with token v.

The partition function $Z(\beta) = \sum_{v \in \mathcal{V}} \exp(-\beta E_v)$ satisfies the fundamental thermodynamic relation:

$$\frac{\partial \log Z(\beta)}{\partial \beta} = -\langle E \rangle = -\sum_{v \in \mathcal{V}} E_v P(v|x_{<t})$$

The entropy of the distribution is given by:

$$S(\beta) = -\sum_{v \in \mathcal{V}} P(v|x_{<t}) \log P(v|x_{<t}) = \beta \langle E \rangle + \log Z(\beta)$$

The following observations are trivial but useful:

(1) As $\beta \to \infty$ ($\tau \to 0$), $P(v|x_{<t}) \to \delta_{v,v^*}$ where $v^* = \arg\max_v z_v$
(2) As $\beta \to 0$ ($\tau \to \infty$), $P(v|x_{<t}) \to 1/V$ (uniform distribution)
(3) Entropy is bounded as $0 \leq S(\beta) \leq \log V$

Next, we discuss some of the notable sampling strategies for next token prediction.

8.3.2.2 *Sampling strategies: mathematical foundations and thermodynamic interpretations*

Different sampling strategies [125,175,288] correspond to different modifications of the Gibbs distribution (Figure 8.3), each with distinct mathematical properties and thermodynamic interpretations.

Greedy sampling ($\tau = 0$): Greedy sampling corresponds to the zero-temperature limit:

$$x_t = \arg\max_{v \in \mathcal{V}} P(v|x_{<t}) = \arg\min_{v \in \mathcal{V}} E_v$$

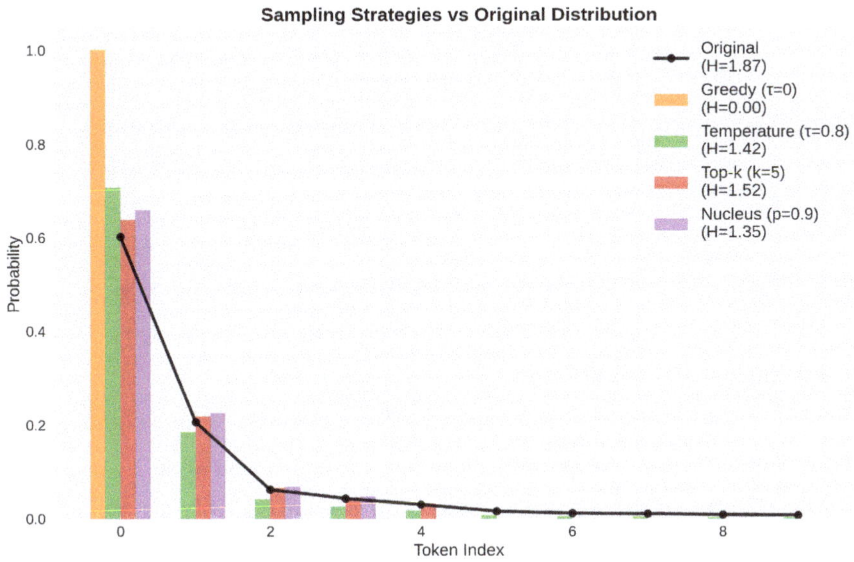

Fig. 8.3: Classical sampling strategies: greedy, temperature, top-k, nucleus

This finds the ground state configuration at each step. The effective entropy is $S = 0$, and the system exhibits no thermal fluctuations.

Temperature sampling: Standard temperature sampling draws from the full Gibbs distribution:

$$P_\tau(v|x_{<t}) = \frac{\exp(z_v/\tau)}{\sum_{v' \in \mathcal{V}} \exp(z_{v'}/\tau)}.$$

The variance of the energy then becomes

$$\mathrm{Var}(E) = \langle E^2 \rangle - \langle E \rangle^2 = \tau^2 \frac{\partial^2 \log Z}{\partial \tau^2}.$$

Top-k sampling: Top-k sampling truncates the distribution to the k most probable tokens. Let $\mathcal{V}_k = \{v_1, v_2, \ldots, v_k\}$ be the set of top-k tokens ordered by probability $P(v_1|x_{<t}) \geq P(v_2|x_{<t}) \geq \ldots \geq P(v_k|x_{<t})$.

The truncated distribution is:

$$P_{\text{top-k}}(v|x_{<t}) = \begin{cases} \frac{P(v|x_{<t})}{\sum_{v' \in \mathcal{V}_k} P(v'|x_{<t})} & \text{if } v \in \mathcal{V}_k \\ 0 & \text{otherwise} \end{cases}$$

The effective partition function becomes:

$$Z_k(\beta) = \sum_{v \in \mathcal{V}_k} \exp(-\beta E_v) \leq Z(\beta)$$

The entropy reduction is:

$$\Delta S = S - S_k = \log\left(\frac{Z(\beta)}{Z_k(\beta)}\right) + \sum_{v \notin \mathcal{V}_k} P(v|x_{<t}) \log P(v|x_{<t})$$

Nucleus sampling (Top-p): Nucleus sampling uses a dynamic cutoff based on cumulative probability. Define the cumulative distribution:

$$F(v_i) = \sum_{j=1}^{i} P(v_j|x_{<t})$$

where tokens are ordered by decreasing probability. The nucleus set is:

$$\mathcal{V}_p = \{v_i : F(v_{i-1}) < p\} = \{v_i : i \leq \min\{j : F(v_j) \geq p\}\}$$

The nucleus distribution is:

$$P_{\text{nucleus}}(v|x_{<t}) = \begin{cases} \frac{P(v|x_{<t})}{\sum_{v' \in \mathcal{V}_p} P(v'|x_{<t})} & \text{if } v \in \mathcal{V}_p \\ 0 & \text{otherwise} \end{cases}$$

The effective vocabulary size adapts to the entropy of the original distribution:

$$|\mathcal{V}_p| = \min\{k : \sum_{i=1}^{k} P(v_i|x_{<t}) \geq p\}$$

Contrastive search: Contrastive search balances model confidence with diversity through a degeneration penalty. The selection criterion is:

$$x_t = \arg\max_{v \in \mathcal{V}_k} \left[\alpha \cdot P(v|x_{<t}) - (1-\alpha) \cdot \max_{x_j \in x_{<t}} \frac{\mathbf{h}_v^T \mathbf{h}_j}{|\mathbf{h}_v||\mathbf{h}_j|}\right]$$

where $\mathbf{h}_v$ is the contextualized embedding (e.g. from a Transformer layer) of token v, and $\alpha \in (0,1)$ controls the balance between model confidence and diversity.

Min-p sampling: Formally, at each time step t, let $\mathcal{V}$ denote the vocabulary, and $P(x_t|x_{1:t-1})$ represent the conditional probability distribution over the vocabulary for the next token x_t. Min-p sampling works as follows:

(1) *Calculate the maximum probability*: Identify the maximum probability token in the distribution, denoted as $p_{\max} = \max_{v \in \mathcal{V}} P(v|x_{1:t-1})$.

(2) *Truncation threshold*: Set a base probability threshold, $p_{\text{base}} \in (0,1]$ Scale it by $p_{\max}$:

$$p_{\text{scaled}} = p_{\text{base}} \times p_{\max}$$

This threshold ensures that tokens with sufficiently high relative probabilities are considered while filtering out less probable tokens in a context-dependent manner.

(3) *Define sampling pool*: Construct pool $\mathcal{V}_{\min}$ of tokens with probabilities $\geq p_{\text{scaled}}$:

$$\mathcal{V}_{\min} = \{v \in \mathcal{V} : P(v|x_{1:t-1}) \geq p_{\text{scaled}}\}$$

(4) *Sample from pool*: Sample the next token x_t from the normalized set $\mathcal{V}_{\min}$:

$$P'(v) = \frac{P(v|x_{1:t-1})}{\sum_{v' \in \mathcal{V}_{\min}} P(v'|x_{1:t-1})} \quad \text{for } v \in \mathcal{V}_{\min}$$

Min-p sampling enables confidence-based filtering of less likely tokens, therefore, reducing an undesirable performance sensitivity to the temperature parameter. Although intuitive, a more principled theoretical framework to explain its benefits remains quite elusive.

8.3.2.3 *Metropolis-Hastings sampling in language models*

The Metropolis-Hastings algorithm provides a principled framework for sampling from complex probability distributions by constructing a Markov chain that converges to the target distribution. In language models, this enables sophisticated sampling strategies that go beyond simple autoregressive generation [126, 274].

Detailed balance and acceptance probability: The acceptance probability in Metropolis-Hastings is designed to satisfy the detailed balance condition, which ensures that the Markov chain converges to the target distribution $P(\mathbf{x})$. Detailed balance requires:

$$P(\mathbf{x}) \cdot T(\mathbf{x} \to \mathbf{x}') = P(\mathbf{x}') \cdot T(\mathbf{x}' \to \mathbf{x})$$

where $T(\mathbf{x} \to \mathbf{x}')$ is the transition probability from state $\mathbf{x}$ to $\mathbf{x}'$.

The transition probability decomposes as:

$$T(\mathbf{x} \to \mathbf{x}') = q(\mathbf{x}'|\mathbf{x}) \cdot \alpha(\mathbf{x} \to \mathbf{x}')$$

where $q(\mathbf{x}'|\mathbf{x})$ is the proposal distribution and $\alpha(\mathbf{x} \to \mathbf{x}')$ is the acceptance probability.

Substituting into the detailed balance condition and solving for α:

$$\alpha(\mathbf{x} \to \mathbf{x}') = \min\left(1, \frac{P(\mathbf{x}') \cdot q(\mathbf{x}|\mathbf{x}')}{P(\mathbf{x}) \cdot q(\mathbf{x}'|\mathbf{x})}\right)$$

This form ensures that detailed balance is satisfied while maximizing the acceptance rate. The $\min(1, \cdot)$ operation prevents acceptance probabilities from exceeding 1, and the ratio form naturally favors transitions to higher-probability states.

Single-token proposals: For language generation, consider proposing modifications at position i:

$$\mathbf{x}' = (x_1, \ldots, x_{i-1}, x_i', x_{i+1}, \ldots, x_T)$$

The acceptance probability becomes:

$$\alpha = \min\left(1, \frac{P(\mathbf{x}') \cdot q(\mathbf{x}|\mathbf{x}')}{P(\mathbf{x}) \cdot q(\mathbf{x}'|\mathbf{x})}\right),$$

For symmetric proposals where $q(\mathbf{x}'|\mathbf{x}) = q(\mathbf{x}|\mathbf{x}')$ (e.g., uniform selection from vocabulary):

$$\alpha = \min\left(1, \frac{P(\mathbf{x}')}{P(\mathbf{x})}\right) = \min\left(1, \exp(-\beta\Delta E)\right),$$

where $\Delta E = H(\mathbf{x}') - H(\mathbf{x})$ is the energy difference between proposed and current states.

This formulation directly connects to statistical mechanics: proposals that lower the energy (increase probability) are always accepted, while energy-increasing proposals are accepted with probability $\exp(-\beta\Delta E)$, following the Boltzmann distribution.

Connection to speculative decoding: Modern speculative decoding schemes [243, 355] leverage similar principles. A fast "draft" model proposes tokens, and a larger "target" model accepts or rejects them based on probability ratios:

$$\alpha_{\text{spec}} = \min\left(1, \frac{P_{\text{target}}(x_t|x_{<t})}{P_{\text{draft}}(x_t|x_{<t})}\right).$$

This maintains the target model's distribution while achieving speedup through parallel proposal generation.

8.3.3 *Information-theoretic quantities in language modeling as thermodynamic observables*

The correspondence between information theory and statistical mechanics provides a rigorous framework for understanding language model behavior through thermodynamic observables. We establish this connection by mapping probability distributions to Gibbs-Boltzmann distributions and information-theoretic quantities to thermodynamic potentials.

8.3.3.1 *Thermodynamic formulation of language models*

Consider a language model defining a probability distribution over sequences $\mathbf{x} = (x_1, \ldots, x_T)$ from vocabulary $\mathcal{V}$. We establish the thermodynamic correspondence by defining an energy function $H(\mathbf{x})$ such that:

$$P_{\text{model}}(\mathbf{x}) = \frac{1}{Z} \exp(-\beta H(\mathbf{x})),$$

where $Z = \sum_{\mathbf{x}} \exp(-\beta H(\mathbf{x}))$ is the partition function and $\beta = 1/T$ is the inverse temperature.

The energy function is related to the model's log-probabilities by:

$$H(\mathbf{x}) = -\frac{1}{\beta} \log P_{\text{model}}(\mathbf{x}) - \frac{1}{\beta} \log Z.$$

8.3.3.2 *Cross-entropy loss and relative entropy*

The cross-entropy loss used in language model training is defined as:

$$\mathcal{L}_{\text{CE}} = -\mathbb{E}_{P_{\text{data}}}[\log P_{\text{model}}(\mathbf{x})] = -\sum_{\mathbf{x}} P_{\text{data}}(\mathbf{x}) \log P_{\text{model}}(\mathbf{x}).$$

Using the information-theoretic identity, this decomposes as:

$$\mathcal{L}_{\text{CE}} = H(P_{\text{data}}) + D_{\text{KL}}(P_{\text{data}} \| P_{\text{model}}),$$

where $H(P_{\text{data}}) = -\sum_{\mathbf{x}} P_{\text{data}}(\mathbf{x}) \log P_{\text{data}}(\mathbf{x})$ is the entropy of the data distribution and

$$D_{\text{KL}}(P_{\text{data}} \| P_{\text{model}}) = \sum_{\mathbf{x}} P_{\text{data}}(\mathbf{x}) \log \frac{P_{\text{data}}(\mathbf{x})}{P_{\text{model}}(\mathbf{x})}$$

is the Kullback-Leibler divergence.

In the thermodynamic formulation, the cross-entropy loss becomes:

$$\mathcal{L}_{\text{CE}} = \beta \langle H(\mathbf{x}) \rangle_{P_{\text{data}}} + \log Z,$$

where $\langle H(\mathbf{x}) \rangle_{P_{\text{data}}} = \sum_{\mathbf{x}} P_{\text{data}}(\mathbf{x}) H(\mathbf{x})$ is the average energy of data configurations under the model's energy function.

8.3.3.3 *Perplexity as effective partition function*

Perplexity is defined as the exponential of cross-entropy loss:

$$\text{PPL} = \exp(\mathcal{L}_{\text{CE}}) = \exp\left(-\mathbb{E}_{P_{\text{data}}}[\log P_{\text{model}}(\mathbf{x})]\right).$$

For autoregressive models, we can decompose this over time steps. The per-token cross-entropy at position t is:

$$\mathcal{L}_t = -\mathbb{E}_{P_{\text{data}}}[\log P_{\text{model}}(x_t | x_{<t})].$$

The corresponding per-token perplexity is:

$$\mathrm{PPL}_t = \exp(\mathcal{L}_t) = \exp\left(-\mathbb{E}_{P_{\text{data}}}[\log P_{\text{model}}(x_t|x_{<t})]\right).$$

In the thermodynamic interpretation, this represents the effective size of the local partition function:

$$\mathrm{PPL}_t = \frac{Z_t(\mathbf{x}_{<t})}{\exp(-\beta E_{x_t}(\mathbf{x}_{<t}))},$$

where $Z_t(\mathbf{x}_{<t}) = \sum_{v\in\mathcal{V}} \exp(-\beta E_v(\mathbf{x}_{<t}))$ is the local partition function at position t given context $\mathbf{x}_{<t}$.

8.3.3.4 *Mutual information and correlation functions*

The mutual information between tokens at positions i and j is defined as:

$$I(X_i; X_j) = \sum_{x_i, x_j} P(x_i, x_j) \log \frac{P(x_i, x_j)}{P(x_i)P(x_j)}.$$

This can be expressed in terms of the connected correlation function. Define the correlation function as:

$$G_{ij}(x_i, x_j) = P(x_i, x_j) - P(x_i)P(x_j).$$

The connected correlation function in the thermodynamic limit becomes:

$$G_{ij}^{(c)}(x_i, x_j) = \frac{\partial^2 \log Z}{\partial J_i(x_i)\partial J_j(x_j)},$$

where $J_i(x_i)$ are external fields coupled to the system.

For language models, mutual information measures long-range dependencies and defines an effective correlation length:

$$\xi = \lim_{|i-j|\to\infty} \frac{|i-j|}{-\log|I(X_i; X_j)|}.$$

Alternatively, using the transfer matrix formalism with eigenvalues $\{\lambda_k\}$:

$$\xi = \frac{1}{\min_{k\neq 0} |\log \lambda_k|}.$$

8.4 Probabilistic Programming and Gibbs Sampling

8.4.1 *Algorithmic foundations*

Gibbs sampling decomposes high-dimensional sampling into conditional updates:

$$x_i^{(t+1)} \sim p\left(x_i \mid x_1^{(t+1)}, \ldots, x_{i-1}^{(t+1)}, x_{i+1}^{(t)}, \ldots, x_n^{(t)}\right)$$

Collapsed Gibbs: Marginalize nuisance variables for faster mixing [253].

Blocked Gibbs: Update groups of correlated variables jointly [316].

Rao-Blackwellization: Reduce variance via conditional expectations [68]

Table 8.1: Gibbs sampling support in various Probabilistic Programming Languages.

System	Discrete vars	Continuous vars	Automatic blocking	GPU support
Stan	Limited	Excellent	Yes	Partial
Pyro	Good	Good	Manual	Yes
Edward2	Good	Excellent	No	Yes
WebPPL	Excellent	Poor	Yes	No

8.4.2 *Hybrid inference engines*

Hybrid inference engines combine multiple sampling strategies to efficiently handle models with both continuous and discrete latent variables. These engines are particularly useful when dealing with complex probabilistic models where no single inference technique performs well across all variable types. One prominent hybrid approach is the Hamiltonian-Gibbs sampler, which leverages the strengths of both Hamiltonian Monte Carlo (HMC) and Gibbs sampling.

Hamiltonian-Gibbs: This hybrid method integrates HMC for efficient sampling of continuous parameters and Gibbs sampling for discrete latent variables. It is particularly well-suited for hierarchical Bayesian models where continuous global parameters interact with discrete local latent states. The main idea is to alternate between:

HMC updates of the continuous variables using Hamiltonian dynamics, which are efficient for high-dimensional and correlated parameter spaces.

Gibbs sampling of the discrete variables conditioned on the current values of the continuous parameters.

This hybridization can significantly improve mixing and convergence over purely discrete or purely continuous samplers, especially in models where discrete and continuous components are tightly coupled.

In practice, a hybrid engine like this is not directly supported in Stan because Stan only supports HMC (and related gradient-based MCMC methods) and does not natively support sampling of discrete parameters. However, in frameworks that allow for more flexible control over the inference process (such as custom PyMC implementations or probabilistic programming languages like Edward2 or Pyro), this hybrid approach can be implemented by interleaving HMC steps for the continuous parameters with exact (or approximate) Gibbs steps for discrete parameters.

A motivating example comes from Bayesian mixture models, where the cluster assignments (discrete latent variables) can be updated via Gibbs sampling, and the cluster means and variances (continuous parameters) via HMC. This approach was explored by [37], who demonstrated that such hybrid samplers can effectively mitigate the pathologies often encountered in purely HMC- or Gibbs-based inference in hierarchical models.

Variational-Gibbs hybrids: Use VI approximations to initialize Gibbs chains [190].

Particle Gibbs: Incorporates particle filtering for state-space models [13].

8.4.3 *Convergence diagnostics*

Convergence diagnostics are essential tools for assessing whether a MCMC algorithm—such as Gibbs sampling—has reached its stationary distribution (see [37]), which is typically a Gibbs measure. Without convergence, samples cannot be trusted to approximate the target distribution.

Since the Gibbs measure $\pi(x) \propto e^{-H(x)}$ is often known only up to a normalization constant, convergence diagnostics rely on empirical statistics over multiple chains to ensure that the Markov chain has sufficiently explored the target space. Below we describe key diagnostics and their statistical foundations.

Gelman-Rubin Statistic ($\hat{R}$). The Gelman-Rubin diagnostic (also called the potential scale reduction factor) is widely used to compare between-chain and within-chain variances across multiple chains started from different initial conditions. It estimates how much variance could be reduced if the chains were continued indefinitely.

$$\hat{R} = \sqrt{\frac{\mathrm{var}^{+}(\psi)}{W}},$$

where W is the average within-chain variance, $\mathrm{var}^{+}(\psi)$ is the pooled variance (weighted average of within- and between-chain variances).

Values of $\hat{R} \approx 1$ indicate convergence, while $\hat{R} > 1.1$ suggests that the chains have not yet mixed properly. The statistic is especially relevant for Gibbs samplers, where slow mixing may occur near phase transitions in the associated Gibbs measure [147].

Effective sample size (ESS). ESS quantifies the number of independent samples that the autocorrelated MCMC chain is equivalent to:

$$\mathrm{ESS} = N \left(1 + 2\sum_{k=1}^{\infty} \rho(k)\right)^{-1},$$

where $\rho(k)$ is the autocorrelation at lag k, and N is the total number of samples. A high autocorrelation reduces ESS, which limits the effective information obtained from the chain.

In machine learning, low ESS is common in models with sharp posterior modes, such as sparse Bayesian regression or variational autoencoders with discrete priors. ESS is therefore critical in understanding how much useful signal has been extracted from complex energy landscapes defined by Gibbs distributions.

Trace-aware and high-dimensional diagnostics. Recent developments have improved diagnostics in high-dimensional models often encountered in machine learning. The Split-$\hat{R}$ and rank-normalized $\hat{R}$ [375] refine the original Gelman–Rubin statistic by: Splitting chains in half to detect non-stationarity, Rank-transforming samples to improve robustness against outliers or non-Gaussian behavior.

These diagnostics are especially important in deep generative models and hierarchical Bayesian structures, where naive assessments of convergence can be misleading due to multimodality or heavy-tailed posteriors.

In modern machine learning, particularly in Bayesian deep learning, probabilistic programming, and energy-based models, convergence diagnostics are crucial for validating inference quality.

Furthermore, convergence diagnostics directly reflect how well the MCMC process explores the configuration space of the underlying Gibbs distribution, providing insights into the geometry of the energy landscape, the presence of metastable states, and computational trade-offs between model expressivity and inference complexity.

Despite widespread adoption, several open challenges remain:

1. *Multimodality awareness*: Standard diagnostics may falsely indicate convergence when chains are trapped in separate modes. Mode-aware diagnostics are an active research area.

2. *Scalability*: Existing diagnostics become computationally expensive or unreliable in high dimensions (e.g., in large neural networks).

3. *Non-Gaussian posteriors*: Many Bayesian machine learning models yield heavy-tailed or skewed posteriors; current diagnostics assume near-normal behavior.

4. *Streaming and online diagnostics*: As models scale and data streams in, there is a growing need for diagnostics that work in real-time or under memory constraints.

5. *Theory-practice gap*: There is a lack of rigorous convergence guarantees for adaptive or heuristic diagnostics used in real-world applications.

8.4.4 *Topic modeling*

Topic modeling refers to a class of unsupervised learning techniques for discovering latent thematic structures in large text corpora [391]. The most well-known model is Latent Dirichlet Allocation (LDA) [47], a generative probabilistic model that assumes each document is a mixture of topics, and each topic is a distribution over words.

LDA defines the following generative process for each document in a corpus:

1. For each topic $k = 1, \ldots, K$, draw a distribution over words $\phi_k \sim \text{Dirichlet}(\beta)$.

2. For each document d, draw topic proportions $\theta_d \sim \text{Dirichlet}(\alpha)$,
For each word w_i in the document:
Sample a topic $z_i \sim \text{Multinomial}(\theta_d)$,
Sample a word $w_i \sim \text{Multinomial}(\phi_{z_i})$.

The latent variables θ, ϕ, z define a hierarchical Bayesian model, and inference focuses on estimating the posterior distribution over the topic assignments z, often using collapsed Gibbs sampling — a form of MCMC that integrates out θ and ϕ analytically.

Collapsed Gibbs sampling for LDA. The collapsed Gibbs sampler samples only the topic assignments z_i for each token w_i, conditioned on all other assignments $\mathbf{z}_{-i}$ and the observed words $\mathbf{w}$. The update rule is:

$$p(z_i = k \mid \mathbf{z}_{-i}, \mathbf{w}) \propto \frac{n_{k,-i}^{(w_i)} + \beta}{n_{k,-i}^{(\cdot)} + V\beta} \cdot \frac{n_{d,-i}^{(k)} + \alpha}{n_{d,-i}^{(\cdot)} + K\alpha},$$

where

$n_{k,-i}^{(w_i)}$ is the number of times word w_i is assigned to topic k, excluding the current token,

$n_{d,-i}^{(k)}$ is the number of words in document d assigned to topic k,

V is the vocabulary size, and K is the number of topics.

This conditional distribution is a Gibbs measure, where the probability of a configuration $z_i = k$ depends exponentially on local sufficient statistics (counts), forming a log-linear energy function. The full posterior over topic assignments is a high-dimensional Gibbs distribution:

$$\pi(\mathbf{z}) \propto \prod_{k=1}^{K} \frac{\Gamma(n_k^{(\cdot)} + V\beta)}{\prod_{v=1}^{V} \Gamma(n_k^{(v)} + \beta)} \prod_{d=1}^{D} \frac{\Gamma(n_d^{(\cdot)} + K\alpha)}{\prod_{k=1}^{K} \Gamma(n_d^{(k)} + \alpha)},$$

which collapsed Gibbs sampling explores via a Markov chain whose stationary distribution is $\pi(\mathbf{z})$.

LDA and its Gibbs sampling implementation have been central to machine learning applications in

Natural language processing: automatic topic discovery, document classification, summarization, information retrieval.

Recommender systems: modeling user-item interactions as latent topics.

Bioinformatics: discovering latent gene or protein groups.

Social network analysis: inferring community structures.

Several open challenges remain:

1. Collapsed Gibbs sampling lacks robust convergence diagnostics in high-dimensional topic models. Existing methods (e.g., $\hat{R}$, ESS) are hard to apply due to the combinatorial nature of the space.

2. LDA struggles with dynamic corpora. Online and streaming variants exist, but maintaining a coherent Gibbs posterior in real-time remains challenging.

3. Integrating deep neural encoders with discrete Gibbs-based latent variables (e.g., in variational autoencoders or transformers) is an emerging direction, raising challenges in amortized inference and differentiability.

8.4.5 *Emerging frontiers*

Recent advances at the intersection of machine learning and statistical physics have opened new directions that extend classical Gibbs-based modeling and sampling to novel computational paradigms. These emerging frontiers aim to overcome limitations of traditional methods in terms of scalability, expressivity, and applicability to real-world data and compute infrastructures.

Differentiable program induction(DPI). DPI seeks to optimize over the space of probabilistic programs using gradient-based techniques. Traditional program induction methods rely on combinatorial search over symbolic structures. In contrast, differentiable approaches embed programs or program traces into continuous spaces amenable to gradient descent [120].

These frameworks often define a Gibbs posterior over program traces or execution paths:

$$\pi(\text{program} \mid \mathcal{D}) \propto \exp\left(-\beta \cdot \mathcal{L}(\text{program}, \mathcal{D})\right),$$

where $\mathcal{L}$ is a differentiable surrogate loss (e.g., negative log-likelihood or behavior discrepancy). Optimization then becomes a form of approximate maximum-a-posteriori (MAP) inference under a Gibbs measure, where the Hamiltonian encodes program fit and complexity.

Enables interpretable, symbolic learning (e.g., equations, grammars, algorithms) using deep learning infrastructure. Applications include program synthesis, automated reasoning, and interpretable reinforcement learning.

Quantum-enhanced sampling: Quantum-enhanced sampling leverages quantum devices to speed up sampling from complex high-dimensional distributions. One approach maps discrete variables (e.g., Ising spins) to quantum bits (qubits), enabling operations like quantum annealing or variational quantum sampling to explore Gibbs-like energy landscapes [28], [75].

Given a Hamiltonian $H(x)$, quantum hardware approximates the Gibbs distribution:

$$\pi(x) \propto e^{-\beta H(x)},$$

by evolving the system toward thermal equilibrium using quantum operations. Quantum techniques can potentially accelerate mixing for multimodal distributions, where classical MCMC struggles due to high-energy barriers.

Relevance to ML: Applies to training of Boltzmann machines, combinatorial optimization, and sampling in generative models. Offers new architectures for energy-based models using variational quantum circuits.

Federated Gibbs Sampling: Federated Gibbs sampling enables collaborative inference across distributed datasets held on different devices or institutions, without exposing raw data. Inspired by federated learning, this approach adapts Gibbs sampling to work in a decentralized setting, where each client samples from local conditional distributions and communicates aggregated statistics (e.g., sufficient statistics or message updates) [64].

In the federated context, the global Gibbs distribution is approximated via coordinated local updates:

$$\pi(x) \propto \prod_{i=1}^{M} \exp(-H_i(x_i)),$$

where each H_i corresponds to a private local energy function. Privacy-preserving mechanisms (e.g., differential privacy, secure aggregation) ensure that shared information does not leak sensitive data.

While promising, these frontiers pose significant theoretical and practical challenges:

1. Differentiable Program Spaces:

How to define smooth, learnable representations of program structure that preserve syntactic correctness?

What are the convergence properties of gradient-based optimization over symbolic programs?

2. Quantum Sampling Limitations:

What classes of distributions benefit from quantum speedup?

How to control noise and decoherence in near-term quantum devices?

3. Scalability of federated Gibbs:

How to ensure convergence of decentralized MCMC chains under communication constraints?

How to design unbiased and differentially private aggregations of sufficient statistics?

4. Theoretical Guarantees: For all three frontiers, formal guarantees of convergence to the true Gibbs distribution (or its approximation) are largely lacking.

5. Benchmarking: Standard benchmarks and evaluation protocols for these new methods are still under development.

8.5 Bayesian Nonparametrics and Infinite Gibbs Measures

8.5.1 *Infinite-dimensional extensions*

Gibbs measures on infinite spaces require projective limits:

$$\mu_\beta(\sigma) = \lim_{\Lambda \uparrow \mathbb{L}} \mu_\beta^\Lambda(\sigma_\Lambda | \sigma_{\partial\Lambda}),$$

where $\mathbb{L}$ is a countable connected graph (e.g., $\mathbb{L} = \mathbb{Z}^d$ or it may be an infinite tree).

Key frameworks:

- Dirichlet-Laplace random measures for sparse priors [38].
- Gaussian process priors: $f \sim \mathcal{GP}(m(\cdot), k(\cdot, \cdot))$ with Gibbs posterior:

$$p(f|\mathcal{D}) \propto \exp\left(-\beta \sum \ell(y_i, f(x_i))\right) \mathcal{GP}(f)$$

8.5.2 *Nonparametric mixture models*

Dirichlet process mixtures:

$$G \sim \mathrm{DP}(\alpha, H), \quad \theta_i | G \sim G, \quad x_i | \theta_i \sim F(\theta_i)$$

Chinese restaurant process representation enables Gibbs sampling [49], [283].

Hierarchical HDPs: Nested DPs for multi-resolution clustering [364], [365].

8.5.3 *Inference techniques*

Probabilistic models, particularly Bayesian nonparametric models such as the Dirichlet Process (DP) and its mixture models, require efficient inference methods due to their infinite-dimensional nature. The following inference techniques are foundational in this context and have connections to Gibbs measures and machine learning.

Slice sampling: This is a class of Markov chain Monte Carlo (MCMC) methods that introduces auxiliary variables to sample from complex or infinite-dimensional distributions [377]. For Bayesian nonparametric models like the Dirichlet Process, slice sampling helps truncate the infinite stick-breaking representation dynamically, making inference computationally tractable. Slice sampling generates samples from a Gibbs measure defined on an extended space, and its ergodic properties are critical for approximating posterior expectations in machine learning tasks such as clustering or density estimation.

Truncated representations: In practice, stick-breaking processes are often approximated using a finite truncation, replacing the infinite mixture with a finite sum [195]. These approximations maintain a variational characterization of the Gibbs posterior while reducing computational complexity. The truncation level implicitly defines an approximation to the underlying Gibbs measure, and these finite representations have been successfully used in topic modeling and deep generative models.

Variational Inference (VI): VI turns Bayesian inference into an optimization problem by approximating the intractable posterior with a simpler distribution (e.g., a mean-field family), minimizing the Kullback-Leibler divergence [48]. In the context of DP mixtures, this technique approximates the Gibbs posterior by projecting it onto a family of factorized distributions. The evidence lower bound (ELBO) that is optimized can be interpreted as a free energy functional, highlighting the link between variational inference and the variational principle in statistical mechanics.

These techniques are essential in modern machine learning, enabling scalable inference in models ranging from Bayesian neural networks to deep latent variable models. Moreover, the underlying probabilistic frameworks often rely on principles from Gibbs measures, such as equilibrium distributions, conditional independence, and energy-based formulations.

8.5.4 *Neural processes*

Neural processes (NPs) are a class of models that combine the expressiveness of neural networks with the probabilistic structure of Gaussian Processes (GPs), allowing for meta-learning over distributions of functions [144]. These models learn to approximate conditional distributions:

$$p_\theta(\mathbf{y}_*|\mathbf{x}_*, \mathcal{D}) = \int p_\theta(\mathbf{y}_*|\mathbf{x}_*, \mathbf{z})\, q_\phi(\mathbf{z}|\mathcal{D})\, d\mathbf{z},$$

where $\mathbf{z}$ is a latent global variable encoding function-level uncertainty, and $\mathcal{D}$ is the context set.

Neural processes perform amortized inference using variational methods. The ELBO (Evidence Lower Bound) objective used in training can be understood in terms of the variational principle underlying Gibbs measures, where the model minimizes the free energy functional:

$$\mathcal{F}(q) = \mathbb{E}_q[\log p(\mathcal{D}, \mathbf{z})] - \mathbb{E}_q[\log q(\mathbf{z})],$$

mirroring the structure of the Gibbs variational principle in statistical mechanics. This deepens the connection between machine learning and thermodynamics-inspired methods, and provides a bridge between stochastic inference and neural computation.

Furthermore, Gibbs-inspired training allows for flexible generalization under uncertainty, which is critical in few-shot learning and meta-learning tasks. By modeling the conditional distributions with neural networks and optimizing a variational objective, NPs provide a scalable, probabilistic alternative to Gaussian processes.

8.6 Graphical models and message passing

Graphical models provide a powerful framework for representing complex probabilistic systems through structured dependencies between variables. They are widely used in machine learning, statistical physics, and signal processing. Message-passing algorithms are central tools for inference in these models, enabling efficient approximation or computation of marginals and MAP configurations.

8.6.1 *Belief propagation algorithms*

Belief propagation (BP), or the sum-product algorithm, is an iterative message-passing procedure for computing marginal distributions in graphical models, especially factor graphs. For a factor graph with variables x_i

and factors $f_a(\mathbf{x}_a)$, the message updates are:

$$m_{i\to a}(x_i) \propto \prod_{b\in\mathcal{N}(i)\setminus a} m_{b\to i}(x_i),$$

$$m_{a\to j}(x_j) \propto \sum_{\mathbf{x}_a\setminus x_j} f_a(\mathbf{x}_a) \prod_{i\in\mathcal{N}(a)\setminus j} m_{i\to a}(x_i),$$

here $\mathcal{N}(i)$ and $\mathcal{N}(a)$ denote the neighborhoods of variable i and factor a, respectively.

On tree-structured graphs (i.e., acyclic), BP converges to the exact marginal distributions [229]. For graphs with loops ("loopy BP"), it often provides good approximations in practice and is closely related to minimizing the Bethe free energy, which connects to variational approximations of Gibbs measures. In machine learning, BP is widely used in applications such as error-correcting codes, image denoising, and structured prediction.

Challenges remain, including improving the convergence of BP on loopy graphs and scaling to high-dimensional datasets. Recent work suggests that reinforcement-based BP variants, such as focused BP (fBP), can mitigate issues with local minima by introducing iterative local fields [258]. Furthermore, the rise of graph neural networks (GNNs) has inspired hybrid BP-GNN models, where message-passing mimics BP updates but incorporates learned features, enhancing performance in tasks like graph-based anomaly detection. These trends highlight BP's evolving role in machine learning, bridging probabilistic inference with modern neural architectures and reinforcing its connection to Gibbs measures.

8.6.2 *Federated learning applications*

In federated learning, data is distributed across multiple clients $k = 1, \ldots, K$, and training is performed without sharing local data. A Gibbs-inspired formulation of global model aggregation uses the following posterior:

$$\pi_g(\theta) \propto \exp\left(-\beta \sum_{k=1}^{K} n_k \hat{\mathcal{R}}_k(\theta)\right) p_0(\theta),$$

where $\hat{\mathcal{R}}_k(\theta)$ is the empirical loss at client k, n_k is the number of data points, $p_0(\theta)$ is a prior, and β serves as an inverse temperature controlling regularization.

This framework corresponds to a Gibbs posterior over model parameters, drawing connections to PAC-Bayesian analysis and Bayesian neural networks. The aggregation step resembles variational inference under distributed constraints.

Moreover, injecting noise into updates for differential privacy can be interpreted as raising the temperature of the Gibbs distribution, thus broadening uncertainty and offering privacy-preserving learning guarantees [204]. This insight bridges statistical physics and privacy-aware machine learning.

8.6.3 *Quantum enhancements*

Quantum computing offers potential advantages for sampling and optimization in graphical models via quantum annealing and QUBO formulations: Quantum annealing: A quantum analogue of simulated annealing, where quantum tunneling helps escape local minima. For Ising models, this can improve sampling from low-temperature Gibbs distributions, useful in combinatorial optimization and unsupervised learning [200].

QUBO (Quadratic Unconstrained Binary Optimization): Many MAP inference problems in graphical models can be reformulated as QUBO problems. These are suitable for quantum solvers like D-Wave and emerging quantum-classical hybrid systems [305]. The connection to Gibbs distributions arises from the equivalence between MAP inference and energy minimization in statistical mechanics. These methods are of growing interest in machine learning for solving NP-hard problems, sparse coding, and generative modeling.

Despite extensive progress, several open challenges remain:

Loopy graphs: Belief propagation lacks convergence guarantees on graphs with cycles. Understanding conditions for convergence or designing robust alternatives remains an open area, especially in deep learning and graphical variational inference [258].

High-dimensional inference: In graphical models with large state spaces or complex latent structure, marginalization becomes intractable. Developing scalable Gibbs-like methods or efficient approximations is a key challenge for high-dimensional Bayesian learning.

Dynamic networks: Many real-world systems, such as social networks or sensor arrays, evolve over time. Message passing and inference on such time-varying graphs require adaptive methods that can track shifting dependencies without retraining from scratch.

These problems lie at the intersection of machine learning, statistical mechanics, and computational mathematics, inviting new theoretical and algorithmic innovations.

8.7 Transfer Learning with Gibbs Priors

Transfer learning seeks to leverage knowledge from related tasks or domains to improve learning efficiency and performance in a target task. A central idea is to encode prior knowledge through probabilistic structures—such as Gibbs priors—thereby providing inductive bias that is shared across tasks. Gibbs measures naturally arise in hierarchical Bayesian modeling, variational learning, and PAC-Bayesian theory, all of which underpin various transfer learning algorithms.

8.7.1 *Hierarchical Bayesian frameworks*

Hierarchical Bayesian models offer a principled approach to multi-task and meta-learning. A shared Gibbs prior across tasks assumes the following form:

$$p(\theta^{(1:D)}|\phi) = \prod_{d=1}^{D} \pi(\theta^{(d)}|\phi), \quad \phi \sim p_0(\phi),$$

where $\theta^{(d)}$ is the model parameter for task d, ϕ is a global latent variable (e.g., a shared hyperparameter), and $\pi(\theta^{(d)}|\phi)$ represents a Gibbs distribution:

$$\pi(\theta^{(d)}|\phi) \propto \exp(-\beta \mathcal{R}_d(\theta^{(d)}|\phi)),$$

with $\mathcal{R}_d$ a task-specific loss or energy function. This framework induces statistical coupling between tasks and enables data-efficient learning by pooling information via the shared prior.

In machine learning, this formulation underpins hierarchical Bayesian neural networks, nonparametric models (e.g., HDPs), and few-shot learning systems. Meta-learning emerges naturally by optimizing ϕ based on empirical performance across tasks [388].

8.7.2 *Model-agnostic meta-learning*

Model-agnostic meta-learning (MAML) is a widely used gradient-based meta-learning algorithm designed to train models that adapt rapidly to new tasks. The adaptation step resembles a Gibbs posterior gradient descent:

$$\theta'_d = \theta - \eta \nabla_\theta \hat{\mathcal{R}}_d(\theta),$$

where $\hat{\mathcal{R}}_d$ is the empirical risk on task d and θ'_d is the adapted parameter. This step can be interpreted as maximizing a local Gibbs posterior centered at θ, linking MAML to the thermodynamic perspective of learning as energy minimization.

Recent PAC-Bayesian interpretations of MAML extend this view by providing generalization guarantees based on Gibbs priors over parameter updates [322]. This connects MAML to the broader framework of stochastic optimization and probabilistic inference in function space.

In machine learning, MAML is widely used for robotics, few-shot classification, and reinforcement learning, especially where rapid adaptation is crucial.

8.7.3 *Challenges in heterogeneous transfer*

Despite its promise, transfer learning—especially in heterogeneous settings—faces several challenges:

Negative transfer: When the source and target distributions differ significantly, shared priors (including Gibbs priors) may bias learning away from optimal solutions. Addressing task mismatch requires robust adaptation and task dissimilarity metrics.

Calibration drift: As the uncertainty landscape varies across domains, Bayesian models may suffer from miscalibrated posteriors. This undermines decision-making in risk-sensitive applications such as medicine or autonomous driving.

Catastrophic forgetting: In sequential learning, updating the model on new tasks often leads to the loss of previously acquired knowledge. Probabilistic approaches based on Gibbs priors help mitigate this via posterior regularization or prior preservation (e.g., EWC, VCL).

Addressing these issues remains an open problem, requiring advances in continual Bayesian learning, meta-regularization, and distributional robustness [180, 302].

8.8 Conclusion

This chapter has surveyed the rapidly evolving landscape of Gibbs measures in modern machine learning, revealing both the maturation of established techniques and the emergence of novel theoretical frameworks that promise to reshape the field. The comprehensive examination across diverse domains—from Bayesian neural networks to large language models, from probabilistic programming to quantum-enhanced graphical models—demonstrates the fundamental role that Gibbs measures continue to play in addressing contemporary challenges in artificial intelligence.

The integration of Gibbs measures into deep learning architectures represents perhaps the most significant development in recent years. The recognition of diffusion models as Gibbs processes has provided new theoretical foundations for generative modeling, while advances in training methodologies for structured Gibbs measures have enabled more principled approaches to uncertainty quantification in neural networks. These developments bridge the gap between the statistical mechanics origins of Gibbs measures and the computational demands of modern deep learning systems.

Particularly noteworthy is the growing sophistication in our understanding of large language models through the lens of statistical mechanics. The interpretation of temperature parameters, sampling strategies, and information-theoretic quantities as thermodynamic observables has opened new avenues for both theoretical analysis and practical optimization of language generation systems. This perspective has proven invaluable in addressing fundamental questions about the emergent properties of autoregressive models and their scaling behavior.

The advances in probabilistic programming and hybrid inference engines have democratized access to sophisticated Gibbs sampling techniques, enabling practitioners to tackle increasingly complex hierarchical models with improved convergence diagnostics and computational efficiency. The emergence of Bayesian nonparametrics and infinite-dimensional Gibbs measures has further expanded the theoretical toolkit, particularly in applications requiring adaptive model complexity and infinite mixture representations.

The extension to federated learning applications and quantum enhancements in graphical models points toward future directions where Gibbs measures will play crucial roles in distributed and quantum computing paradigms. These developments suggest that the theoretical foundations established in statistical mechanics continue to provide robust frameworks for emerging computational architectures.

However, significant challenges remain. The computational complexity of exact inference in high-dimensional Gibbs measures continues to drive research into approximate methods and variational techniques. There are ongoing difficulties in leveraging Gibbs priors across heterogeneous domains, particularly in model-agnostic meta-learning frameworks where theoretical guarantees remain elusive.

Looking forward, the convergence of multiple research streams—statistical mechanics, information theory, quantum computation, and distributed systems—suggests that Gibbs measures will continue to serve as a unifying theoretical framework for understanding and developing machine learning algorithms. The mathematical elegance and computational tractability of Gibbs measures, combined with their natural capacity to encode uncertainty and structure, position them as essential tools for addressing the next generation of challenges in artificial intelligence.

For future reading, see [41, 79, 80, 173, 226–228, 354, 374].

Bibliography

[1] L. U. Abdullaev, T. M. Nguyen, *Transformer Meets Twicing: Harnessing Unattended Residual Information*, International Conference on Learning Representations, 2025.

[2] M. Ackerman, S. Ben-David, *Measures of Clustering Quality: A Working Set of Axioms for Clustering.* In NIPS (2008), 8 pages.

[3] D. H. Ackley, G. E. Hinton, T. J. Sejnowski, A learning algorithm for Boltzmann machines, *Cognitive Science*, **9**(1) (1985), 147–169.

[4] C. O. Acuna, Texture modeling using Gibbs distributions. *CVGIP: Graphical Models and Image Processing*, **54**(3) (1992), 210–222.

[5] C. Acuna, *Parameter Estimation for Stochastic Texture Models*, Ph.D. thesis, Department of Mathematics and Statistics, University of Massachusetts, 1988.

[6] G. Allenby, P. Rossi, *Bayesian Applications in Marketing*, The Oxford Handbook of Bayesian Econometrics, 2009.

[7] E. Alpaydin, *Introduction to Machine Learning.* The MIT Press, 2014.

[8] S. Alpert, *Spectral partitioning: The More Eigenvectors, the Better.* In 32nd Design Automation Conference (1995), pp. 195–200.

[9] P. Alquier, User-friendly introduction to PAC-Bayes bounds, 17(2), 2024, 174–303. and arXiv:2110.11216v6

[10] M. A. Álvarez, L. Rosasco, N. D. Lawrence, Kernels for vector-valued functions: A review, *Foundations and Trends in Machine Learning* **4**(3) (2012), 195–266.

[11] D.J. Amit, H. Gutfreund, H. Sompolinsky, Statistical mechanics of neural networks near saturation. *Annals of Physics*, **173**(1) (1987), 30–67.

[12] C. Andrieu, A. Lee, M. Vihola, Uniform ergodicity of the iterated conditional SMC and geometric ergodicity of particle Gibbs samplers. *Bernoulli*, **24**(2) (2018), 842–872.

[13] C. Andrieu, A. Doucet, R. Holenstein, Particle Markov chain Monte Carlo methods, *Journal of the Royal Statistical Society Series B*, 72(3) (2010), 269–342.

[14] K. Appel, W. Haken, Every planar map is four colorable. *Part I: Discharging. Illinois Journal of Mathematics*, **21**(3) (1977), 429–490.

[15] K. Appel, W. Haken, Every planar map is four colorable. *Part II: Reducibility. Illinois Journal of Mathematics*, **21**(3) (1977), 491–567.

[16] É.O. Aquino, F.W. Lima, A.D. Araújo, F.R.N. Costa, Potts model in one-dimension on directed small-world networks. *Journal of Statistical Physics,* **171**(6) (2018), 1112–1121.

[17] S. Ariosto, *Statistical Physics of Deep Neural Networks: Generalization Capability, Beyond the Infinite Width, and Feature Learning,* arXiv:2501.19281v1, 2025.

[18] M. Arjovsky, S. Chintala, L. Bottou, *Wasserstein Generative Adversarial Networks*, Proceedings of the 34-th International Conference on Machine Learning, Sydney, Australia, PMLR 70, 2017.

[19] M. S. Arulampalam, S. Maskell, N. Gordon, T. Clapp, *A Tutorial on Particle Filters for Online Nonlinear/Non-Gaussian Bayesian Tracking,* in IEEE Transactions on Signal Processing, **50**(2) (2002), 174–188.

[20] D. Bahdanau, K. Cho, Y. Bengio. *Neural Machine Translation by Jointly Learning to Align and Translate*, arXiv:1409.0473, 2014.

[21] Y. Bahri1, J. Kadmon, J. Pennington, S. S. Schoenholz, J. Sohl-Dickstein, S. Ganguli, Statistical Mechanics of Deep Learning, *Annual Review of Condensed Matter Physics*, **11** (2020), 501–528.

[22] J. Barbier, D. Panchenko, Strong replica symmetry in high-dimensional optimal Bayesian inference. *Communications in Mathematical Physics,* **393** (2022), 1199–1239.

[23] R. Bardenet, A. Doucet, C. Holmes. On Markov chain Monte Carlo methods for tall data. *Journal of Machine Learning Research,* **18**(47) (2017), 1–43.

[24] R. Barney, M. Winer, V. Galitski, *Neural Networks as Spin Models: From Glass to Hidden Order Through Training.* arXiv:2408.06421.

[25] R.J. Baxter, *Exactly Solved Models in Statistical Mechanics.* Academic Press. London. 1982.

[26] F. Behrens, N. Mainali, C. Marullo, S. Lee, B. Sorscher, H. Sompolinsky, Statistical mechanics of deep learning, *Journal of Statistical Mechanics: Theory and Experiment* (2024) 104007.

[27] S. Ben-David, U. von Luxburg, *Relating Clustering Stability to Properties of Cluster Boundaries.* In COLT (2008), 12 pages.

[28] M. Benedetti, J. Realpe-Gómez, R. Biswas, A. Perdomo-Ortiz, Quantum-assisted learning of hardware-embedded probabilistic graphical models, *Physical Review X* **7**, 041052 (2017), 17 pages.

[29] Y. Bengio, A. C. Courville, P. Vincent, Representation learning: A review and new perspectives. *IEEE Transactions on Pattern Analysis and Machine Intelligence*, **35** (2012), 1798–1828.

[30] Yoshua Bengio, Réjean Ducharme, Pascal Vincent, Christian Jauvin, A Neural probabilistic language model, *Journal of Machine Learning Research*, 2003.

[31] A. Benveniste, M. Métivier, P. Priouret, *Adaptive Algorithms and Stochastic Approximations*, Springer-Verlag, 1990.

[32] M. Belkin, D. Hsu, S. Ma, S. Mandal, Reconciling modern machine-learning practice and the classical bias–variance trade-off. *PNAS* **116**, 32 (2019), 15849–15854.

[33] I. Beltagy, M. E. Peters, A. Cohan, *Longformer: The Long-Document Transformer*, arXiv:2004.05150, 2020.

[34] M. Beraha, R. Argiento, J. Møller, A. Guglielmi, MCMC Computations for Bayesian Mixture Models Using Repulsive Point Processes. *Journal of Computational and Graphical Statistics*, **31** (2020), 422–435.

[35] M. Beraha, R. Corradin, Bayesian Nonparametric Model-based Clustering with Intractable Distributions: An ABC Approach, *Bayesian Analysis Advance Publication* 1–28, 2024.

[36] J. Besag, On the statistical analysis of dirty pictures. *Journal of the Royal Statistical Society: Series B* **48**(3) (1986), 259–302.

[37] M. Betancourt, *A Conceptual Introduction to Hamiltonian Monte Carlo*, arXiv:1701.02434v2.

[38] A. Bhattacharya, D. Pati, N.S. Pillai, D.B. Dunson, Dirichlet-Laplace priors for optimal shrinkage. *Journal of the American Statistical Association* 2015 **110**(512) (2015), 1479–1490.

[39] I. Bhattacharya, R. Martin, Gibbs posterior inference on multivariate quantiles. *Journal of Statistical Planning and Inference* **218** (2022), 106–121.

[40] P. Billingsley, *Convergence of Probability Measures*, Wiley-Interscience, 1999.

[41] C.M. Bishop, *Pattern Recognition and Machine Learning.* Springer. 2006.

[42] C.M. Bishop, H. Bishop, *Deep Learning: Foundations and Concepts.* Springer. 2024.

[43] P.G. Bissiri, C.C. Holmes, S.G. Walker, A general framework for updating belief distributions, *Journal of the Royal Statistical Society. Series B (Statistical Methodology)* **78**(5) (2016), 1103–1130.

[44] A. Blake, P. Kohli, C. Rother (Eds.) *Markov Random Fields for Vision and Image Processing.* The MIT Press., 2011.

[45] P.M. Bleher, J.G. Sinai, Investigation of the critical point in models of the type of Dyson's hierarchical models. *Communications in Mathematical Physics* **33** (1973), 23–42.

[46] D.M. Blei, A. Kucukelbir, J.D. McAuliffe, Variational Inference: A Review for Statisticians. *Journal of the American Statistical Association,* **112**(518) (2017), 859–877.

[47] D.M. Blei, A. Y. Ng, M. I. Jordan, Latent Dirichlet Allocation. *Journal of Machine Learning Research,* **3** (2003), 993–1022.

[48] D. M. Blei, M. I. Jordan, Variational inference for Dirichlet process mixtures. *Bayesian Anal.* **1**(1) (2006), 2006121–143.

[49] D.M. Blei, M. David P.I. Frazier, Distance dependent Chinese restaurant processes, *Journal of Machine Learning Research.* **12** (2011), 2461–2488.

[50] A. Blum, C. Dan, S. Seddighin, *Learning Complexity of Simulated Annealing*, Proceedings of The 24th International Conference on Artificial Intelligence and Statistics, 2021.

[51] R. Bommasani, D. Hudson, *et al. On the Opportunities and Risks of Foundation Models*, arXiv:2108.07258v3. 2022.

[52] L. Bottou, *Large-scale Machine Learning with Stochastic Gradient Descent.* In: Lechevallier, Y., Saporta, G. (eds) Proceedings of COMPSTAT'2010.

[53] L. Bottou, F. E. Curtis, J. Nocedal, Optimization methods for large-scale machine learning. *SIAM Review.* **60**(2) (2018), 223–311.

[54] A. Bovier, *Statistical Mechanics of Disordered Systems: A Mathematical Perspective.* Cambridge University Press, 2006.

[55] S. Boyd, L. Vandenberghe, *Convex Optimization.* Cambridge University Press. 2004.

[56] Y. Boykov, O. Veksler, R. Zabih, Fast approximate energy minimization via graph cuts. *IEEE Transactions on Pattern Analysis and Machine Intelligence* **23**(11) (2001), 1222–1239.

[57] M. Bramson, R. Durrett, R.H. Schonmann, The contact process on a finite set. *The Annals of Probability,* **19**(3) (1991), 960–983.

[58] P. Brémaud, *Markov Chains: Gibbs Fields, Monte Carlo Simulation, and Queues.* Springer New York, NY. 1999.

[59] T.B. Brown, B. Mann, *et al. Language Models are Few-Shot Learners,* arXiv:2005.14165v4, 2020.

[60] M. J. Brusco, J. D. Cradit, A variable-selection heuristic for k-means clustering, *Psychometrika,* **66** (2001), 249–270.

[61] A. Buades, B. Coll, and J-M Morel. *A Non-local Algorithm for Image Denoising.* In Computer Vision and Pattern Recognition, 2005. CVPR 2005. IEEE Computer Society Conference on, volume 2, pages 60–65. IEEE, 2005.

[62] J. Buolamwini, T. Gebru, *Gender Shades: Intersectional Accuracy Disparities in Commercial Gender Classification.* Proceedings of the 1st Conference on Fairness, Accountability and Transparency, PMLR **81** (2018), 77–91.

[63] B. Cakmak, M. Opper, A Dynamical Mean-Field Theory for Learning in Restricted Boltzmann Machines, *Journal of Statistical Mechanics* (2020) 103303.

[64] L. Cao, H. Chen, X. Fan, J. Gama, Y.-S. Ong, V. Kumar, *Bayesian Federated Learning: A Survey.* Proceedings of the Thirty-Second International Joint Conference on Artificial Intelligence Survey Track. 2023, 7233–7242.

[65] O. Cappé, E. Moulines, T. Rydén, *Inference in Hidden Markov Models.* Springer. 2005.

[66] R. Caruana, Multitask Learning. *Machine Learning,* **28** (1997), 41–75.

[67] G. Casella, R.L. Berger, Statistical inference. *Cengage Learning.* 2001.

[68] G. Casella, C. P. Robert, Rao-Blackwellisation of sampling schemes, *Biometrika,* **83**(1) (1996), 81–94.

[69] M. Cassandro, G. Jona-Lasinio, Critical multiscale phenomena in hierarchical systems. *Advances in Physics,* **28**(4) (1979), 465–522.

[70] M. Cassandro, T. Merola, P. Picco, U.A. Rozikov, One-dimensional Ising models with long range interactions: Cluster expansion, Phase-separating point. *Communications in Mathematical Physics* **327**(3) (2014), 951–991.

[71] M. Cassandro, P.A. Ferrari, I. Merola, E. Presutti, Geometry of contours and Peierls estimates in $d = 1$ Ising models with long range interactions. *Journal of Mathematical Physics* **46**(5) (2005), 053305, 22 pp.

[72] C. Castellano, S. Fortunato, V. Loreto, Statistical physics of social dynamics. *Reviews of Modern Physics,* **81**(2) (2009), 591–646.

[73] O. Catoni, *PAC-Bayesian Supervised Classification The Thermodynamics of Statistical Learning*, IMS Lecture Notes Monograph Series, **56**(i–xii) (2007), 1–163.

[74] G. Celeux, F. Forbes, N. Peyrard. *EM-based Image Segmentation Using Potts Models with External Field.* [Research Report] RR-4456, INRIA. 2002. inria-00072132.

[75] M. Cerezo, A. Arrasmith, R. Babbush, *et al.* Variational quantum algorithms. *Nature Reviews Physics* **3** (2021), 625–644.

[76] A. Chakraborty, A. Bhattacharya, D. Pati, A Gibbs Posterior Framework for Fair Clustering. *Entropy*, **26** (2024), 63, 18 pages.

[77] V. Chandola, A. Banerjee, V. Kumar, Anomaly detection: A survey. *ACM Computing Surveys (CSUR)*, **41**(3) (2009), 1–58.

[78] P. Chaudhari, A. Choromanska, S. Soatto, Y. LeCun, C. Baldassi, C. Borgs, J. Chayes, L. Sagun, R. Zecchina, *Entropy-SGD: Biasing Gradient Descent Into Wide Valleys*, International Conference on Learning Representations, 2017.

[79] S. Chen, Z. Lin, Y. Polyanskiy, P. Rigollet, *Quantitative Clustering in Mean-Field Transformer Models*, arXiv preprint arXiv:2504.14697, 2025

[80] S. Chen, Z. Lin, Y. Polyanskiy, P. Rigollet, *Residual Connections Provably Mitigate Oversmoothing in Graph Neural Networks*, arXiv preprint arXiv:2501.00762, 2025

[81] V. Chernozhukov, H. Hong, *An MCMC Approach to Classical Estimation.* SSRN 2002.

[82] F. Chierichetti, R. Kumar, S. Lattanzi, S. Vassilvitskii, *Fair Clustering Through Fairlets.* In Proceedings of the Advances in Neural Information Processing Systems; Guyon, I., Luxburg, U.V., Bengio, S., Wallach, H., Fergus, R., Vishwanathan, S., Garnett, R., Eds.; Curran Associates, Inc.: Red Hook, NY, USA, 2017; Volume 30.

[83] A. Choromanska, M. Henaff, M. Mathieu, G. Ben Arous, Y. LeCun, *The Loss Surfaces of Multilayer Networks*, arXiv:1412.0233v3, 2015.

[84] P. Clifford, A. Sudbury, A model for spatial conflict. *Biometrika*, **60**(3) (1973), 581–588.

[85] S. Cocco, G. Croce, F. Zamponi, Adaptive cluster expansion for Ising spin models. *The European Physical Journal B* **92**, 259 (2019), 17 pages.

[86] A. Coja-Oghlan, A. Galanis, L.A. Goldberg, J. B. Ravelomanana, D. Štefanković, E. Vigoda, Metastability of the Potts Ferromagnet on Random Regular Graphs. *Communications in Mathematical Physics* **401** (2023), 185–225.

[87] A. Coja-Oghlan, W. Perkins, Spin Systems on Bethe Lattices. *Communications in Mathematical Physics* **372** (2019), 441–523.

[88] M. Collins, Discriminative training methods for hidden Markov models: Theory and experiments with perceptron algorithms. *Proc. EMNLP.* **10**. 2002.

[89] J.-B. Cordonnier, A. Loukas, M. Jaggi, *Multi-Head Attention: Collaborate Instead of Concatenate*, arXiv:2006.16362v2.

[90] G. Cybenko, Approximation by superpositions of a sigmoidal function. *Mathematics of Control, Signals, and Systems*, **2** (1989), 303–314.

[91] Y. Dauphin, R. Pascanu, C. Gülcehre, K. Cho, S. Ganguli, Y. Bengio, *Identifying and Attacking the Saddle Point Problem in High-Dimensional Non-convex Optimization*, arXiv:1406.2572v1. 2014.

[92] A. Decelle, An introduction to machine learning: A perspective from statistical physics, *Physica A: Statistical Mechanics and its Applications*, **631**(1) (2023), 128154, 31 pages.

[93] A. Decelle, F. Krzakala, C. Moore, L. Zdeborová, Inference and phase transitions in the detection of modules in sparse networks. *Physical Review Letters* **107**(6) (2011), 065701.

[94] M. P. Deisenroth, A. A. Faisal, C. S. Ong. *Mathematics for Machine Learning*. Cambridge University Press. 2020.

[95] A. Dempster, N. Laird, D. Rubin, Maximum likelihood for incomplete data via the EM algorithm. *Journal of the Royal Statistical Society: Series B (Statistical Methodology)* **39**(1) (1977), 1–38.

[96] M. Demircigil, J. Heusel, M. Löwe, S. Uǵur Üngör, F. Vermet, On a model of associative memory with huge storage capacity, *Journal of Statistical Physics*, **168**(2) (2017), 288–299.

[97] H. Derin and H. Elliott, Modeling and segmentation of noisy and textured images using Gibbs random fields, *IEEE Transactions on Pattern Analysis and Machine Intelligence*, **PAMI-9**(1), (1987), 39–55.

[98] X. Descombes, J.F. Mangin, E. Pechersky, M. Sigelle, *Fine Structures Preserving Model for Image Processing*. Proc. 9th SCIA 95, Uppsala, Sweden, 1995, 349–356.

[99] X. Descombes, R. Minlos, E. Zhizhina, *Object Extraction Using a Stochastic Birth-and-Death Dynamics in Continuum*, Research Report 6135, INRIA, 2007.

[100] X. Descombes, M. Sigelle, F. Preteux, GMRF Parameter Estimation in a non-stationary Framework by a Renormalization Technique: Application to Remote Sensing Imaging. *IEEE Transactions on Image Processing*, **8**(4) (1999), 490–503.

[101] X. Descombes, E. Zhizhina, Applications of Gibbs fields methods to image processing problems. *Problems of Information Transmission*, **40**(3) (2004), 108–125.

[102] X. Descombes, E. Zhizhina, The Gibbs fields approach and related dynamics in image processing, *Condensed Matter Physics*, **11**(2(54)) (2008), 293–312.

[103] R.L. Dobrushin, S. B. Shlosman, Gibbsian description of Gibbs random fields. *Russian Mathematical Surveys*, **40**(2) (1985), 149–212.

[104] E. Domany, R. Meir, *Layered Neural Networks*. Springer Berlin Heidelberg, 1991, pp. 307–334.

[105] D. L. Donoho, *High-Dimensional Data Analysis: The Curses and Blessings of Dimensionality*. Conference "Mathematical Challenges of the 21st Century", American Math. Society, Los Angeles. 2000.

[106] A. Doucet, E. Moulines, A. Thin, Differentiable samplers for deep latent variable models, *Philosophical Transactions of the Royal Society A: Mathematical, Physical and Engineering Sciences*, 38120220147 (2023), 13 pages.

[107] Y. Du, I. Mordatch, *Implicit Generation and Modeling with Energy-Based Models.* Proceedings of the 33rd International Conference on Neural Information Processing Systems. Curran Associates Inc., Red Hook, NY, USA, Article 324 (2019), 3608–3618.

[108] S. Dudoit, J. Fridlyand, A rediction-based resampling method for estimating the number of clusters in a dataset. *Genome Biology,* **2**(7) (2002), 00361–21.

[109] B. Dupuis, P. Viallard, G. Deligiannidis, U. Simsekli, Uniform generalization bounds on data-dependent hypothesis sets via PAC-Bayesian theory on random sets, *Journal of Machine Learning Research* **25** (2024), 1–55.

[110] R. Durrett, *Lecture Notes on Particle Systems and Percolation,* Wadsworth and Brooks/Cole Advanced Books, Software, 1988.

[111] R. Durrett, D. Griffeath, Supercritical contact processes on $\mathbb{Z}^d$. *The Annals of Probability,* **11**(1) (1983), 1–26.

[112] M. Dyer, A. Frieze, R. Kannan, A random polynomial-time algorithm for approximating the volume of convex bodies. *Journal of the ACM (JACM),* **38**(1) (1991), 1–17.

[113] F. J. Dyson, Existence of a phase-transition in a one-dimensional Ising ferromagnet. *Physical Review Letters* **12**(2) (1969), 91–107.

[114] G. K. Dziugaite and D. M. Roy, *Computing Nonvacuous Generalization Bounds for Deep (Stochastic) Neural Networks with Many More Parameters than Training Data,* Proceedings of the 33-Conference on Uncertainty in Artificial Intelligence, UAI 2016, August 11–15, 2017, Sydney, NSW, Australia.

[115] D. J. Earl, M. W. Deem, *Parallel Tempering: Theory, Applications, and New Perspectives,* arXiv:physics/0508111v2.

[116] G. Egri, X. Han, *Attention is Kernel Trick Reloaded,* Conference on Neural Information Processing Systems, 2021.

[117] M. Ekeberg, C. Lø"vkvist, Y. Lan, M. Weigt, E. Aurell, Improved contact prediction in proteins: Using pseudolikelihoods to infer Potts models, *Physical Review E* **87** (2013), 012707.

[118] M. Elad, M. Aharon, Image denoising via sparse and redundant representations over learned dictionaries. *IEEE Transactions on Image Processing,* **15**(12) (2006), 3736–3745.

[119] R.S. Ellis, *Entropy, Large Deviations, and Statistical Mechanics.* Springer. 2006.

[120] K. Ellis, L. Wong, M. Nye, M. Sablé-Meyer, L. Cary, L. Anaya Pozo, L. Hewitt, A. Solar-Lezama, J.B. Tenenbaum, DreamCoder: Growing generalizable, interpretable knowledge with wake–sleep Bayesian program learning. *Philosophical Transactions of the Royal Society A: Mathematical, Physical and Engineering Sciences* **381** (2023), 20220050.

[121] A. Engel, C. Van den Broeck, *Statistical Mechanics of Learning,* Cambridge University Press, 2001.

[122] A.C.D. van Enter, V.N. Ermolaev, G. Iacobelli, C. Külske, Gibbs-non-Gibbs properties for evolving Ising models on trees. *Annales de l'I.H.P. Probabilités et statistiques,* **48**(3) (2012), 774–791.

[123] S. Esmaeili, B. Brubach, L. Tsepenekas, J. Dickerson, *Probabilistic Fair Clustering. In Proceedings of the Advances in Neural Information Processing Systems*; Larochelle, H., Ranzato, M., Hadsell, R., Balcan, M., Lin, H., Eds.; Curran Associates, Inc.: Red Hook, NY, USA, **33** (2020), 12743–12755.

[124] A. Esteva, A. Robicquet, B. Ramsundar, V. Kuleshov, M. DePristo, K. Chou, C. Cui, G. Corrado, S. Thrun, J. Dean, A guide to deep learning in healthcare. *Nature Medicine* **25** (2019), 24–29.

[125] A. Fan, M. Lewis, Y. Dauphin, *Hierarchical Neural Story Generation*, Proceedings of the 56th Annual Meeting of the Association for Computational Linguistics **1** (2018), 889–898.

[126] G. Faria, S. Agrawal, A. Farinhas, R. Rei, J. G. C. de Souza, A. Martins, *QUEST: Quality-Aware Metropolis-Hastings Sampling for Machine Translation*, Conference on Neural Information Processing Systems, 2024.

[127] C. Finn, P. Abbeel, S. Levine, *Model-Agnostic Meta-Learning for Fast Adaptation of Deep Networks.* Proceedings of the 34th International Conference on Machine Learning, PMLR **70** (2017), 1126–1135.

[128] D. Firth. Bias reduction of maximum likelihood estimates. *Biometrika*, **80**(1) (1993), 27–38.

[129] K. Fisher, Y. Marzouk, *Can Bayesian Neural Networks Make Confident Predictions?* arXiv preprint arXiv:2501.11773, 2025.

[130] A. Fischer, C. Igel, *An Introduction to Restricted Boltzmann Machines.* Progress in Pattern Recognition, Image Analysis, Computer Vision, and Applications, 2012, 14–36.

[131] S. Fortunato, Community detection in graphs, *Physics Reports*, **486**(3–5) (2010), 75–174.

[132] S. Fort, H. Hu, B. Lakshminarayanan, *Deep Ensembles: A Loss Landscape Perspective*, arXiv preprint arXiv:1912.02757, 2019.

[133] V. Francois-Lavet, P. Henderson, R. Islam, M. G. Bellemare, J. Pineau, An introduction to deep reinforcement learning, *Foundations and Trends in Machine Learning*, **11**(3-4) (2018), 219–354.

[134] C. Fraley, A. Raftery, How many clusters? Which clustering method? Answer via model-based clustering. *The Computer Journal*, **41**(8) (1998), 578–588.

[135] B. Frey, D. Dueck, Clustering by passing messages between data points. *Science* **315** (2007), 972–976.

[136] S. Friedli, Y. Velenik, *Statistical Mechanics of Lattice Systems. A Concrete Mathematical Introduction.* Cambridge University Press, Cambridge, 2018.

[137] J. H. Friedman, J. J. Meulman, Clustering objects on subsets of attributes (with discussion), *Journal of the Royal Statistical Society Series B: Statistical Methodology*, **66**(4) (2004): 815–849.

[138] F. Friedrich, A. Kempe, V. Liebscher, G. Winkler, Complexity penalized M-estimation: Fast computation. *Journal of Computational and Graphical Statistics* **17**(1) (2008), 201–224.

[139] C. Furtlehner, *Statistical Physics Methods for Machine Learning and Traffic Forecasting*, Statistical Mechanics. University Paris Saclay, 2020.

[140] H. Fu, C. Li, X. Liu, J. Gao, A. Celikyilmaz, L. Carin, 2019. *Cyclical Annealing Schedule: A Simple Approach to Mitigating KL Vanishing.* In Proceedings of the 2019 Conference of the North American Chapter of the Association for Computational Linguistics: Human Language Technologies, Association for Computational Linguistics, 2019.

[141] Y. Gal, Z. Ghahramani, *Dropout as a Bayesian Approximation: Representing Model Uncertainty in Deep Learning.* Proceedings of the 33rd International Conference on Machine Learning. 2016.

[142] G. Gallavotti, *Statistical Mechanics: A Short Treatise*, Springer, 1999.

[143] E. Gardner, B. Derrida, Three unfinished works on the optimal storage capacity of networks. *Journal of Physics A: Mathematical and General* **22**(12) (1989), 1983–1994.

[144] M. Garnelo, D. Rosenbaum, C. Maddison, T. Ramalho, D. Saxton, M. Shanahan, Y. W. Teh, D. Rezende, S. M. A. Eslami, *Conditional Neural Processes*, Proceedings of the 35th International Conference on Machine Learning, PMLR, **80** (2018), 1704–1713.

[145] M. Geiger, *et al.* Scaling description of generalization with number of parameters in deep learning. *Journal of Statistical Mechanics: Theory and Experiment* 2020, 2 (2020), 023401.

[146] A. Gelman, J.B. Carlin, H. S. Stern, D. B. Dunson, A. Vehtari, D.B. Rubin, *Bayesian Data Analysis* (3rd ed.). CRC Press. 2013.

[147] A. Gelman, D. B. Rubin. Inference from iterative simulation using multiple sequences. *Statistical Science* **7**(4) (1992), 457–472.

[148] S. Geman, D. Geman, Stochastic relaxation, Gibbs distributions, and the Bayesian restoration of images. *IEEE Transactions on Pattern Analysis and Machine Intelligence*, **6**(6) (1984), 721–741.

[149] H.O. Georgii, *Gibbs Measures and Phase Transitions*, Second edition. de Gruyter Studies in Mathematics, 9. Walter de Gruyter, Berlin, 2011.

[150] B. Geshkovski, C. Letrouit, Y. Polyanskiy, P. Rigollet, A mathematical perspective on transformers, *Bulletin of the American Mathematical Society* **62**(3) (2025), 427–479.

[151] Z. Ghahramani, *Unsupervised Learning.* In: Bousquet, O., von Luxburg, U., Rätsch, G. (eds) Advanced Lectures on Machine Learning. ML 2003. Lecture Notes in Computer Science, vol 3176. 2004, Springer, Berlin, Heidelberg.

[152] Z. Ghahramani, Probabilistic machine learning and artificial intelligence. *Nature*, **521** (2015), 452–459.

[153] L. Ghilotti, M. Beraha, A. Guglielmi, Bayesian clustering of high-dimensional data via latent repulsive mixtures, *Biometrika*, **112**(2) (2025), asae059 (17 pages).

[154] B. Ghojogh, M. Crowley, F. Karray, A. Ghodsi, *Elements of Dimensionality Reduction and Manifold Learning*, Springer Nature. Switzerland 2023.

[155] S. Ghosal, A. W. van der Vaart, *Fundamentals of Nonparametric Bayesian Inference*, Cambridge University Press. 2017.

[156] W. R. Gilks, S. Richardson, D. J. Spiegelhalter, *Markov Chain Monte Carlo in Practice.* Chapman and Hall/CRC. (1995).

[157] A. Giorlandino, S. Goldt, *Two Failure Modes of Deep Transformers and how to Avoid Them: A Unified Theory of Signal Propagation at Initialisation*, arXiv:2505.24333v1. 2025.

[158] G.H. Givens, J. A. Hoeting, *Computational Statistics* (2nd ed.). John Wiley and Sons. 2012.

[159] Y. Goldberg, O. Levy, *Word2vec Explained: Deriving Mikolov et al.'s Negative-Sampling Word-Embedding Method*, arXiv:1402.3722 (2014).

[160] R. Gonzalez, R. Woods, Digital image processing. *Pearson*, 2017.

[161] I. Goodfellow, J. Pouget-Abadie, M. Mirza, B. Xu, D. Warde-Farley, S. Ozair, A. Courville, Y. Bengio, Generative adversarial nets. *Advances in Neural Information Processing Systems.* 2014.

[162] I. Goodfellow, Y. Bengio, A. Courville, *Deep Learning.* MIT Press. (2016).

[163] I. Goodfellow, J. Shlens, C. Szegedy, *Explaining and Harnessing Adversarial Examples.* International Conference on Learning Representations. 2015.

[164] W. Grathwohl, K.-C. Wang, J.-H. Jacobsen, D. Duvenaud, M. Norouzi, K. Swersky, *Your Classifier is Secretly an Energy Based Model and You Should Treat it Like One*, ICLR 2020.

[165] T.L. Griffiths, M. Steyvers, Finding scientific topics. *Proceedings of the National Academy of Sciences*, **101**(Suppl 1) (2004), 5228–5235.

[166] G. Grimmett, D. Stirzaker, *Probability and Random Processes* (3rd ed.). Oxford University Press. 2001.

[167] C. Guo, G. Pleiss, Y. Sun, K. Weinberger, *On Calibration of Modern Neural Networks*, International Conference on Machine Learning, 2017.

[168] P. Grünwald, T. van Ommen, Inconsistency of Bayesian inference for misspecified linear models, and a proposal for repairing it. *Bayesian Analysis*, **12**(4) (2017), 1069–1103.

[169] B. Guedj, J. Shawe-Taylor, *A Primer on PAC-Bayesian Learning.* ICML 2019 - Thirty-sixth International Conference on Machine Learning, Jun 2019, Long Beach, United States. 2019. hal-02537094.

[170] L. Hamilton, F. Koehler, A. Moitra, *Information Theoretic Properties of Markov Random Fields, and their Algorithmic Applications*, arXiv:1705.11107 and Advances in Neural Information Processing Systems 30 (NIPS 2017).

[171] D. Hassabis, D. Kumaran, C. Summerfield, M. Botvinick, Neuroscience-inspired artificial intelligence. *Neuron.* **19**, 95(2) (2017), 245–258.

[172] F.H. Haydarov, B.A. Omirov, U.A. Rozikov, *Coupled Ising-Potts Model: Rich Sets of Critical Temperatures and Translation-Invariant Gibbs Measures*, arXiv:2502.12014. 2025.

[173] S. Hayou, On the infinite-depth limit of finite-width neural networks, *Transactions on Machine Learning Research*, 2023.

[174] B. Hoover, D. H. Chau, H. Strobelt, P. Ram, D. Krotov, *Dense Associative Memory Through the Lens of Random Features*, The Thirty-eighth Annual Conference on Neural Information Processing Systems, 2024.

[175] A. Holtzman, J. Buys, L. Du, M. Forbes, Y. Choi, *The Curious Case of Neural Text Degeneration*, International Conference on Learning Representations, 2020.

[176] D.O. Hebb, *The Organization of Behavior*, Psychology Press., 2002.
[177] G. Heinrich, *Parameter Estimation for Text Analysis.* Technical Report, Fraunhofer Institute for Autonomous Intelligent Systems. 2005.
[178] F. Henning, C. Kuelske, A. Le Ny, U.A. Rozikov, Gradient Gibbs measures for the SOS model with countable values on a Cayley tree. *Electronic Journal of Probability.* **24**(104) (2019), 1–23.
[179] F. Herrera, U. A. Rozikov, M. V. Velasco, Ising models with hidden Markov structure: Applications to probabilistic inference in machine learning. *Journal of Statistical Mechanics: Theory and Experiment* 2025. 073201, 21 pages.
[180] J. Hickey, J. Williams, E. C. Hector, Transfer learning with uncertainty quantification: Random effect calibration of source to target (RECaST), *Journal of Machine Learning Research* **25** (2024), 1–40.
[181] G. E. Hinton, T. J. Sejnowski, *Learning and Relearning in Boltzmann Machines.* In D. E. Rumelhart, J. L. McClelland, and PDP Research Group (Eds.), Parallel Distributed Processing: Explorations in the Microstructure of Cognition: Foundations (Vol. 1). MIT Press. (1986).
[182] G. E. Hinton, P. Dayan, B. J. Frey, R. M. Neal. The wake-sleep algorithm for unsupervised neural networks. *Science*, **268**(5214) (1995), 1158–1161.
[183] G.E. Hinton, A practical guide to training restricted Boltzmann machines. *Momentum*, **9**(1) (2010), 599–619.
[184] G.E. Hinton, S. Osindero, Y.W. Teh, A fast learning algorithm for deep belief nets. *Neural Computation*, **18**(7) (2006), 1527–1554.
[185] G.E. Hinton, R. R. Salakhutdinov, Reducing the dimensionality of data with neural networks. *Science*, **313**(5786) (2006), 504–507.
[186] G.E. Hinton, Training products of experts by minimizing contrastive divergence. *Neural Computation* **14**(8) 2002, 1771–1800.
[187] M. Hirt, V. Kreouzis, P. Dellaportas, Learning variational autoencoders via MCMC speed measures. *Statistical Computing* **34**, 164, 2024.
[188] J. Ho, A. Jain, P. Abbeel, *Denoising Diffusion Probabilistic models.* Advances in Neural Information Processing Systems (NeurIPS). 2020.
[189] S. Hochreiter, J. Schmidhuber, Flat minima. *Neural Computation*, **9**(1) (1997), 1–42.
[190] M. Hoffman, D. Blei, *Stochastic Structured Variational Inference.* Proceedings of the Eighteenth International Conference on Artificial Intelligence and Statistics, PMLR **38** (2015), 361–369.
[191] C.C. Holmes, S.G. Walker, Assigning a value to a power likelihood in a general Bayesian model. *Biometrika*, **104** (2017), 497–503.
[192] L. Hong, R. Martin, Model misspecification, Bayesian versus credibility estimation, and Gibbs posteriors. *Scandinavian Actuarial Journal* **2020** (2020), 634–649.
[193] J.J. Hopfield, Neural network and physical systems with emergent collective computational abilities. *PNAS* **79** (1982), 2554–2558.
[194] A. Hyvärinen, Estimation of non-normalized statistical models by score matching, *Journal of Machine Learning Research*, **6** (2005), 695–709.

[195] H. Ishwaran, J.F. Lancelot, Gibbs sampling methods for stick-breaking priors, *Journal of the American Statistical Association* **96**(453) (2001), 161–173.

[196] T.S. Jaakkola, M.I. Jordan, Variational Probabilistic Inference and the QMR-DTNetwork, *Journal of Artificial Intelligence Research,* **10** (1999), 291–322.

[197] A. Jacot, F. Gabriel, C. Hongler, *Neural Tangent Kernel: Convergence and Generalization in Neural Networks.* In NIPS **31** (2018), 8571–8580.

[198] A. Jafari, M. Rezagholizadeh, P. Sharma, A. Ghodsi, *Annealing Knowledge Distillation.* In Proceedings of the 16th Conference of the European Chapter of the Association for Computational Linguistics, Association for Computational Linguistics, 2021.

[199] E. T. Jaynes, Information theory and statistical mechanics. *Physical Review,* 1957.

[200] M. Johnson, M. Amin, S. Gildert, *et al.* Quantum annealing with manufactured spins. *Nature* **473** (2011), 194–198.

[201] M.I. Jordan, Z. Ghahramani, T.S. Jaakkola, *et al.* An introduction to variational methods for graphical models. *Machine Learning.* **37** (1999), 183–233.

[202] M. I. Jordan, T. M. Mitchell, Machine learning: Trends, perspectives, and prospects. *Science,* **349** (2015), 255–260.

[203] W. Jiang, M.A. Tanner, Gibbs posterior for variable selection in high-dimensional classification and data mining. *The Annals of Statistics* **36** (2008), 2207–2231.

[204] P. Kairouz, *et al.* Advances and open problems in federated learning, Foundations and Trends in Machine Learning, **14**(1–2) (2021), 1–210.

[205] I. Kanter, Potts-glass models of neural networks. *Physical Review A* (3) **37**(7) (1988), 2739–2742.

[206] J. Kaplan, S. McCandlish, T. Henighan, T.B. Brown, B. Chess, R. Child, S. Gray, A. Radford, J. Wu, D. Amodei, *Scaling Laws for Neural Language Models.* CoRR abs/2001.08361 (2020).

[207] R. Karakida, M. Okada, S.-I. Amari, Dynamical analysis of contrastive divergence learning: Restricted Boltzmann machines with Gaussian visible units. *Neural Networks,* **79** (2016), 78–87.

[208] T. Karras, S. Laine, T. Aila, *A Style-based Generator Architecture for Generative Adversarial Networks.* In Proceedings of CVPR (2019), 4401–4410.

[209] Z. Kato, J. Zerubia, Markov random fields in image segmentation. *Foundations and Trends in Signal Processing,* **5**(12) (2011), 1–155.

[210] S.M. Kay, *Fundamentals of Statistical Signal Processing: Estimation Theory.* Pearson. 1993.

[211] J. Kennedy and R. Eberhart, *Particle Swarm Optimization,* Proceedings of ICNN'95 - International Conference on Neural Networks, Perth, WA, Australia, **4** (1995), 1942–1948.

[212] N.S. Keskar, J. Nocedal, P.T.P. Tang, D. Mudigere, M. Smelyanskiy, *On Large-Batch Training for Deep Learning: Generalization Gap and Sharp*

minima. Paper presented at 5th International Conference on Learning Representations, ICLR 2017, Toulon, France.

[213] R. Kindermann, J.L. Snell, *Markov Random Fields and Their Applications.* American Mathematical Society. (1980).

[214] D. P. Kingma, J. Ba, *Adam: A Method for Stochastic Optimization,* CoRR, 2014, https://api.semanticscholar.org/CorpusID:6628106.

[215] S. Kirkpatrick, C. D. Gelatt, M. P. Vecchi, Optimization by simulated annealing. *Science,* **220**(4598) (1983), 671–680.

[216] T. Kim, Y. Bengio, *Deep Directed Generative Models with Energy-Based Probability Estimation,* ICLR Workshop Track (2016).

[217] J. Kleinberg, *An Impossibility Theorem for Clustering.* In NIPS (2002), 8 pages.

[218] F. Knoll, J. Daly, J. Meyer, Solving Boltzmann optimization problems with deep learning. *Unconventional Computing* **1**(5) (2024), 7 pages.

[219] D. Koller, N. Friedman, *Probabilistic Graphical Models: Principles and Techniques.* MIT Press. 2009.

[220] A. N. Kolmogorov, S. V. Fomin, *Elements of the Theory of Functions and Functional Analysis.* Dover Publications, 1999.

[221] C. Krapu, R. Stewart, A. Rose. *A Review of Bayesian Networks for Spatial Data. ACM Transactions on Spatial Algorithms and Systems* **9**(1), Article 7 (2023), 21 pages.

[222] A. Krogh, J.A. Hertz, A simple weight decay can improve generalization. *Advances in Neural Information Processing Systems (NeurIPS),* 1991.

[223] D. Krotov, J. J. Hopfield, *Large Associative Memory Problem in Neurobiology and Machine Learning,* In *International Conference on Learning Representations,* 2021.

[224] D. Krotov, J. J. Hopfield, *Dense Associative Memory for Pattern Recognition,* arXiv:1606.01164, 2016.

[225] E. Kropff, A. Treves, The storage capacity of Potts models for semantic memory retrieval. *Journal of Statistical Mechanics: Theory and Experiment* **8** (2005), P08010, 19 pp.

[226] F. Krzakala, M. Mézard, F. Sausset, Y. Sun, L. Zdeborová, Statistical-physics-based reconstruction in compressed sensing. *Physical Review X* **2**(2), 2012, 021005.

[227] F. Krzakala, L. Zdeborová, *Statistical Physics Methods in Optimization and Machine Learning,* https://sphinxteam.github.io/EPFLDoctoralLecture 2021/Notes.pdf.

[228] F. Krzakala, C. Moore, E. Mossel, J. Neeman, A. Sly, L. Zdeborová, P. Zhang, Spectral redemption in clustering sparse networks. *PNAS* **110**(52) (2013), 20935–20940.

[229] F. R. Kschischang, B. J. Frey, H. Loeliger, Factor graphs and the sum-product algorithm. *IEEE Transactions on Information Theory,* **47**(2) (2001), 498–519.

[230] J. Lafferty, A. McCallum, F. Pereira, *Conditional Random Fields: Probabilistic Models for Segmenting and Labeling Sequence Data.* Proc. 18th International Conf. on Machine Learning. (2001), 282–289.

[231] B. Lakshminarayanan, A. Pritzel, C. Blundell, Simple and scalable predictive uncertainty estimation using deep ensembles, *Advances in Neural Information Processing Systems*, 2017.

[232] S.L. Lauritzen, *Graphical Models.* Oxford University Press. 1996.

[233] Y. LeCun, S. Chopra, R. Hadsell, M. A. Ranzato, F. J. Huang, A tutorial on energy-based learning. *Predicting Structured Data.* 2006.

[234] Y. LeCun, Y. Bengio, G.E. Hinton, Deep learning. *Nature* **521** (2015), 436–444.

[235] Y. LeCun, B. Boser, J. S. Denker, D. Henderson, R. E. Howard, W. Hubbard, L. D. Jackel, Backpropagation applied to handwritten zip code recognition, *Neural Computation,* **1**(4) (1989), 541–551.

[236] S.Y. Lee, B. Lei, B. Mallick, Estimation of COVID-19 spread curves integrating global data and borrowing information. *PLoS One.* 2020 Jul 29;15(7):e0236860.

[237] J. Lee, Y. Bahri, R. Novak, S. S. Schoenholz, J. Pennington, J. Sohl-Dickstein, Deep neural networks as Gaussian processes, *International Conference on Learning Representations*, 2018.

[238] S.Y. Lee, B. Mallick, Bayesian hierarchical modeling: Application towards production results in the Eagle Ford Shale of South Texas. *Sankhya B.* **84** (2021), 1–43.

[239] A. Lenci, M. Sahlgren, *Distributional Semantics.* Cambridge University Press, 2023.

[240] M. Leone, Sumedha and M. Weigt, Clustering by soft-constraint affinity propagation: Applications to gene expression data. *Bioinformatics*, **23** (2007), 2708–2715.

[241] Y. Leviathan, M. Kalman, Y. Matias, Fast Inference from Transformers via Speculative Decoding, *International Conference on Machine Learning*, 2023.

[242] M. Leone, Sumedha and M. Weigt, Unsupervised and semi supervised clustering by message passing: Soft-constraint affinity propagation. *Eur. Phys. J. B*, (2008), 125–135.

[243] D.A. Levin, Y. Peres, *Markov Chains and Mixing Times.* AMS. 2017.

[244] H. Li, Z. Xu, G. Taylor, C. Studer, T. Goldstein, *Visualizing the Loss Landscape of Neural Nets*, arXiv:1712.09913v3, 2018.

[245] S. Z. Li, *Markov Random Field Modeling in Image Analysis*, Springer-Verlag London, 2009.

[246] Q. Li, H. Sompolinsky, Statistical mechanics of deep linear neural networks: The backpropagating Kernel renormalization. *Physical Review X*, **11**(3), 031059 (2021), 32 pages.

[247] Y. Lipman, R. T. Q. Chen, H. Ben-Hamu, M. Nickel, M. Le, *Flow Matching for Generative Modeling*, The Eleventh International Conference on Learning Representations, 2023.

[248] Y. Li, *Deep Reinforcement Learning: An Overview,* arXiv:1701.07274, 2017.

[249] T.M. Liggett, *Interacting Particle Systems.* Springer Berlin, Heidelberg., 2005.

[250] H. Lin, M. Tegmark, D. Rolnick, Why does deep and cheap learning work so well? *Journal of Statistical Physics* **168**(6) (2017), 1223–1247.

[251] A.D. Linkevich, A sequential pseudo-inverse learning rule for networks of formal neurons. *Journal of Physics A* **25**(15) (1992), 4139–4146.

[252] Q. Liu, D. Wang, Stein variational gradient descent: A general purpose Bayesian inference algorithm, *Advances in Neural Information Processing Systems* 29, 2016.

[253] J. S. Liu, The collapsed Gibbs sampler in Bayesian computations with applications to a gene regulation problem, *Journal of the American Statistical Association*, **89**(427) (1994), 958–966.

[254] H. A. Loeliger, An introduction to factor graphs, *IEEE Signal Processing Magazine*, **21** (2004), 28–41.

[255] M. Löwe, F. Vermet, The capacity of q-state Potts neural networks with parallel retrieval dynamics. *Statistics & Probability Letters* **77**(14) (2007), 1505–1514.

[256] Y. Lu, Z. Li, Di He, Z. Sun, B. Dong, T. Qin, L. Wang, T.-Y. Liu, *Understanding and Improving Transformer From a Multi-Particle Dynamic System Point of View*, arXiv:1906.02762v1 [cs.LG], 2019.

[257] C. Lucibello, M. Mezard, Exponential capacity of dense associative memories, *Physical Review Letters*, 2024.

[258] C. Lucibello, F. Pittorino, G. Perugini, R. Zecchina, Deep learning via message passing algorithms based on belief propagation, *Machine Learning: Science and Technology* **3**3 (2022), 035005.

[259] W. Maass, T. Natschläger H. Markram, Real-time computing without stable states: A new framework for neural computation based on perturbations. *Neural Computation*, **14**(11) (2002), 2531–60.

[260] D. J. MacKay. A practical Bayesian framework for backpropagation networks, *Neural Computation*, **4**(3) (1992), 448–472.

[261] A. Madry, A. Makelov, L. Schmidt, D. Tsipras, A. Vladu. *Towards Deep Learning Models Resistant to Adversarial Attacks.* International Conference on Learning Representations. 2018.

[262] S. Mandt, M.D. Hoffman, D.M. Blei, Stochastic gradient descent as approximate Bayesian inference. *Journal of Machine Learning Research.* **18** (2017), 1–35.

[263] S. Mallat, Understanding deep convolutional networks. *Philosophical Transactions of the Royal Society A: Mathematical, Physical and Engineering Sciences* **374**, 2065 (2016), 20150203.

[264] F. Martinelli, *Lectures on Glauber Dynamics for Discrete Spin Models.* In Lectures on Probability Theory and Statistics (pp. 93–191). Springer. 1999.

[265] A. De Masi, E. Presutti, *Mathematical Methods for Hydrodynamic Limits.* Springer Berlin, Heidelberg, 1991.

[266] C. Maugis, Model-based clustering of high-dimensional data: A review. *Computational Statistics & Data Analysis* **71** (2014), 52–78.

[267] D.A. McAllester, PAC-Bayesian stochastic model selection. *Machine Learning*, **51** (2003), 5–21.

[268] J.-W. van de Meent, B. Paige, H. Yang, F. Wood, *An Introduction to Probabilistic Programming*, arXiv:1809.10756.

[269] P. Mehta, M. Bukov, C. Wang, A. Day, C. Richardson, C. Fisher, D. Schwab, A high-bias, low-variance introduction to machine learning for physicists. *Physics Reports*, **810** (2019), 1–124.

[270] N. Mehrabi, F. Morstatter, N. Saxena, K. Lerman, A. Galstyan, A survey on bias and fairness in machine learning. *ACM Computing Surveys (CSUR)* **54** (2019), 1–35.

[271] N. Meinshausen, P. Bühlmann. High-dimensional graphs and variable selection with the Lasso. *The Annals of Statistics* **34**(3) (2006), 1436–1462.

[272] M. Mézard, G. Parisi, M. Virasoro, *Spin Glass Theory and Beyond.* World Scientific, Singapore, 1987.

[273] M. Mézard, J.P. Nadal, Learning in feedforward layered networks: The tiling algorithm. *Journal of Physics A: Mathematical and General* **22**(12) (1989), 2191–2203.

[274] N. Miao, H. Zhou, L. Mou, R. Yan, L. Li, *CGMH: Constrained Sentence Generation by Metropolis-Hastings Sampling*, Proceedings of the Thirty-Third AAAI Conference on Artificial Intelligence and Thirty-First Innovative Applications of Artificial Intelligence Conference and Ninth AAAI Symposium on Educational Advances in Artificial Intelligence, 2019.

[275] T. Mikolov, K. Chen, G. Corrado, J. Dean, *Efficient Estimation of Word Representations in Vector Space.* International Conference on Learning Representations (2013).

[276] B. Millidge, T. Salvatori, Y. Song, T. Lukasiewicz, R. Bogacz, *Universal Hopfield networks: A General Framework for Single-Shot Associative Memory Models.* In International Conference on Machine Learning, PMLR (2022), 15561–15583.

[277] J. Møller, A. N. Pettitt, R. Reeves, K. K. Berthelsen, An efficient Markov chain Monte Carlo method for distributions with intractable normalising constants, *Biometrika* **93**(2) (2006), 451–458.

[278] K.P. Murphy, *Machine Learning: A Probabilistic Perspective.* MIT Press. 2012.

[279] P. Nakkiran, G. Kaplun, Y. Bansal, T. Yang, B. Barak, I. Sutskever, Deep double descent: Where bigger models and more data hurt. *Journal of Statistical Mechanics* 124003 (2021), 32 pages.

[280] M. Naim, V. Boboeva, C.J. Kang, A. Treves, Reducing a cortical network to a Potts model yields storage capacity estimates. *Journal of Statistical Mechanics: Theory and Experiment* **4** (2018), 043304, 35 pp.

[281] R.M. Neal, *Probabilistic Inference Using Markov Chain Monte Carlo Methods.* Technical Report, Department of Computer Science, University of Toronto. 1993.

[282] R.M. Neal, *Bayesian Learning for Neural Networks.* Springer Science and Business Media, 2012.

[283] R.M. Neal, Markov chain sampling methods for Dirichlet process mixture models, *Journal of Computational and Graphical Statistics* **9**(2) (2000), 249–265.

[284] B. Neyshabur, R. Tomioka, M. Srebro, *Norm-Based Capacity Control in Neural Networks*, Proceedings of The 28th Conference on Learning Theory, PMLR **40** (2015), 1376–1401.

[285] A. Ng, M. Jordan, W. Yair, On spectral clustering: Analysis and an algorithm. *In NIPS* **14** (2002), 849–856.

[286] Tam M. Nguyen, C. A. Uribe, Tan M. Nguyen, R. Baraniuk, *PIDformer: Transformer Meets Control Theory*, In Forty-first International Conference on Machine Learning, 2024.

[287] Tam. M. Nguyen, M. Pham, Tan. M. Nguyen, K. Nguyen, S. Osher, N. Ho, FourierFormer: Transformer meets generalized Fourier integral theorem, *Advances in Neural Information Processing Systems*, 2022.

[288] N. M. Nguyen, A. Baker, C. Neo, A. G. Roush, A. Kirsch, R. Shwartz-Ziv, *Turning Up the Heat: Min-p Sampling for Creative and Coherent LLM Outputs*, International Conference on Learning Representations, 2025.

[289] S. Nielsen, L. Abdullaev, R. Teo, T. M. Nguyen, *Elliptical Attention*, The Thirty-eighth Annual Conference on Neural Information Processing Systems, 2024.

[290] S. K. Nielsen, R. S.Y. Teo, L. U. Abdullaev, T. M. Nguyen, *Tight Clusters Make Specialized Experts*, International Conference on Learning Representations, 2025.

[291] H. Nishimori, *Statistical Physics of Spin Glasses and Information Processing: An Introduction.* Oxford University Press, 2001.

[292] E. Nichani, J. D. Lee, A. Bietti, *Understanding Factual Recall in Transformers via Associative Memories*, The Thirteenth International Conference on Learning Representations, 2025.

[293] S. Nowozin, C. H. Lampert, *Structured Learning and Prediction in Computer Vision*, 2011.

[294] A. van den Oord, N. Kalchbrenner, K. Kavukcuoglu, *Pixel Recurrent Neural Networks.* Proceedings of The 33rd International Conference on Machine Learning, PMLR **48** (2016), 1747–1756.

[295] L. Panait, S. Luke, Cooperative multi-agent learning: The state of the art. *Autonomous Agents and Multi-Agent Systems*, **11**(3) (2005), 387–434.

[296] B. Pang, T. Han, E. Nijkamp, S.-C. Zhu, Y. N. Wu. *Learning Latent Space Energy-based Prior Model.* In Proceedings of the 34th International Conference on Neural Information Processing Systems. Curran Associates Inc., Red Hook, NY, USA, Article 1845 (2020), 21994–22008.

[297] G. Papamakarios, E. Nalisnick, D. J. Rezende, S. Mohamed, B. Lakshminarayanan, Normalizing flows for probabilistic modeling and inference, *Journal of Machine Learning Research*, **22** (2021), 1–64.

[298] T. Papamarkou, Approximate blocked Gibbs sampling for Bayesian neural networks. *Statistics and Computing* **33**, 119 (2023), 19 pages.

[299] G. A. Pavliotis, *Stochastic Processes and Applications: Diffusion Processes, the Fokker-Planck and Langevin Equations.* Springer, 2014.

[300] J. Pearl, *Probabilistic Reasoning in Intelligent Systems: Networks of Plausible Inference.* Morgan Kaufmann. 1988.

[301] R. Peierls, On Ising model of ferro magnetism, *Proceedings of the Cambridge Philosophical Society* **32** (1936), 477–481.

[302] B. Peng, S. Parthasarathy, X. Ning, *Multi-modality Meets Relearning: Mitigating Negative Transfer in Sequential Recommendation*, arXiv:2309.10195, 2023.

[303] V.V. Prasolov, *Polynomials.* (Springer-Verlag Berlin Heidelberg, 2004)

[304] C. Preston, *Gibbs States on Countable Sets*, Cambridge University Press, London (1974).

[305] A. P. Punnen (editor), *Quadratic Unconstrained Binary Optimization Problem*, Theory, Algorithms, and Applications, Springer, 2022.

[306] L.R. Rabiner, A tutorial on hidden Markov models and selected applications in speech recognition. *Proceedings of the IEEE*, **77**(2) (1989), 257–286.

[307] H. Ramsauer, B. Schäfl, J. Lehner, P. Seidl, M. Widrich, L. Gruber, M. Holzleitner, T. Adler, D. Kreil, M. K. Kopp, G. Klambauer, J. Brandstetter, S. Hochreiter, *Hopfield Networks is All You Need.* In *International Conference on Learning Representations*, 2021. URL: `https://openreview.net/forum?id=tL89Rnz1Ccd`.

[308] M. Ranzato, Y-L. Boureau, S. Chopra, Y. LeCun, A unified energy-based framework for unsupervised learning, *Proceedings of the 11th International Conference on Artificial Intelligence and Statistics, PMLR* (2007), 371–379.

[309] Z. Razaee, A. Amini, *The Potts-Ising Model for Discrete Multivariate Data*, 34th Conference on Neural Information Processing Systems (NeurIPS 2020), Vancouver, Canada.

[310] D.J. Rezende, S. Mohamed, D. Wierstra, *Stochastic Backpropagation and Approximate Inference in Deep Generative Models.* ICML 2014, arXiv:1401.4082.

[311] D.J. Rezende, S. Mohamed, *Variational Inference with Normalizing Flows.* Proceedings of the 32nd International Conference on Machine Learning, PMLR **37** (2015), 1530–1538.

[312] T. Rigon, A.H. Herring, D.B. Dunson, A generalized Bayes framework for probabilistic clustering. *Biometrika*, **110** (2023), 559–578.

[313] O. Rivasplata, I. Kuzborskij, C. Szepesvari, J. Shawe-Taylor, PAC-Bayes analysis beyond the usual bounds, *Advances in Neural Information Processing Systems* 33, 2020.

[314] C.P. Robert, G. Casella, *Introducing Monte Carlo Methods with R.* Springer. (2010).

[315] C.P. Robert, G. Casella, *Monte Carlo Statistical Methods.* Springer Science and Business Media. 2004.

[316] G. O. Roberts, S. K. Sahu, Updating schemes, correlation structure, blocking and parameterization for the Gibbs sampler, *Journal of the Royal Statistical Society Series B, Royal Statistical Society,* **59(2)** (1997), 291–317.

[317] G. O. Roberts, J. S. Rosenthal, Minimising MCMC variance via diffusion limits, with an application to simulated tempering, *The Annals of Applied Probability* **24**(1) (2014), 131–149.

[318] B. Rodríguez-Gálvez, R. Thobaben, M. Skoglund. More PAC-Bayes bounds: from bounded losses, to losses with general tail behaviors, to anytime validity. *Journal of Machine Learning Research* **25**(1), Article 110 (2024), 43 pages.
[319] R. Rojas, *Neural Networks: A Systematic Introduction.* Springer. 1996.
[320] X. Rong, *Word2vec Parameter Learning Explained*, arXiv:1411.2738, 2016.
[321] S.Roth, M.J. Black, *Fields of Experts: A Framework for Learning Image Priors.* Proceedings of the IEEE Conference on Computer Vision and Pattern Recognition. 2005.
[322] J. Rothfuss, V. Fortuin, M. Josifoski, A. Krause, *PACOH: Bayes-Optimal Meta-Learning with PAC-Guarantees*, Proceedings of the 38th International Conference on Machine Learning, PMLR, **139** (2021), 9116–9126.
[323] U.A. Rozikov, *Gibbs Measures on Cayley Trees. World Scientific Publishing* Singapore. 2013.
[324] U.A. Rozikov, *Gibbs Measures in Biology and Physics: The Potts Model. World Scientific Publishing* Singapore. 2023.
[325] U.A. Rozikov, A contour method on Cayley trees, *Journal of Statistical Physics* **130**(4) (2008), 801–813.
[326] U.A. Rozikov, On q-component models on Cayley tree: Contour method, *Letters in Mathematical Physics* **71**(1) (2005), 27–38.
[327] U.A. Rozikov, What are p-adic numbers? What are they used for? *Asia Pacific Mathematics Newsletter* **3**(4) (2013), 1–6.
[328] U.A. Rozikov, Bubble coalescence in interacting system of DNA molecules. *International Journal of Biomathematics* **17**(4) (2024), 22 pages.
[329] H. Rue, L. Held, *Gaussian Markov Random Fields: Theory and Applications.* Chapman and Hall/CRC. 2005.
[330] R. Salakhutdinov, A. Mnih, G. Hinton, *Restricted Boltzmann Machines for Collaborative Filtering.* Proceedings of the 24th International Conference on Machine Learning. 2007.
[331] R. Salakhutdinov, I. Murray, *On the Quantitative Analysis of Deep Belief Networks.* Proceedings of the 25th International Conference on Machine Learning (2008), 872–879.
[332] R. Salakhutdinov, G. Hinton, *Deep Boltzmann Machines.* Proceedings of the Twelfth International Conference on Artificial Intelligence and Statistics (AISTATS), **5** (2009), 448–455.
[333] A. Saxe, J. L. McClelland, S. Ganguli, *Exact Solutions to the Nonlinear Dynamics of Learning in Deep Linear Neural Networks.* CoRR abs/1312.6120 (2013).
[334] J. Schmidhuber, Deep learning in neural networks: An overview, *Neural Networks*, **61** (2015), 85–117.
[335] E. Seneta, *Non-negative Matrices and Markov Chains.* Springer. 2006.
[336] H.S. Seung, H. Sompolinsky, N. Tishby, Statistical mechanics of learning from examples. *Physical Review A* **45** (1992), 6056–6091.
[337] S. Shalev-Shwartz, S. Ben-David, *Understanding Machine Learning: From Theory to Algorithms*, Cambridge University Press, 2014.

[338] J. Shi, J. Malik, Normalized cuts and image segmentation, *IEEE Transactions on Pattern Analysis and Machine Intelligence*, **22**(8) (2000), 888–905.

[339] A. N. Shiryaev, *Probability*, 2nd Ed. Springer, 1996.

[340] M.A. Shwe, B. Middleton, D.E. Heckerman, M. Henrion, E.J. Horvitz, H. P. Lehmann, G.F. Cooper, Probabilistic diagnosis using a reformulation of the INTERNIST-1/QMR knowledge base. *Methods of Information in Medicine* **30** (1991), 241–255.

[341] K. F. Siburg and A. B. Nobel, A Gibbs sampler for the disk packing problem, *Journal of Statistical Physics*, **110**(3–6) (2003), 755–762.

[342] V. Simkus, B. Rhodes, M. U. Gutmann, Variational Gibbs inference for statistical model estimation from incomplete data, *Journal of Machine Learning Research*, **24** (2023), 1–72.

[343] Ya.G. Sinai, *Theory of Phase Transitions: Rigorous Results*, Pergamon, Oxford, 1982.

[344] Ya.G. Sinai, *Probability Theory – An Introductory Course*. Springer, 1992.

[345] D.A. Smith, J. Eisner, *Dependency Parsing by Belief Propagation*. Proceedings of the 46th Annual Meeting of the Association for Computational Linguistics (ACL). (2008), 145–156.

[346] J. Sohl-Dickstein, E.A. Weiss, N. Maheswaranathan, S. Ganguli. *Deep Unsupervised Learning Using Nonequilibrium Thermodynamics*. In Proceedings of the 32nd International Conference on International Conference on Machine Learning. **37** (2015), 2256–2265.

[347] Y. Song, S. Ermon, Generative modeling by estimating gradients of the data distribution. *Advances in Neural Information Processing Systems (NeurIPS)*. (2019).

[348] Y. Song, P.K. Diederik, *How to Train your Energy-based Models*, arXiv:2101.03288v2 (2021).

[349] P. Spirtes, C. Glymour, R. Scheines, *Causation, Prediction, and Search*, Springer, New York, NY. 1993.

[350] S. Still, W. Bialek, How many clusters?: An information-theoretic perspective. *Neural computation*, **16** (2004), 2483–2506.

[351] Yu.M. Suhov, U.A. Rozikov, A hard-core model on a Cayley tree: An example of a loss network, *Queueing Systems* **46**(1/2) (2004), 197–212.

[352] T. Sutter, A. Ganguly, H. Koeppl, A variational approach to path estimation and parameter inference of hidden diffusion processes, *Journal of Machine Learning Research*, **17**(190) (2016), 1–37.

[353] R.S. Sutton, A. G. Barto, *Reinforcement Learning: An Introduction* (2nd ed.). MIT Press. 2018.

[354] I. Sutskever, O. Vinyals, Q. V. Le, *Sequence to Sequence Learning with Neural Networks*, NeurIPS, 2014.

[355] Z. Sun, U. Mendlovic, Y. Leviathan, A. Aharoni, J. H. Ro, A. Beirami, A. T. Suresh, *Block Verification Accelerates Speculative Decoding*, International Conference on Learning Representations, 2025.

[356] R.H. Swendsen, J.S. Wang, Nonuniversal critical dynamics in Monte Carlo simulations. *Physical Review Letters* **58**(2) (1987), 86–88.

[357] N. Syring, R. Martin, Robust and rate-optimal Gibbs posterior inference on the boundary of a noisy image. *The Annals of Statistics* **48** (2020), 1498–1513.
[358] N. Syring, L. Hong, R. Martin, Gibbs posterior inference on value-at-risk. *Scandinavian Actuarial Journal* **2019** (2019), 548–557.
[359] N. Syring, M. Ryan, *Gibbs Posterior Concentration Rates Under Sub-Exponential Type Losses*, arXiv:2012.04505v6, 2022.
[360] R. Szeliski, R. Zabih, *An Experimental Comparison of Stereo Algorithms.* Vision Algorithms: Theory and Practice. 1999.
[361] C. Szegedy, W. Zaremba, I. Sutskever, *et al. Intriguing properties of neural networks.* 2nd International Conference on Learning Representations, Banff, 14–16 April 2014, 1–10.
[362] D.R. Tan, C.D. Li, W.P. Zhu, F.J. Jiang, A comprehensive neural networks study of the phase transitions of Potts model. *New Journal of Physics* **22** (2020), 063016 (18 pages).
[363] T. Tao, *An Introduction to Measure Theory.* Graduate Studies in Mathematics, vol. 126, American Mathematical Society, 2011.
[364] Y.W. Teh, *Dirichlet Process.* Encyclopedia of Machine Learning, Springer. 2010.
[365] Y. W. Teh, M. I. Jordan, M. J. Beal, D. M. Blei, Hierarchical Dirichlet processes, *Journal of the American Statistical Association* **101**(476) (2006), 1566–1581.
[366] D.J. Thouless, P. W. Anderson, R. G. Palmer, Solution of 'solvable model of a spin glass'. *The Philosophical Magazine: A Journal of Theoretical Experimental and Applied Physics*, **35**(3) (1977), 593–601.
[367] T. Tieleman. *Training Restricted Boltzmann Machines Using Approximations to the Likelihood Gradient.* In Proceedings of the 25th international conference on Machine learning (ICML '08). Association for Computing Machinery, New York, NY, USA (2008), 1064–1071.
[368] A. Tirelli, D. O. Carvalho, L. A. Oliveira, J.P. Lima, N. C. Costa, R. R. dos Santos, Unsupervised machine learning approaches to the q-state Potts model, *The European Physical Journal* **B 95**, 189 (2022).
[369] N. Tishby, N. Zaslavsky, *Deep Learning and the Information Bottleneck Principle*, IEEE Information Theory Workshop (ITW), Jerusalem, Israel (2015), 1–5.
[370] P. Turchin, *Complex Population Dynamics: A Theoretical/Empirical Synthesis.* Princeton University Press. 2003.
[371] Y.-H. H. Tsai, S. Bai, M. Yamada, L.-P. Morency, R. Salakhutdinov, *Transformer Dissection: An Unified Understanding for Transformer's Attention via the Lens of Kernel.* In Proceedings of the 2019 Conference on Empirical Methods in Natural Language Processing and the 9th International Joint Conference on Natural Language Processing, 2019, 4344–4353.
[372] L. Valiant, *A Theory of the Learnable.* In Proceedings of the Sixteenth Annual ACM Symposium on Theory of Computing (1984), STOC '84, pp. 436–445.

[373] V. Vapnik, *The Nature of Statistical Learning Theory*. Springer-Verlag, Berlin, Heidelberg, 1995.

[374] A. Vaswani, N. Shazeer, N. Parmar, J. Uszkoreit, L. Jones, A. N. Gomez, L. Kaiser, I. Polosukhin. Attention is all you need. In Proceedings of the 31st International Conference on Neural Information Processing Systems (NIPS'17). Curran Associates Inc., Red Hook, NY, USA (2017), 6000–6010.

[375] A. Vehtari, A. Gelman, D. Simpson, B. Carpenter, P.-C. Bürkner, Rank-normalization, folding, and localization: An improved $\hat{R}$ for assessing convergence of MCMC. *Bayesian Analysis* **16**(2) (2021), 667–718.

[376] J. Vuckovic, A. Baratin, R. T. des Combes, *A Mathematical Theory of Attention*, arXiv:2007.02876v2 [stat.ML], 2020.

[377] S. G. Walker, Sampling the Dirichlet mixture model with slices. *Communications in Statistics — Simulation and Computation*, **36**(1) (2007), 45–54.

[378] M.J. Wainwright, M. I. Jordan, Graphical models, exponential families, and variational inference. *Foundations and Trends in Machine Learning*, **1**(1-2) (2008), 1–305.

[379] K. Wang, J. Zhang, D. Li, X. Zhang, T. Guo, Adaptive affinity propagation clustering. *Acta Automatica Sinica*, **33**(12) (2007), 1242–1246.

[380] W. Wang, C. Wu, J. Deng, Piecewise constant signal and image denoising using a selective averaging method with multiple neighbors. *Inverse Problems and Imaging*, **13**(5) (2019), 903–930.

[381] J. Wei *et al. Emergent Abilities of Large Language Models.* arXiv:2206.07682v2, 2022.

[382] A.Weinmann, M. Storath, L. Demaret, The L^1-Potts functional for robust jump-sparse reconstruction. *SIAM Journal on Numerical Analysis* **53**(1) (2015), 644–673.

[383] Y. Weiss, W T. Freeman, *Correctness of Belief Propagation in Gaussian Graphical Models of Arbitrary Topology*, Advances in Neural Information Processing Systems 12 (NIPS 1999), 673–679.

[384] M. Welling, Y. W. Teh, *Bayesian Learning via Stochastic Gradient Langevin Dynamics*, ICML 2011.

[385] F. Wenzel, K. Roth, B. S. Veeling, J. Swiatkowski, L. Tran, S. Mandt, J. Snoek, T. Salimans, R. Jenatton, S. Nowozin. *How Good is the Bayes Posterior in Deep Neural Networks Really?* In Proceedings of the 37th International Conference on Machine Learning. **119**, Article 949 (2020), 10248–10259.

[386] J. Weston, C. Watkins, *Support vector machines for multi-class pattern recognition.* ESANN'1999 proceedings - European Symposium on Artificial Neural Networks Bruges (Belgium), 21–23 April 1999, D-Facto public., ISBN 2-600049-9-X, pp. 219–222.

[387] M. Winer, B. Hanin, *Deep Nets as Hamiltonians*, arXiv:2503.23982, 2025.

[388] A. G. Wilson, Z. Hu, R. Salakhutdinov, E. P. Xing, D. Kernel, *Deep kernel learning*, Proceedings of the 19th International Conference on Artificial Intelligence and Statistics, PMLR, **51** (2016), 370–378.

[389] S. Wiseman, M. Blatt, E. Domany, Super-paramagnetic clustering of data. *Physical Review E* **57** (1998), 3767–3787.

[390] D. Xiong, H. Zhao, q-state Potts-glass neural network based on pseudoinverse rule, *Journal of Physics A* **43**(44) (2010), 445001, 9 pp.

[391] L. Yao, D. Mimno, A. McCallum, Efficient methods for topic model inference on streaming document collections, *Knowledge Discovery and Data Mining.* 2009.

[392] D. Yamins, J. DiCarlo, Using goal-driven deep learning models to understand sensory cortex. *Nature Neuroscience* **19** (2016), 356–365.

[393] L. Yang, Z. Zhang, Y. Song, H. Shenda, R. Xu, Y. Zhao, W. Zhang, B. Cui, M. Yang, Diffusion models: A comprehensive survey of methods and applications, *ACM Computing Surveys* (2023) **56**(4), 1–39.

[394] J. S. Yedidia, W. T. Freeman and Y. Weiss, Constructing free-energy approximations and generalized belief propagation algorithms, *IEEE Transactions on Information Theory,* **51**(7) (2005), 2282–2312.

[395] T. Ye, L. Dong, Y. Xia, Y. Sun, Y. Zhu, G. Huang, F. Wei, *Differential Transformer*, The Thirteenth International Conference on Learning Representations, 2025.

[396] M. Zahradnik, An alternate version of Pirogov-Sinai theory, *Communications in Mathematical Physics* **93** (1984), 559–581.

[397] L. Zdeborová, F. Krzakala, Statistical physics of inference: Thresholds and algorithms. *Advances in Physics,* **65**(5) (2016), 453–552.

[398] M.D. Zeiler, R. Fergus, *Visualizing and Understanding Convolutional Networks.* In: Fleet, D., Pajdla, T., Schiele, B., Tuytelaars, T. (eds) Computer Vision - ECCV 2014. Lecture Notes in Computer Science, vol 8689. Springer, Cham.

[399] C. Zhang, S. Bengio, M. Hardt, B. Recht, O. Vinyals, *Understanding deep learning requires rethinking generalization*, arXiv:1611.03530v2. 2017.

[400] Z. H. Zhou, A brief introduction to weakly supervised learning, *National Science Review,* **5**(1) (2018), 44–53.

[401] J. Zhao, M. Mathieu, Y. LeCun, *Energy-based Generative Adversarial Networks* International Conference on Learning Representations (2017). OpenReview: `https://openreview.net/forum?id=ryh9pmcee`.

[402] S. Zhong, M. Xu, T. Ao, G. Shi, *Understanding Transformer from the Perspective of Associative Memory*, arXiv:2505.19488, 2025.

[403] W. Zucchini, I.L. MacDonald, R. Langrock, *Hidden Markov Models for Time Series: An Introduction Using R.* CRC Press. (2016).

Index

www.ingramcontent.com/pod-product-compliance
Ingram Content Group UK Ltd.
Pitfield, Milton Keynes, MK11 3LW, UK
UKHW020351150926
8624IPUK00004B/103

9 789819 814565